Sensors Fusion in Digital Healthcare Applications

Sensors Fusion in Digital Healthcare Applications

Guest Editor

Boon Giin Lee

MDPI

Basel • Beijing • Wuhan • Barcelona • Belgrade • Novi Sad • Cluj • Manchester

Guest Editor
Boon Giin Lee
School of Computer Science
University of Nottingham Ningbo China
Ningbo
China

Editorial Office
MDPI AG
Grosspeteranlage 5
4052 Basel, Switzerland

This is a reprint of the Special Issue, published open access by the journal *Sensors* (ISSN 1424-8220), freely accessible at: https://www.mdpi.com/journal/sensors/special_issues/sensors_fusion_digital_healthcare.

For citation purposes, cite each article independently as indicated on the article page online and as indicated below:

Lastname, A.A.; Lastname, B.B. Article Title. *Journal Name* **Year**, *Volume Number*, Page Range.

ISBN 978-3-7258-7571-9 (Hbk)
ISBN 978-3-7258-7572-6 (PDF)
https://doi.org/10.3390/books978-3-7258-7572-6

Contents

About the Editor

Boon Giin Lee

Boon Giin Lee is an Associate Professor in the School of Computer Science at the University of Nottingham Ningbo China, where he also serves as the School Director of Research and Knowledge Exchange and Deputy Director of the In-the-Wild Human–Computer Interaction Laboratory. His research interests lie at the intersection of human-AI interaction, affective and cognitive computing, and interactive wearable systems, with a focus on developing sensor-fusion-driven solutions for digital health and human–machine interfaces. He has authored over 100 papers in high-impact journals and has filed more than 30 patents, with 13 granted in China and others in the USA and Korea. He is recognized on Stanford University's list of the Top 2% of World Scientists since 2022 and has led several funded projects on wearable sensors for safety assessment and indoor localization. He is IEEE Senior Member and actively serves on committee members for international conferences and has received multiple awards for both his research and his mentorship of award-winning student innovation teams.

sensors

MDPI

Editorial

Sensors Fusion in Digital Healthcare Applications

Boon Giin Lee

School of Computer Science, University of Nottingham Ningbo China, Ningbo 315100, China; boon-giin.lee@nottingham.edu.cn

1. Introduction

Sensors constitute a fundamental element of modern digital healthcare, acting as the primary interface for capturing physiological, environmental, and biomechanical data from patients and clinical settings. The convergence of multiple sensing modalities, including wearable physiological monitors, environmental sensors, depth cameras, and inertial measurement units, has enabled a more comprehensive understanding of human health and disease processes [1–3]. This sensor fusion framework, coupled with advances in artificial intelligence, has opened new possibilities for personalized medicine, remote patient monitoring, and clinical decision support. However, the growing complexity and volume of healthcare data, together with the demands for real-time processing, privacy preservation, and model interpretability, present ongoing challenges that require innovative methodological approaches [4–6].

The intersection of sensor technologies and AI has accelerated the development of intelligent healthcare systems capable of automated detection, prediction, and intervention in various clinical domains. Sensor fusion approaches are transforming healthcare delivery and monitoring approaches, from assessing mental workload and detecting stress to analyzing performances in rehabilitation [7–9]. However, the main challenges remain, including ensuring data quality, addressing class imbalance, achieving hardware efficiency for wearable platforms, and maintaining the clinical interpretability of complex models [10–12]. The generation of high-quality synthetic data to augment limited real-world datasets, together with the benchmarking of wearable systems against gold-standard references, further highlights the multi-dimensional character of this research area.

This Special Issue, "Sensors Fusion in Digital Healthcare Applications", brings together 11 original research articles and a comprehensive systematic review that address these challenges through novel sensor-integration strategies, advanced signal processing techniques, and clinically validated AI models. The collected contributions showcase the wide scope of sensor fusion applications, spanning the multimodal physiological monitoring in autism spectrum disorders, biomechanical assessments of stair descent, the synthetic generation of medical time series, and hardware acceleration for EEG processing. Together, they highlight the transformative promise of integrated sensing technologies for advancing digital healthcare throughout life.

2. Overview of the Published Papers

In the first contribution, the authors present a sensor- and machine learning-based sensory management recommendation system (SMRS) designed to assist children with autism spectrum disorder (ASD) in managing atypical sensory responses within classroom settings. The SMRS integrates multi-modal data from wearable and environmental sensors, including physiological monitors, accelerometers, and ambient light, noise, and temperature

https://doi.org/10.3390/s26061870

sensors, with machine learning models for the real-time detection of attention and stress levels. A novel aspect is the integration of a standardized sensory profile questionnaire to personalize the SMRS's understanding of each child's unique sensory processing patterns, alongside a fuzzy logic-based decision-making module that generates context-aware, real-time recommendations for caregivers and teachers. Through a comprehensive evaluation involving 30 children with ASD and 30 typically developing peers, the study demonstrates that SMRS can effectively identify distractions and stressors and that the implementation of its recommended strategies leads to statistically significant improvements in children's attention and stress regulation. The results of a System Usability Scale questionnaire further indicate that the SMRS is perceived as user-friendly. The authors conclude that while the proposed SMRS represents a significant step toward low-cost, personalized, and real-time support for children with ASD.

In the second contribution, the authors explore the application of sensor fusion to optimize the trade-off between airborne pandemic control and energy consumption in digital indoor healthcare settings. The study uses a multi-sensor array consisting of air-quality monitors, temperature and humidity sensors, energy consumption meters, and 3D LiDAR scanners to characterize the physical and environmental dynamics of hospital inpatient rooms, classrooms, and conference rooms under different ventilation conditions. Through combining these heterogeneous sensor data, the authors construct high-fidelity input parameters for a computational fluid dynamics model coupled with an epidemiological simulation, allowing for detailed analysis of how open windows, pedestal fans, and air-conditioning systems influence airborne pathogen transmission. A Pareto front analysis using fused sensor datasets reveals that while all three ventilation solutions can be epidemiologically effective, certain configurations, such as low-speed air-conditioning, are highly energy-inefficient compared to high-speed fans. This work demonstrates how sensor fusion can inform evidence-based, multi-objective decision-making in digital public health infrastructure, balancing infection control with sustainability goals. Future research directions include the integration of real-time thermal comfort sensors to include human factors in the optimization framework and the development of room-level adaptive ventilation control systems driven by continuous sensor data streams.

In the third contribution, the authors present a sensor fusion-based methodology to estimate patient length of stay in intensive care units using data collected during the first 24 h after admission. The work integrates heterogeneous data from the MIMIC-III database, including monitors of vital signs at the bedside, laboratory test results, and demographic information, transforming raw variables into 139 statistical indicators. Using XGBoost as the core estimator, a novel two-step Bayesian optimization approach on GPU architecture automatically identifies the optimal hyperparameter configuration to minimize prediction error. Experimental results from 33,847 stays in the ICU demonstrate that the methodology achieves a mean absolute error of 2.529 days, significantly exceeding both the default XGBoost model (3.040 days) and the single-step optimization baseline (2.539 days), while also outperforming the current state-of-the-art generic LoS prediction models. The GPU-accelerated implementation reduces execution time by approximately 85% compared to multi-core CPU processing. The authors conclude that such automated machine learning systems driven by sensor fusion can serve as powerful digital healthcare tools for managing ICU occupancy and supporting clinical decision-making.

In the fourth contribution, the authors present a sensor fusion-based deep learning framework to assess stress levels in university students under ecologically valid problem-solving conditions. The study addresses a gap in stress detection research by designing an experimental protocol centered on Sudoku tasks in three distracting scenarios: exposure to discordant audio and horror videos, silent observation, and comforting stimuli. The

https://doi.org/10.3390/s26061870

study integrates multi-modal physiological data from wearable sensors, including PPG, ECG, and EEG, capturing responses associated with cognitive load and environmental stressors. The authors evaluated three advanced deep learning architectures, including StressNeXt, a long-term recurrent convolutional network (LRCN), and a self-supervised CNN, for stress classification. The findings demonstrate that sensor fusion, particularly the combination of ECG and EEG, enables the LRCN model to achieve high proficiency in discriminating stress levels, while StressNeXt shows exceptional performance under comforting conditions. The study further reveals that the ECG alone serves as a highly effective signal, whereas the EEG data are more susceptible to motion artifacts. The authors conclude that such sensor fusion-driven systems can provide valuable digital healthcare tools to monitor student well-being in academic settings.

In the fifth contribution, the authors present a sensor fusion-based data augmentation method to improve indoor localization in nursing care facilities using Bluetooth Low-Energy (BLE) beacons. The work addresses data imbalances in real-world healthcare settings, where certain areas are underrepresented due to variability in caregiving routines. Through combining RSSI data from 25 stationary beacons, the authors analyze signal patterns across semantic locations using standard deviation and KL divergence as similarity metrics. A novel relabeling approach identifies matching signal patterns between underrepresented (minority) and well-represented (majority) room classes, enabling beacon data to be reassigned to augment minority class samples. Two variations are implemented: full matching, requiring complete surrounding beacon configurations, and partial matching for incomplete sensor coverage. The augmented data are used within a Random Forest classifier for room-level localization. The authors demonstrate that this relabeling strategy, particularly KL divergence with full matching, achieves better minority class classification and improves overall performance compared to conventional techniques. The study concludes that the use of existing sensor data through intelligent signal pattern analysis offers a practical and data-efficient approach to improving indoor positioning in nursing settings.

In the sixth contribution, the authors present a sensor fusion-based comparative study examining the impact of two cockpit display interface designs, which are the traditional Steam Gauge panel and the modern G1000 Glass panel, on novice pilots' mental workload and situational awareness within a flight simulator environment. The system integrates multi-modal physiological sensing, combining EEG to capture the brain activity patterns associated with cognitive load, and HRV from electrocardiogram sensors to assess autonomic nervous system responses. Through integrating these objective physiological measures with subjective NASA-TLX assessments, the authors evaluated how different interface designs influence pilots' cognitive states during simulated flight tasks with varying secondary task demands. The findings demonstrate that the G1000 Glass panel is more effective in supporting situational awareness and reducing mental workload compared to the distributed analog gauges, as evidenced by its significantly better flight performance and increased alpha band activity in frontal EEG channels. The study concludes that sensor fusion-driven cognitive state assessment provides valuable insight into human–computer interaction research in aviation, influencing the design of future cockpit interfaces.

In the seventh contribution, the authors present a systematic review of hardware acceleration techniques for CNNs applied to EEG signal analysis in digital healthcare applications, including emotion classification, motor imagery, epilepsy detection, and sleep monitoring. Unlike previous reviews that focused on software-based solutions, this work addresses the challenge of deploying CNN-based EEG analysis on resource-constrained wearable devices, where small form factor, low power consumption, real-time processing, and data security are critical. The review classifies the EEG feature extraction methods into five categories, including frequency-based, time-based, time–frequency, spatial features, and raw signals, and analyzes their suitability for hardware implementation on FPGAs, ASICs, and GPUs. The authors also examine efficient CNN design techniques adapted for EEG signals, including pruning, quantization, tensor decomposition, knowledge distillation, and neural architecture searches. Through synthesizing current research and identifying critical gaps, such as the need for standardized metrics, multimodal datasets, and embedded artifact preprocessing, this review offers a foundational roadmap for the development of hardware-accelerated EEG systems for real-world healthcare settings.

In the eighth contribution, the authors present QAMT, a novel LLM framework for generating quality-assured medical time-series data to address the challenges of limited data volume, poor quality, and privacy concerns in digital healthcare. The study integrates heterogeneous data from hospital sensors, including continuous vital signs and laboratory measurements, with static patient information. Through the construction of a health knowledge graph that combines domain expertise from open medical databases with instance-level information from MIMIC-III and eICU, QAMT uses a modular architecture combining a GAN-based module for generating continuous temporal data and an LLM-based retrieval-augmented generation module for discrete static events. A quality-assurance module uses the knowledge graph through chain-of-thought reasoning to enforce clinical constraints and validate variable dependencies. The authors demonstrate that this approach produces higher-fidelity synthetic data that outperform existing GAN- and LLM-based methods in fidelity, utility, and privacy metrics while preserving interpretability throughout the generation pipeline.

In the ninth contribution, the authors present a sensor fusion-based biomechanical study investigating the effects of lower-limb muscle fatigue on dynamic postural control during a descent down stairs. The study integrates kinetic data from four stair-embedded force plates with kinematic data from a ten-camera 3D motion capture system, allowing for the synchronized collection of ground reaction forces and joint angle trajectories. Through inducing multi-joint fatigue through a standardized sit-to-stand task with heel raises, the authors use one-dimensional statistical parametric mapping to analyze continuous waveform changes in joint kinematics and kinetics. The sensor fusion approach reveals that fatigue significantly impairs dynamic stability, as evidenced by the increased margin of stability and required coefficient of friction values in critical gait events. Compensatory adaptations include increased knee flexion angles, increased step width, reduced swing phase proportion, and enhanced ankle plantar flexion moments, suggesting neuromuscular redistribution in which the distal joints compensate for proximal muscle fatigue. The authors conclude that this integrated approach provides quantitative biomechanical evidence for the assessment of fall risk and can inform joint-specific interventions in rehabilitation and prevention programs.

In the tenth contribution, the authors present a sensor fusion-based approach to modernize the Fukuda Stepping Test for objective postural control assessment in older adults living in the community. The study uses a Microsoft Kinect v2 depth sensor to capture three-dimensional skeletal kinematics during stepping with the eyes closed, combining temporal, linear displacement and segmental angular parameters with established clinical measures of balance, mobility, cognitive function, physical activity and quality of life. Through replacing subjective visual estimation with automated kinematic tracking, the authors demonstrate that depth sensor-derived parameters, particularly trunk flexion and upper trunk rotation, exhibit clinically meaningful associations with cognitive performance and functional mobility that substantially exceed traditional displacement-only metrics. The sensor fusion approach reveals that excessive trunk flexion may reflect impaired cognitive–motor integration, while increased head mediolateral displacement correlates with frailty and fear of falling. The authors conclude that integrating markerless

https://doi.org/10.3390/s26061870

motion capture technology into geriatric assessment protocols provides objective, multidimensional information on postural control mechanisms, enabling more precise fall-risk screening and individualized rehabilitation targeting in aging populations.

In the eleventh contribution, the authors present a validation study of an IMU-based method to estimate the kinematics of three-dimensional lower limb in asymptomatic individuals and patients with gait disorders, including cerebral palsy. The sensor fusion approach integrates data from seven IMUs positioned on the pelvis, thighs, shanks, and feet, synchronizing accelerometer and gyroscope signals with a reference optoelectronic motion capture system. Through combining inertial data through advanced orientation estimation and sensor-to-segment alignment algorithms, the authors demonstrate that IMU-derived kinematics achieve good to excellent correlation with the reference measures, with mean root mean square errors below 10° and centered errors below 3.2° after offset removal. Reliability analysis reveals moderate to excellent intra- and inter-operator agreement for sagittal- and frontal plane kinematics. A comparison of the gait profile scores of both methods shows moderate agreement, with no statistically significant differences in pathological populations. The authors conclude that this methodology provides a clinically viable and accessible alternative to traditional gait laboratories to characterize impairments and support therapeutic decision-making.

3. Conclusions

This Special Issue presents 10 original research articles and one comprehensive review that collectively showcase the transformative potential of sensor fusion in digital healthcare. Contributions span various domains, including neurodevelopmental disorders, assessment of mental workload, indoor localization, biomechanical analysis, and synthetic data generation, demonstrating how multimodal sensor integration combined with advanced AI enables personalized monitoring, clinical decision support, and improved patient outcomes. Several recurring themes appear throughout the collection: the significance of context-aware systems that adapt to individual characteristics, the importance of addressing data quality and imbalance through innovative methodologies, and the central role of hardware efficiency and interpretability in enabling practical, real-world deployment.

We hope these selected papers offer valuable insight into the methodological innovations and practical applications shaping the future of sensor-based digital healthcare, serving as both a reference for current best practices and an inspiration for future research in this rapidly evolving field.

We extend our sincere gratitude to all authors who contributed their outstanding research to this Special Issue, as well as to the dedicated reviewers whose rigorous and constructive feedback ensured the high quality of the published works. Their collective efforts have made this comprehensive overview of sensor fusion in digital healthcare possible.

Conflicts of Interest: The author declares no conflicts of interest.

List of Contributions

1. Deng, L.; Rattadilok, P. A Sensor and Machine Learning-Based Sensory Management Recommendation System for Children with Autism Spectrum Disorders. *Sensors* **2022**, *22*, 5803. https://doi.org/10.3390/s22155803.
2. Alexi, A.; Rosenfeld, A.; Lazebnik, T. The Trade-Off between Airborne Pandemic Control and Energy Consumption Using Air Ventilation Solutions. *Sensors* **2022**, *22*, 8594. https://doi.org/10.3390/s22228594.
3. González-Nóvoa, J.A.; Busto, L.; Campanioni, S.; Fariña, J.; Rodríguez-Andina, J.J.; Vila, D.; Veiga, C. Two-Step Approach for Occupancy Estimation in Intensive Care Units Based on Bayesian Optimization Techniques. *Sensors* **2023**, *23*, 1162. https://doi.org/10.3390/s23031162.

https://doi.org/10.3390/s26061870

4. Chen, Q.; Lee, B.G. Deep Learning Models for Stress Analysis in University Students: A Sudoku-Based Study. *Sensors* **2023**, *23*, 6099. https://doi.org/10.3390/s23136099.
5. Garcia, C.; Inoue, S. Relabeling for Indoor Localization Using Stationary Beacons in Nursing Care Facilities. *Sensors* **2024**, *24*, 319. https://doi.org/10.3390/s24020319.
6. Tang, H.; Lee, B.G.; Towey, D.; Pike, M. The Impact of Various Cockpit Display Interfaces on Novice Pilots' Mental Workload and Situational Awareness: A Comparative Study. *Sensors* **2024**, *24*, 2835. https://doi.org/10.3390/s24092835.
7. Xie, Y.; Oniga, S. A Comprehensive Review of Hardware Acceleration Techniques and Convolutional Neural Networks for EEG Signals. *Sensors* **2024**, *24*, 5813. https://doi.org/10.3390/s24175813.
8. Luo, Y.; Zhang, Y.; Xing, C.; Ren, P.; Liu, X. QAMT: An LLM-Based Framework for Quality-Assured Medical Time-Series Data Generation. *Sensors* **2025**, *25*, 5482. https://doi.org/10.3390/s25175482.
9. Wang, L.; Ma, W.; Zhu, W.; Xie, Q.; Sun, Y. Impact of Lower-Limb Muscle Fatigue on Dynamic Postural Control During Stair Descent: A Study Using Stair-Embedded Force Plates. *Sensors* **2025**, *25*, 5570. https://doi.org/10.3390/s25175570.
10. Ünal, H.T.; Koçak, M.; Çetin, S.Y.; Kara, Ö.K.; Doğan, M. Depth Sensor-Based Instrumentation of the Fukuda Stepping Test: Reliability and Clinical Associations in Older Adults. *Sensors* **2026**, *26*, 1623. https://doi.org/10.3390/s26051623.
11. Carcreff, L.; Payen, G.; Grouvel, G.; Cardoso-Fonseca, M.; Massé, F.; Armand, S. Validity, Reliability and Interpretability of an IMU-Based System to Measure 3D Lower Limb Kinematics of Patients with Heterogeneous Gait Disorders. *Sensors* **2026**, *26*, 1746. https://doi.org/10.3390/s26061746.

References

1. Gutruf, P. Towards A Digitally Connected Body for Holistic and Continuous Health Insight. *Commun. Mater.* **2024**, *5*, 2. [CrossRef]
2. Clarke, L.; Liau, C.; Wei, V.W.; Liao, C. Healthcare Monitoring with Earables: A Paradigm Shift Towards Digital Health. *Biomed. Anal.* **2025**, *2*, 207–222. [CrossRef]
3. Mahato, K.; Saha, T.; Ding, S.; Sandhu, S.S.; Chang, A.; Wang, J. Hybrid Multimodal Wearable Sensors for Comprehensive Health Monitoring. *Nat. Electron.* **2024**, *7*, 735–750. [CrossRef]
4. Khoo, L.S.; Lim, M.K.; Chong, C.Y.; McNaney, R. Machine Learning for Multimodal Mental Health Detection: A Systematic Review of Passive Sensing Approaches. *Sensors* **2024**, *24*, 348. [CrossRef] [PubMed]
5. Acosta, J.N.; Falcone, G.J.; Rajpurkar, P.; Topol, E.J. Multimodal Biomedical AI. *Nat. Med.* **2022**, *28*, 1773–1784. [CrossRef] [PubMed]
6. Ates, H.C.; Nguyen, P.Q.; Gonzalez-Macia, L.; Morales-Narváez, E.; Güder, F.; Collins, J.J.; Dincer, C. End-to-End Design of Wearable Sensors. *Nat. Rev. Mater.* **2022**, *7*, 887–907. [CrossRef] [PubMed]
7. Assaad, R.H.; Mohammadi, M.; Poudel, O. Developing an Intelligent IoT-Enabled Wearable Multimodal Biosensing Device and Cloud-Based Digital Dashboard for Real-Time and Comprehensive Health, Physiological, Emotional, and Cognitive Monitoring using Multi-Sensor Fusion Technologies. *Sens. Actuators A Phys.* **2025**, *381*, 116074. [CrossRef]
8. Soenksen, L.R.; Ma, Y.; Zeng, C.; Boussioux, L.; Carballo, K.V.; Na, L.; Wiberg, H.M.; Li, M.L.; Fuentes, I.; Bertsimas, D. Integrated Multimodal Artificial Intelligence Framework for Healthcare Applications. *npj Digit. Med.* **2022**, *5*, 149. [CrossRef] [PubMed]
9. Akre, S.; Seok, D.; Douglas, C.; Aguilera, A.; Carini, S.; Dunn, J.; Hotopf, M.; Mohr, D.C.; Bui, A.A.T.; Freimer, N.B. Advancing Digital Sensing in Mental Health Research. *npj Digit. Med.* **2024**, *7*, 362. [CrossRef] [PubMed]
10. Kargarandehkordi, A.; Li, S.; Lin, K.; Phillips, K.T.; Benzo, R.M.; Washington, P. Fusing Wearable Biosensors with Artificial Intelligence for Mental Health Monitoring: A Systematic Review. *Biosensors* **2025**, *15*, 202. [CrossRef] [PubMed]
11. Zhao, L.; Liang, C.; Huang, Y.; Zhou, G.; Xiao, Y.; Ji, N.; Zhang, Y.-T.; Zhao, N. Emerging Sensing and Modeling Technologies for Wearable and Cuffless Blood Pressure Monitoring. *npj Digit. Med.* **2023**, *6*, 93. [CrossRef] [PubMed]
12. Wall, C.; Hetherington, V.; Godfrey, A. Beyond the Clinic: The Rise of Wearables and Smartphones in Decentralising Healthcare. *npj Digit. Med.* **2023**, *6*, 219. [CrossRef] [PubMed]

https://doi.org/10.3390/s26061870

sensors

MDPI

Article

Validity, Reliability and Interpretability of an IMU-Based System to Measure 3D Lower Limb Kinematics of Patients with Heterogeneous Gait Disorders

Lena Carcreff [1,2,*], Gabriel Payen [3], Gautier Grouvel [1,4], Mickael Cardoso-Fonseca [1], Fabien Massé [3] and Stéphane Armand [1,4]

1 Kinesiology Laboratory, Geneva University Hospitals and Geneva University, 1205 Geneva, Switzerland; gautier.grouvel@unige.ch (G.G.); mickael.cfonseca@gmail.com (M.C.-F.); stephane.armand@unige.ch (S.A.)
2 Department of Physical Medicine and Rehabilitation, University Hospital of Angers, 49100 Angers, France
3 MindMaze, 1006 Lausanne, Switzerland; gabriel.payen.gp@gmail.com (G.P.); masse.fabien@gmail.com (F.M.)
4 Center of Research on Skeletal Muscle and Movement, Geneva University Hospitals and Geneva University, 1205 Geneva, Switzerland
* Correspondence: lena.carcreff@gmail.com

Abstract

Inertial measurement units (IMUs) represent a promising alternative to optoelectronic systems for estimating gait kinematics in less resource-intensive laboratories. However, evidence regarding the clinical interpretability of IMU-based gait analysis in pathological populations remains limited. This study aimed to evaluate the concurrent validity, reliability, and interpretability of an IMU-based method for computing 3D lower limb kinematics in asymptomatic and pathological populations. Fifty-five participants, including asymptomatic individuals (AS, n = 15), patients with cerebral palsy (CP, n = 15), and individuals with various motor disorders (OMD, n = 25), were assessed using a 7-IMU system and an optoelectronic system. Validity was quantified using root mean square error (RMSE), centered RMSE, and Pearson correlation coefficients (CCs) across 11 commonly reported lower limb kinematic outcomes. Reliability was assessed using intraclass correlation coefficients (ICCs), and interpretability was examined by comparing Gait Profile Scores (GPS) derived from both systems. Mean RMSE values were 7.1° (AS), 9.8° (CP), and 9.3° (OMD), with centered RMSE values below 3.2°. The correlation between IMU- and optoelectronic-based kinematics was good to excellent (mean CC = 0.76). Reliability ranged from moderate to excellent, and GPS showed moderate agreement between systems (CC = 0.42). These findings support the clinical validity, reliability, and interpretability of IMU-based gait kinematics across heterogeneous gait disorders.

Keywords: inertial measurement units; clinical gait analysis; lower limb kinematics; concurrent validity; reliability; interpretability

1. Introduction

Tridimensional gait kinematics evaluation is crucial in some clinical situations and therapeutic decision processes. Gait kinematics is indeed an indicator of motor function and is highly valuable to assess the progression after treatment or rehabilitation over time. These quantified gait deviations are commonly assessed during a standardized examination called clinical gait analysis (CGA) [1,2] using an optoelectronic system, seen as the clinical reference and silver standard for kinematics computation. However, CGA requires expensive equipment, highly qualified personnel, and is time consuming which makes it inaccessible for some patients.

The widespread availability of inertial measurement units (IMUs) has permitted the emergence of methods to cheaply and simply estimate gait kinematics. IMU-based gait kinematics computation could indeed be possible outside of cutting edge laboratories and be performed by the clinicians themselves as the devices are simple to use and not restricted to a laboratory setting [3]. Even if the large majority of methods have been developed for asymptomatic populations, a growing interest in inertial sensors in clinical practice is observed [4,5]. Existing research has primarily focused on methodological aspects and the overall validity of the IMU-based kinematics computation [6]. Heterogeneous results were found among the studies given the diversity of methods for sensor configuration [7], sensor orientation estimation (sensor fusion process) [8] and sensor-to-segment alignment (pose estimation) [9,10]. In addition, various results were found according to the joint angle evaluated. According to Poitras et al. [11] in their review of 42 studies (including asymptomatic populations only), the root mean square errors (RMSEs) of IMUs against traditional motion capture ranged between 0.4° and 18.8° among the 12 commonly reported joint kinematic variables: pelvis, hip, knee, and ankle flexion/extension, abduction/adduction and internal/external rotation. Similarly, correlation coefficients between IMU-based and optoelectronic system-based kinematics ranged between 0.33 and 1.00 [11]. The reliability was found fair to excellent across all variables [11]. Overall, joint kinematics generally prove good to excellent validity and reliability in the sagittal and frontal planes, but this must be formally demonstrated for every new method proposed as no consensus exists on each methodological step.

Just like any other health related-patient reported outcomes (HR-PROs), biomechanical outcomes resulting from gait analysis must be reliable (free from measurement error) and valid (measures the construct(s) it purports to measure) [12]. As previously mentioned, reliability and validity aspects are commonly reported in studies aiming at proposing a new assessment tool. Notably, concurrent validity (extent of the agreement between two measures taken at the same time) is always reported. However, to be incorporated into clinical practice, gait analysis outcomes need to guarantee a high level of interpretability (degree to which one can assign qualitative meaning [12]). To the best of our knowledge, no study has assessed the interpretability of IMU-based kinematic outputs.

Therefore, the objectives of this study were to assess the concurrent validity (A), reliability (B) and interpretability (C), according to COSMIN recommendations [12], of an IMU-based method for computing 3D lower limb kinematics in asymptomatic and pathological populations with heterogeneous gait disorders. We hypothesized that the proposed method would demonstrate acceptable agreement with optoelectronic reference measures, moderate-to-excellent reliability, and clinically meaningful interpretability across populations.

The remainder of this manuscript is organized as follows: Section 2 describes participant characteristics, instrumentation, data collection, data processing, outcome selection and statistical analyses; Section 3 reports the validity, reliability, and interpretability outcomes; and Section 4 interprets findings, addresses limitations, and outlines implications for clinical practice.

2. Materials and Methods

2.1. Design of Experiment

The present study is a prospective study of concurrent validity, reliability, and interpretability The participants were all assessed once for validity and interpretability purposes. The participants who agreed to come back for a second measurement session one week

https://doi.org/10.3390/s26061746

later were included in the reliability part of the study. The protocol was approved by and carried out in accordance with the hospital's institutional ethical committee (Cantonal Commission for Research Ethics of Geneva—CCER-2020-00358).

2.2. Participants

Three groups of participants were included in this study (from the 'Kinemagics' project): an asymptomatic group (AS), a cerebral palsy (CP) group, and a group of patients with various other motor disorders (OMD) (i.e., stroke survivors, patients with myopathy, with idiopathic toe walking, club foot, etc.). In each group, three age categories were targeted: children (5–13 years), teenagers (14–17 years) and adults (>18 years). For the two pathological populations (CP and OMD), a representative sample of the patients generally coming to the laboratory for CGA was included. The general and group-specific inclusion and exclusion criteria are detailed in the Supplementary Materials (Table S1).

Sample size justification:

A minimum of 12 participants in the AS group, 10 participants in the CP group and 24 in the OMD group were determined based on the concurrent validity objective for this study. The sample sizes were calculated using the means and standard deviations of the RMSE found in the literature assessing an IMU system against an optoelectronic system. The power was set to 0.95, and α to 0.05. The calculation was based on the following nine outcomes that were reported by previous studies: the hip, knee and ankle angles in flexion/extension, abduction/adduction and internal/external rotations. Two reference studies [13,14] were used to compute the sample size for the AS group (maximal sample size computed = 12 among all outcomes and studies), and a reference study [15] was used to compute the sample size of the CP group (maximal sample size computed = 10 among all outcomes). Since no reference values were found for other pathological populations, we decided to double the sample size of the AS group for the OMD group.

2.3. Equipment

The equipment was installed by one operator. Participants were simultaneously measured by a 7-IMU (Physilog5, GaitUp, Renens, Switzerland) inertial system at 256 Hz and a twelve-camera (Oqus7+) optoelectronic system (Qualisys, Göteborg, Sweden) at 100 Hz. The IMUs were positioned on the lower back (at sacrum level, in the middle and below the posterior superior iliac spines), the thighs (on the frontal side), the shanks (on the distal lateral side) and the feet (on the top) (Figure 1). Tridimensional acceleration and angular velocity were acquired with ranges of $\pm$16 g and $\pm$2000°/s respectively. Reflective markers were placed according to the Conventional Gait Model 2.4 (CGM 2.4) [16] on the lower limbs (Figure 1).

2.4. Protocol

At the beginning and the end of each measurement session, a synchronization ('sync') trial was performed. During this trial, the acceleration signal from one Physilog IMU and the trajectory of a reflective marker, both attached to a rigid wand, were recorded simultaneously while the wand was dropped onto the ground. The resulting impact generated a distinct acceleration peak in the IMU signal and a clear event in the optoelectronic trajectory data. These 'sync' trials were used to temporally align the inertial and optoelectronic systems as no hardware synchronization was available.

Each participant was asked to stand still in a neutral pose with the legs as vertical as possible and parallel feet, then to walk back and forth along the 10 m walkway at spontaneous speed like during a classical CGA session.

https://doi.org/10.3390/s26061746

For the first session, the protocol was led by one operator (A). For the second session, the protocol was led first by one operator (A) and then, after equipment removal and 15 min rest, the protocol was repeated and led by another operator (B).

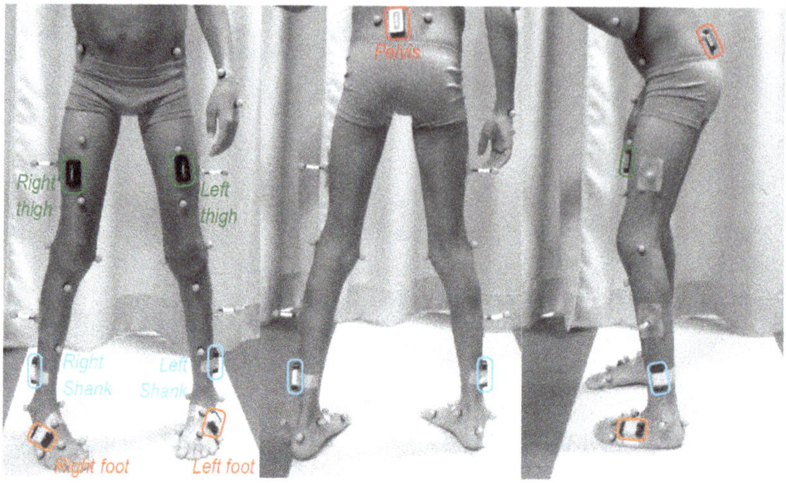

Figure 1. Equipment on the participants including 7 IMUs on the pelvis (red), thighs (green), shanks (blue) and feet (orange), and 34 reflective markers placed following the Conventional Gait Model 1.1.

2.5. Data Processing

Data processing and analysis were performed using Matlab R2018b software (Mathworks, Natick, MA, USA) and R v.4.1.1 and the RStudio interface (v.1.4.1717, Rstudio Team) following the workflow presented in Figure 2.

Synchronization between the optoelectronic and IMU systems was performed offline during post-processing using the 'sync' trials described above. The impact event generated during the wand drop was identified in both the IMU acceleration signal and the optoelectronic marker trajectory. The absolute timestamps of these events were then used to compute the temporal offset between systems. Following synchronization, inertial data were segmented into individual gait trials according to the optoelectronic trial timestamps, ensuring temporal consistency between both data streams for all subsequent kinematic analyses. The gaps in the marker trajectories were automatically filled using information of inter-correlated markers obtained from a principal component analysis [17]. Gait events (foot strikes and foot offs) were computed from the feet and pelvis marker trajectories as proposed by Zeni et al. [18] and checked manually.

Sensor orientation estimation, sensor-to-segment alignment and kinematics computation were performed as described in Carcreff et al. [19], with the exception of the sitting trial which, in the current study, was replaced by a lying trial for the same purpose (mediolateral axis correction). This lying trial was performed at the beginning of each measurement session, asking the participant to lie still on a medical table with legs extended and feet facing the ceiling.

The resulting kinematics data was cut into gait cycles, using the events detected on the raw inertial data as proposed by Mariani et al. [20] for the IMU data. The reference kinematic data was computed using the CGM 1.1 [16]. The marker set used during acquisition was compatible with the PyCGM2.x framework including CGM 2.4 (Figure 1); however, joint kinematics were computed using the CGM 1.1 model, chosen for its widespread clinical use and status as the reference standard. The additional markers were included to enhance trajectory reconstruction and ensure complete data for CGM 1.1 calculations.

https://doi.org/10.3390/s26061746

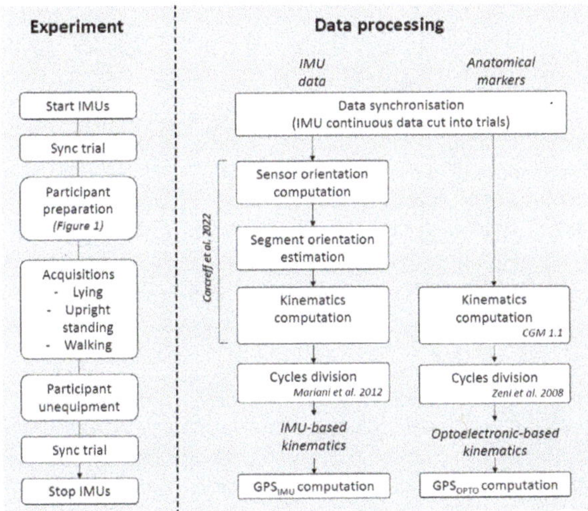

Figure 2. Flowchart of the experimental procedure and data processing main steps. IMUs: inertial measurement units, CGM: Conventional Gait Model; GPS: Gait Profile Score; OPTO: optoelectronic. References: Carcreff et al., 2022 [19], CGM1.1 [16], Mariani et al., 2012 [20], Zeni et al., 2008 [18].

2.6. Outcomes

The 11 most commonly reported gait kinematic outcomes [21] were computed: pelvis ante/retroversion, pelvis obliquity, pelvis in/external rotations, hip flex/extension, hip ab/adduction, hip in/external rotations, knee flex/extension, knee ab/adduction, knee in/external rotations, ankle dorsi/plantar flexion, and foot progression angle (Figure 3).

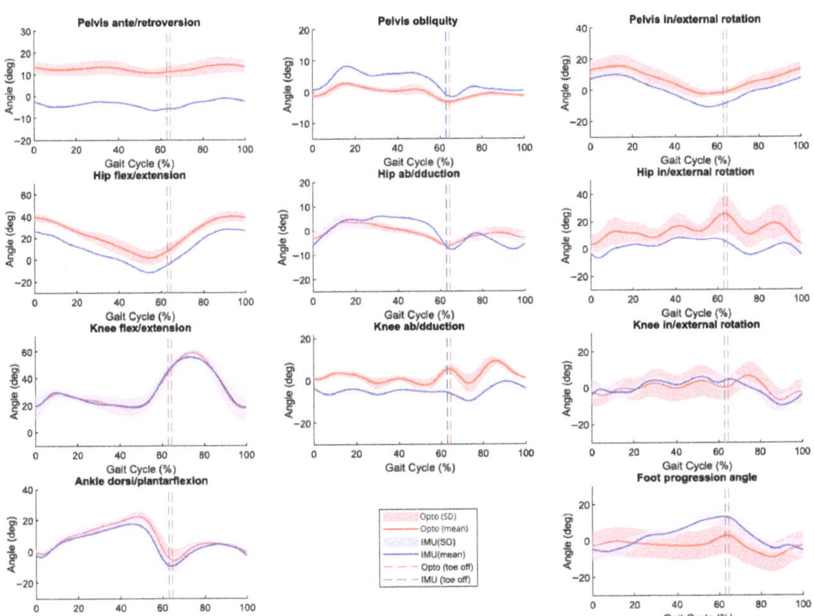

Figure 3. Mean (and standard deviation (SD)) kinematics of the right side of one study participant with cerebral palsy computed from the optoelectronic (red) and IMU (blue) systems across several gait trials.

https://doi.org/10.3390/s26061746

The Gait Profile Scores (GPS) were also computed in order to have an index of the gait pattern quality [22]. It consists of the direct root mean square distance between an individual's data and the mean normal data calculated across the gait cycle for nine joint angles (pelvis and hip angles in three planes, knee and ankle angles in the sagittal plane and foot progression) of one side [22]. The higher the GPS (°), the more affected the gait pattern gets, with a normal value approaching 5° [22]. The GPS computed from optoelectronic data (GPS_{OPTO}) and the GPS computed from the IMU data (GPS_{IMU}) were reported separately. For the GPS_{OPTO}, the mean normal data was the mean kinematics of the AS group computed with the optoelectronic system. For the GPS_{IMU}, the mean normal data was the mean kinematics of the AS group computed with the IMU system.

2.7. *Data Analysis*

2.7.1. A-Concurrent Validity

The concurrent validity was assessed for the first session only, through the RMSE, the RMSE centered at the mean (as proposed in Nuesch et al. [23]), the Pearson's correlation coefficient (CC) and the absolute difference in range of motion (ΔROM) between the optoelectronic-based and the IMU-based kinematics. These validity metrics were computed cycle by cycle and then averaged across sides and gait trials. The results were separated by groups. According to CGA standards, errors below 5° were considered as clinically acceptable, in line with the intrinsic accuracy limits of optoelectronic motion capture systems [24].

Furthermore, the link between the IMU system's errors and the level of disability was assessed through Pearson's CCs between the RMSEs and the mean of right and left GPS_{OPTO} for all the study participants. Altman's guidelines were used to interpret the correlations: poor, if $CC < 0.20$; fair, if $0.20 \leq CC < 0.40$; moderate, if $0.40 \leq CC < 0.60$; good, if $0.60 \leq CC < 0.80$; and very good, if $CC \geq 0.80$ [25].

CCs were also computed between the RMSE and the GPS_{OPTO} to investigate whether the errors were linked to the participant's level of gait disorder.

2.7.2. B-Reliability

The reliability was assessed through intraclass correlation coefficient (ICC). For its calculation, the variance of 'class' components [26], associated with the session and the operator, were computed from a linear two-way mixed effects model [27] with the lme4 package of R [28]. The intra-operator reliability was assessed between the first and the second sessions, one week apart, led by one operator (A). ICCs were computed for discrete parameters including the maximal, minimal and mean values of the kinematic curves which correspond to clinically relevant events [29]. The inter-operator reliability was assessed between the two operators (A and B) during the second session. In line with COSMIN recommendations, Koo and Li's guidelines were used to read ICC scores: poor reliability for $ICC < 0.50$, moderate reliability for $0.50 \leq ICC < 0.75$, good reliability for $0.75 \leq ICC < 0.90$ and excellent reliability for $ICC \geq 0.90$ [30].

2.7.3. C-Interpretability

The interpretability was assessed through the comparison between the GPS_{OPTO} and the GPS_{IMU} (averaged by sides) using a paired student t-test and Pearson correlation for normally distributed data, and a paired Wilcoxon test and Spearman correlation for non-normally distributed data. Comparisons between systems (OPTO-IMU) for each group were conducted using paired Wilcoxon tests with Bonferroni correction. Group comparisons were performed using an appropriate analysis of variance with pairwise comparisons.

Interpretation was based on previously reported minimal detectable changes (MDCs): 1.7° for the CP and OMD groups [31,32], and 0.7° for the AS group [33].

https://doi.org/10.3390/s26061746

3. Results

Fifty-five participants were included in total in this study: 15 in the AS group, 15 in the CP group and 25 in the OMD group. Among the OMD group, the main clinical profiles were club foot, idiopathic toe walking, drop foot, muscular dystrophia, torsional troubles, poliomyelitis, ataxia, and foot deformities. The distribution across age subgroups is presented in Table 1.

Table 1. Participants' characteristics.

Group		Asymptomatic (n = 15)			Cerebral Palsy (n = 15)			Other Motor Disorders (n = 25)		
Age Group		**Children (n = 4)**	**Teenagers (n = 5)**	**Adults (n = 6)**	**Children (n = 8)**	**Teenagers (n = 5)**	**Adults (n = 2)**	**Children (n = 13)**	**Teenagers (n = 3)**	**Adults (n = 9)**
Age	mean	9.0	16.0	32.4	10.1	15.0	18.5	9.2	15.3	42.2
(years)	SD	1.4	1.0	5.7	1.5	0.9	0.5	2.2	0.5	20.2
Height	mean	135.1	169.9	165.3	141.8	166.8	167.8	140.3	163.7	166.3
(cm)	SD	10.4	8.6	10.6	7.2	8.4	0.8	14.8	7.6	11.3
Weight	mean	30.4	58.7	65.6	36.1	56.1	68.3	39.0	55.5	71.1
(kg)	SD	7.1	13.8	17.4	5.7	16.2	10.3	14.7	4.8	12.7
BMI	mean	16.4	20.1	23.5	17.9	19.8	24.2	19.0	20.8	25.6
(kg/m^2)	SD	2.0	3.4	3.2	2.2	3.8	3.4	4.0	1.5	2.7
GMFCS	n				I—5 II—2 III—1	I—4 III—1	I—2			
GPS (°)	mean	5.0	4.0	4.4	8.1	8.0	7.4	6.6	8.6	6.6
	SD	0.4	0.8	0.5	2.4	1.7	1.8	0.6	2.6	1.2

BMI: Body Mass Index; GMFCS: Gross Motor Classification System; GPS: Gait Profile Score.

A table with all the computed kinematic outcomes is available in an open access repository (Zenodo) [34].

3.1. A-Concurrent Validity

The validity metrics per kinematic outputs and groups are depicted in Figure 4 and the corresponding numeric values are available in the Supplementary Materials (Table S2).

The mean RMSEs across all kinematic outcomes were 7.1° for the AS group, 9.8° for the CP group and 9.3° for the OMD group, with maximal errors found at the pelvis ante/retroversion level and minimal errors found at the pelvis rotation level for the three groups. These errors corresponded on average to about 0.8–0.9% (with a maximum of 5.5%) of the kinematic curve amplitude ('Relative RMSE' in Table S2).

The mean centered RMSEs, with offset removal, were 3.1° for the AS group, 3.1° for the CP group and 3.0° for the OMD group. The highest errors were then found at the foot progression angle and hip rotation levels.

Overall good correlations were found across all outcomes and groups (mean CC between 0.72 and 0.73). Very good correlations were found for the three groups in the sagittal plane (hip, knee and ankle), as well as in the pelvis rotation and hip ab/adduction (CC > 0.8). The poorest correlations were found for the hip rotation (CC: [0.32–0.45]), knee ab/adduction (CC: [0.46–0.59]) and rotation (CC: [0.43–0.52]), and foot progression angle (CC: [0.22–0.36]).

The means ΔROM across all outcomes were 3.5° for the AS group, 3.8° for the CP group and 3.6° for the OMD group.

Correlations between the RMSE and the GPS_{OPTO} (Table 2) ranged from poor (CC = 0.15 for knee ab/adduction) to moderate (CC = 0.51 for pelvis rotation), indicating a poor link between measurement error and the level of disability.

https://doi.org/10.3390/s26061746

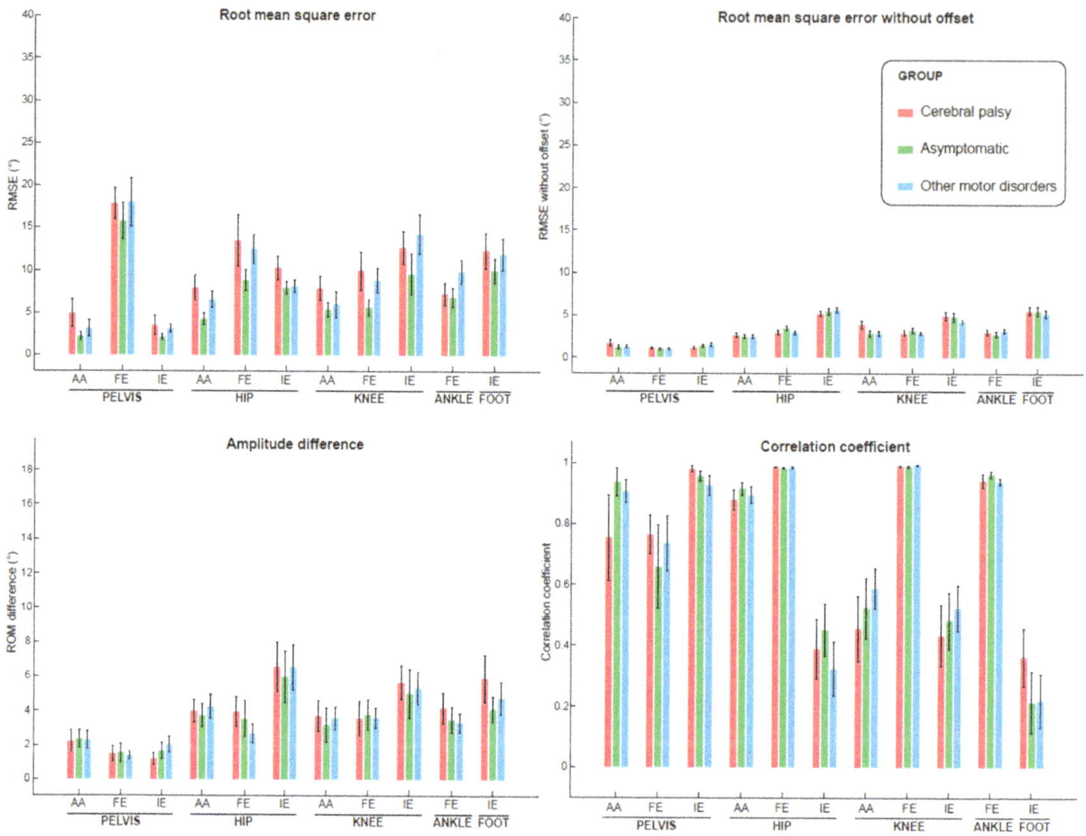

Figure 4. Metrics (root mean square error (RMSE), RMSE centered at the mean, correlation coefficient, and difference in range of motion (ROM)) of validity evaluation of the IMU method against the opto-electronic reference for each group. Histograms represent the mean values (and interval confidence bars) for all participants, for each joint/segment, and for each plane. AA: obliquity–ab/adduction; FE: ante/retroversion–flex/extension; IE: in/external rotation–progression angle.

Table 2. Correlations between the root mean square error (RMSE) and the Gait Profile Score (GPS).

Kinematic Outcomes	Pearson's Correlation Coefficient	*p*-Value
Pelvis ante/retroversion	0.26	0.053
Pelvis obliquity	0.17	0.204
Pelvis rotation	0.51	<0.001
Hip flexion/extension	0.38	<0.001
Hip ab/adduction	0.32	0.001
Hip rotation	0.21	0.025
Knee flexion/extension	0.46	<0.001
Knee ab/adduction	0.15	0.108
Knee rotation	0.23	0.014
Ankle flexion/extension	0.19	0.047
Foot progression angle	0.23	0.014

https://doi.org/10.3390/s26061746

3.2. B-Reliability

Thirty-one individuals participated in the second measurement session: fifteen from the AS group (same as the first session), eight from the CP group (five children, one teenager, and two adults), and eight from the OMD group (four children, two teenagers, and two adults). Due to technical issues with the sensors (e.g., synchronization between inertial sensors, incorrect handling, battery loss), only 27 participants (13 AS, 7 CP, 7 OMD) were included in the reliability analysis.

Table 3 reports the intra- and inter-operator ICCs of the IMU system computed for four discrete parameters (maximal, minimal values, range of motion (ROM), and the mean over gait cycles), as well as those of the reference system. A more complete table (with more parameters tested and with more statistical outcomes reported) is available in the Supplementary Materials (Table S3). From poor to good intra-operator reliability was found for the IMU-based kinematics in the sagittal and frontal planes (minimal ICC = 0.24 for the knee ab/adduction ROM; maximal ICC = 0.82 for the mean pelvis obliquity). Mainly poor intra-operator reliability was found in the transverse plane (ICCs ranging from 0.13 for the mean pelvis rotation to 0.53 for the foot progression angle ROM). From moderate to good inter-operator reliability was found in the sagittal and frontal planes (from 0.53 for the minimum hip flexion/extension to 0.85 for the maximum knee flexion/extension) and poor to moderate in the transverse plane (from 0.08 for the minimum pelvis rotation to 0.68 for the hip rotation ROM) (Table 3).

Intra-operator ICCs for kinematic variables obtained from the reference system across joints and planes ranged from 0.30 to 0.84, and inter-operator ICCs ranged from 0.46 to 0.85.

Table 3. Intra- and inter-operator intraclass correlation coefficients (ICCs) for both inertial (IMU) and reference optoelectronic (OPTO) systems for 4 discrete parameters (maximal, minimal values, range of motion (ROM) and the mean over gait cycles). Color gradient from excellent (green) to poor (red) reliability according to Koo & Li [30].

Plane	Segment/Joint	Parameter	Intra-Operator ICC		Inter-Operator ICC	
			IMU	OPTO	IMU	OPTO
Sagittal	Pelvis	Max	0.40	0.82	0.57	0.85
		Min	0.49	0.67	0.56	0.77
		ROM	0.76	0.78	0.84	0.80
		Mean	0.39	0.77	0.50	0.82
	Hip	Max	0.63	0.71	0.59	0.81
		Min	0.37	0.78	0.53	0.85
		ROM	0.71	0.77	0.68	0.79
		Mean	0.56	0.69	0.57	0.82
	Knee	Max	0.82	0.62	0.85	0.62
		Min	0.67	0.49	0.72	0.39
		ROM	0.60	0.75	0.74	0.76
		Mean	0.69	0.49	0.77	0.44
	Ankle	Max	0.51	0.73	0.61	0.75
		Min	0.50	0.65	0.66	0.76
		ROM	0.68	0.73	0.72	0.76
		Mean	0.45	0.71	0.57	0.75

https://doi.org/10.3390/s26061746

Table 3. *Cont.*

Plane	Segment/Joint	Parameter	Intra-Operator ICC		Inter-Operator ICC	
			IMU	OPTO	IMU	OPTO
Frontal	Pelvis	Max	0.81	0.68	0.77	0.65
		Min	0.72	0.63	0.70	0.65
		ROM	0.69	0.79	0.70	0.82
		Mean	0.82	0.49	0.80	0.51
	Hip	Max	0.63	0.70	0.69	0.74
		Min	0.69	0.55	0.70	0.72
		ROM	0.63	0.67	0.66	0.76
		Mean	0.69	0.63	0.75	0.69
	Knee	Max	0.58	0.28	0.73	0.60
		Min	0.70	0.48	0.71	0.58
		ROM	0.24	0.32	0.55	0.46
		Mean	0.68	0.50	0.71	0.78
Transverse	Pelvis	Max	0.35	0.68	0.39	0.68
		Min	0.40	0.46	0.08	0.49
		ROM	0.50	0.56	0.17	0.50
		Mean	0.13	0.59	0.06	0.60
	Hip	Max	0.22	0.35	0.44	0.64
		Min	0.49	0.30	0.70	0.59
		ROM	0.47	0.37	0.68	0.63
		Mean	0.36	0.31	0.54	0.55
	Knee	Max	0.51	0.49	0.60	0.58
		Min	0.39	0.50	0.42	0.66
		ROM	0.50	0.39	0.56	0.65
		Mean	0.42	0.48	0.58	0.58
	Foot Progression Angle	Max	0.52	0.81	0.60	0.84
		Min	0.29	0.73	0.28	0.83
		ROM	0.53	0.67	0.57	0.69
		Mean	0.29	0.82	0.37	0.88

3.3. C-Interpretability

The normality of GPS_{OPTO} and GPS_{IMU} distributions was visually assessed using histograms. Non-parametric tests were applied due to the uneven distribution of GPS_{OPTO} across the two pathological groups. No statistical difference was found between both scores for the pathological groups ($p = 0.84$), and a significant moderate correlation (CC = 0.42, p-value = 0.007) was found. Figure 3 and Table 4 detail the scores for the groups.

Table 4. Mean (and standard deviation) of the mean left and right Gait Profile Scores (GPS) (°) computed from inertial (IMU) and optoelectronic (OPTO) data for the 3 groups.

	GPS_{OPTO}	GPS_{IMU}	Systems Comparisons [c]
Asymptomatic (n = 15)	4.4 (0.7)	4.9 (0.7)	$p = 0.015$

https://doi.org/10.3390/s26061746

Table 4. *Cont.*

	GPS_{OPTO}	GPS_{IMU}	Systems Comparisons [c]
Cerebral palsy (n = 15)	8.0 (1.9)	7.7 (1.7)	$p = 0.720$
Other motor disorders (n = 25)	6.8 (1.4)	7.0 (1.5)	$p = 0.491$
Groups comparisons [a]	$p < 0.001$	$p < 0.001$	
Pairwise comparisons with asymptomatic group [b]	CP vs. TD: $p < 0.001$ OMD vs. TD: $p < 0.001$	CP vs. TD: $p < 0.001$ OMD vs. TD: $p < 0.001$	

[a]: Kruskall–Wallis ANOVA; [b]: Wilcoxon unpaired tests with Bonferroni adjustment; [c]: Wilcoxon paired tests.

4. Discussion

The aim of this study was to evaluate the concurrent validity (A), reliability (B) and interpretability (C) of a previously described IMU-based 3D lower limb kinematics computational method [19] on asymptomatic individuals and patients with heterogeneous gait disorders. Concurrent validity was generally found acceptable (RMSE < 10°, good correlations and ΔROM < 5°) for all groups of subjects in the three planes. However, important exceptions must warrant attention such as the pelvis ante/retroversion which resulted in a noteworthy offset (constant error throughout the gait cycle) of about 15°. The errors were not associated with the severity of the gait deviations. Intra- and inter-operator reliability were highly dependent on the chosen discrete parameter and the plane assessed (from poor to good reliability). Finally, the proposed IMU-based system could be used to compute gait quality scores such as the GPS since it was not found different from the one computed from the optoelectronic reference.

4.1. A-Concurrent Validity

The results obtained in the current study were similar to and even better than the results obtained in our previous paper discussing the concurrent validity of the IMU system against the optoelectronic reference for asymptomatic individuals only [19]. Indeed, for the AS group, we found a mean RMSE of 7.1° across all joints and planes, while 7.5° was previously found, a mean centered RMSE of 3° while 4° previously, a mean correlation coefficient of 0.73, while 0.63 previously, and a mean ΔROM of 3.5° while 6.8° previously. A possible explanation for these slight differences between both studies is the different locations of the IMUs on the body segments (e.g., frontal side of the thighs instead of lateral, lateral and distal side of the shank instead of medial and proximal). Just like reflective markers, IMUs suffer from soft tissue artifacts. A recent study has demonstrated for instance that an IMU placed on the frontal side of the thigh gives better stability than an IMU located on the lateral side of the thigh [7]. The better results found with the current sensor placement might confirm this statement.

Except for the pelvis ante/retroversion, the present results seemed comparable to existing studies. In the meta-analysis of Kobsar et al. [35], RMSEs ranged between 0.7° and 11.2° across 14 studies assessing the validity of 3D kinematic outcomes in healthy individuals. In pathological populations, there are few existing studies, and making comparisons is challenging due to disparities in sensor placement, reference systems, methodologies employed for kinematic calculations, and the diversity of reported outcomes. Prisco et al. recently reviewed the literature for the concurrent validity of wearables for gait analysis [36]. In this review, none of the studies which included pathological populations (10/32) studied kinematics.

The difference between the asymptomatic and the pathological (CP and OMD) individuals regarding the concurrent validity of the IMU system was only revealed by the RMSE values. A higher mean RMSE of about 2.5° was indeed found in the CP and OMD

https://doi.org/10.3390/s26061746

groups, but centered RMSE, correlation coefficients and ROM differences were very close to the AS group, meaning that the kinematic waveforms were well respected in the three groups. The main difference was thus provided by the kinematic offset, i.e., constant angle separating the IMU-based kinematic curve from the optoelectronic-based kinematic curve. This offset is a well-known issue while computing gait kinematics from IMUs [13,23,37–39] and is often related to the difference in biomechanical models, with different definitions of the anatomical framework [38,39]. While the optoelectronic system determines the segment frames thanks to the anatomical points located by cutaneous markers, the IMU system constructs the segment frames from functional axes determined by calibration postures or movements [38]. To address this issue, Ortigas Vàsquez et al. [40] used a robotic simulator driven by real-world data to evaluate the accuracy of their IMU system in computing 3D knee kinematics. The resulting errors of their tested IMU system were significantly smaller (RMSE $\leq 0.9°$, maximum absolute error $\leq 3.2°$). The kinematic offset can also be attributed to the lack of knowledge of the initial position of the IMU sensors. In general, participants perform predefined calibration postures and movements to enrich the knowledge of sensor positions related to the reference frame and/or the underlying body segments. In our method, these calibration postures and movements were kept to a minimum to be feasible for patients with motor disorders. This is the main reason why a substantial constant offset was observed at the pelvis level. No specific calibration posture or movement was required to align the pelvic-mounted sensor with the anatomical pelvis in the sagittal plane. As a result, the natural sacral inclination was implicitly considered as the neutral pelvic orientation, which differs from the definition used in the Conventional Gait Model. This methodological choice primarily affects absolute pelvic angle values, while preserving the overall waveform shape and temporal characteristics.

From a clinical perspective, this limitation may restrict the use of absolute pelvic ante/retroversion values for diagnostic interpretation or inter-subject comparison. However, the preserved waveform consistency suggests that IMU-based pelvic kinematics remain relevant for within-subject analyses, longitudinal follow-up, and the assessment of changes induced by therapeutic interventions.

4.2. B-Reliability

The intra- and inter-operator reliability provided by our IMU system were poor to good depending on the chosen discrete kinematic parameter and the plane. Overall, the reliability of the system was found better for the sagittal and frontal kinematics, which is in accordance with previous studies [11,35]. However, our ICC values appeared lower than previously published values. This may be explained by the inclusion of pathological subjects instead of only asymptomatic individuals [29,38,41]. Inter-operator reliability was found higher than intra-operator reliability. This could be explained by the intrinsic gait variability between two visits (one week apart) [42]. Finally, higher ICCs were mostly found for the reference optoelectronic system, especially at the hip and knee levels in the sagittal plane, and the pelvis and the foot level in the transverse plane. Intra-operator ICCs for kinematic variables obtained from the IMU system ranged from 0.13 to 0.82 across joints and planes, and ICCs obtained from the reference system ranged from 0.30 to 0.84. Inter-operator ICCs obtained from the IMU system ranged from 0.06 to 0.85, and those obtained from the reference ranged from 0.46 to 0.85. Lower reliability values were predominantly observed for transverse-plane kinematics, particularly at the pelvis and foot levels. From a clinical standpoint, this finding indicates that the use of an IMU system for transverse-plane kinematics may be more appropriate for descriptive or exploratory analyses rather than precise clinical decision-making.

https://doi.org/10.3390/s26061746

4.3. C-Interpretability

To be incorporated into clinical practice, gait analysis outcomes need to guarantee a high level of interpretability (the degree to which one can assign qualitative meaning [12]). We found that one of the most commonly used gait scores, the GPS, used to summarize the gait deviation level, was correctly estimated from the IMU system as compared to the optoelectronic system for both pathological groups. Surprisingly, this was not the case for the AS group. This discrepancy may be explained by the difference in heterogeneity between the pathological and the asymptomatic groups. Pathological participants show larger and more heterogeneous gait deviations, making the GPS less sensitive to small measurement differences, whereas asymptomatic participants have low and homogeneous GPS values (centered around 5°), where minor methodological biases can lead to statistically significant differences. These findings suggest that while IMU-based GPS estimates are appropriate for summarizing gait deviations in populations with motor disorders, their interpretability may be reduced in asymptomatic individuals.

Additionally, individuals identified as outliers based on optoelectronic GPS values (GPS_{OPTO}; Figure 5) were not identified as outliers when using the IMU-based method, suggesting an attenuation of large kinematic deviations relative to the reference system. This finding highlights a potential limitation of our IMU-based kinematic analysis in capturing extreme gait deviations. To the best of our knowledge, no study evaluated a similar interpretability metric. Further analysis is needed to explore the interpretability in greater depth, including a larger sample size.

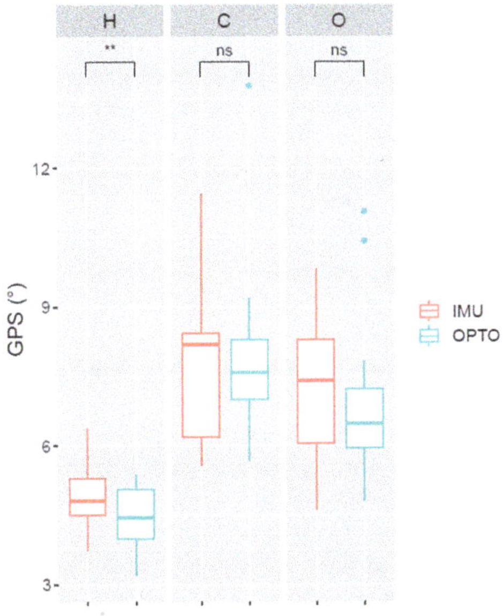

Figure 5. Mean left and right Gait Profile Scores (GPS) (°) computed from inertial (IMU) and optoelectronic (OPTO) data for the 3 groups (H: asymptomatic, C: cerebral palsy, O: other motor disorders); ns: non-significant difference; **: significant difference. The three outliers highlighted in Figure 5 based on optoelectronic data correspond to participants with severe gait impairments (bilateral CP with crouch gait, unilateral drop foot, and excessive external rotation of feet and hips). They were not identified as outliers in the IMU-based kinematic outcomes.

4.4. Limitations

The patients included in this study were not selected based on their walking capacities or motor disorders. They were included as they came for a CGA, provided they were willing

https://doi.org/10.3390/s26061746

to participate and met the inclusion criteria. This may explain the relatively low impairment degree in the two pathological subgroups, and may partly explain the equivalence in the method's performance for pathological groups as compared to the healthy group. Another limitation of this study was the use of different gait events (foot strikes and foot offs) for the two compared methods (IMU and OPTO). Errors in gait event detection could thus have led to errors in the kinematics; however, this choice was made to be the closest to the future conditions and processes (IMUs alone).

4.5. Perspectives

Our study highlighted the potential of wearable technology in gait analysis, offering promising applications for low-resource laboratories and clinics. However, several challenges have been identified and discussed in the preceding sections. Computing gait kinematics with IMUs remains complex, and integrating AI into algorithms should enhance the accuracy and reliability of the results, as demonstrated by several studies [43,44].

Three-dimensional kinematic computations, similar to those performed by optoelectronic systems in laboratories, may not be the optimal application for IMU-based systems. Instead, their greatest value lies in their usability outside laboratory settings, particularly for long-term and real-world monitoring [45]. Future efforts should focus on reducing the number of sensors required without compromising accuracy, making these systems less obtrusive and more practical.

Finally, given the challenges of 3D kinematics computation with IMUs, markerless motion capture methods, bolstered by advancements in AI, could surpass wearable sensors in specific applications [46]. In fact, rather than viewing these technologies as competitors, they should be seen as complementary. Optoelectronic systems excel in providing highly detailed and accurate biomechanical data in controlled environments. IMU-based systems are ideal for real-world, long-duration measurements, while markerless methods offer significant potential for efficient screening and assessment. The choice of technology should therefore be guided by the specific context and objectives of the analysis [46].

5. Conclusions

This study demonstrates the validity, reliability, and interpretability of IMU-based systems for estimating 3D lower limb kinematics in both asymptomatic individuals and populations with gait disorders. The findings highlight the potential of IMU-based methods as a viable alternative to optoelectronic systems in clinical and resource-limited settings. Despite the observed limitations in specific parameters, such as pelvis ante/retroversion, the good agreement between the kinematic waveforms, strong agreement in Gait Profile Scores and acceptable reliability across most metrics underscore the relevance of this approach. The acquired kinematic data can be used to objectively characterize gait impairments, support clinical decision-making, and monitor changes in gait patterns over time, particularly in contexts where conventional gait laboratories are not accessible. Such data may also facilitate large-scale or longitudinal studies by enabling repeated gait assessments in ecological environments. Future research should focus on optimizing sensor alignment and expanding applications to enhance the precision and versatility of IMU-based gait analysis.

Supplementary Materials: The following supporting information can be downloaded at https://www.mdpi.com/article/10.3390/s26061746/s1, Table S1: Inclusion and exclusion criteria; Table S2: IMU-based kinematics validity against the optoelectronic-based kinematics.; Table S3: Reliability results including intraclass correlation coefficients (ICCs), standard errors of measurement (SEMs) and minimal detectable changes (MDCs).

Author Contributions: Conceptualization, L.C. and S.A.; Methodology, L.C. and G.P.; Software, L.C. and G.P.; Validation, L.C., G.P., G.G. and S.A.; Formal Analysis, L.C. and G.P.; Investigation, L.C. and M.C.-F.; Resources, F.M. and S.A.; Data Curation, L.C. and G.P.; Writing—Original Draft Preparation, L.C.; Writing—Review and Editing, G.P., G.G., M.C.-F., F.M. and S.A.; Visualization, L.C.; Supervision, F.M. and S.A.; Project Administration, S.A.; Funding Acquisition, L.C. and S.A. All authors have read and agreed to the published version of the manuscript.

Funding: This research was funded by INNOSUISSE—Swiss Innovation Agency (Bern, Switzerland), grant number 38215.1 IP-LS—'KINEMAGICS: 3D kinematics for clinical gait analysis using wearable sensors'.

Institutional Review Board Statement: The study was conducted in accordance with the Declaration of Helsinki, and approved by the Ethics Committee of Geneva University Hospital (Cantonal Commission for Research Ethics of Geneva—CCER-2020-00358, 17 April 2020).

Informed Consent Statement: Informed consent was obtained from all subjects involved in the study.

Data Availability Statement: The data presented in the study are openly available in Zenodo at https://doi.org/10.5281/zenodo.18068953.

Acknowledgments: We wish to thank the participants and their families. We thank Xavier Gasparutto for his help with statistical analysis. We thank Cléo Moulin, Cyntia Duc and Benoit Mariani for their help in the protocol setting and the algorithm implementation.

Conflicts of Interest: F. Massé and G. Payen were employed by Mindmaze SA when the study was conducted. The remaining authors declare that the research was conducted in the absence of any commercial or financial relationships that could be construed as a potential conflict of interest.

Abbreviations

The following abbreviations are used in this manuscript:

AS	Asymptomatic
CC	Correlation coefficient
CGA	Clinical gait analysis
CGM	Conventional Gait Model
CP	Cerebral palsy
GMFCS	Gross motor function classification system
GPS	Gait Profile Score
HR-PRO	Health related-patient reported outcomes
ICC	Intraclass correlation coefficient
IMU	Inertial measurement unit
OMD	Other motor disorders
OPTO	Optoelectronic
RMSE	Root mean square error
ROM	Range of motion

References

1. Kirtley, C. *Clinical Gait Analysis—Theory and Practice*; Churchill Livingstone-Elsevier, Ed.; Elsevier: Philadelphia, PA, USA, 2006.
2. Baker, R.; Esquenazi, A.; Benedetti, M.G.; Desloovere, K. Gait analysis: Clinical facts. *Eur. J. Phys. Rehabil. Med.* **2016**, *52*, 560–574.
3. Leineweber, M.J.; Orozco, M.D.G.; Andrysek, J.; Gomez Orozco, M.D.; Andrysek, J. Evaluating the feasibility of two post-hoc correction techniques for mitigating posture-induced measurement errors associated with wearable motion capture. *Med. Eng. Phys.* **2019**, *71*, 38–44. [CrossRef]
4. Weygers, I.; Kok, M.; Konings, M.; Hallez, H.; De Vroey, H.; Claeys, K. Inertial sensor-based lower limb joint kinematics: A methodological systematic review. *Sensors* **2020**, *20*, 673. [CrossRef]
5. Kataoka, Y.; Ishida, T.; Osuka, S.; Takeda, R.; Tadano, S.; Yamada, S.; Tohyama, H. Validity of Wearable Gait Analysis System for Measuring Lower-Limb Kinematics during Timed Up and Go Test. *Sensors* **2024**, *24*, 6296. [CrossRef]

https://doi.org/10.3390/s26061746

6. Hindle, B.R.; Keogh, J.W.L.; Lorimer, A.V. Inertial-based human motion capture: A technical summary of current processing methodologies for spatiotemporal and kinematic measures. *Appl. Bionics Biomech.* **2021**, *2021*, 6628320. [CrossRef] [PubMed]
7. Scalera, G.M.; Ferrarin, M.; Marzegan, A.; Rabuffetti, M. Assessment of Stability of MIMU Probes to Skin-Marker-Based Anatomical Reference Frames During Locomotion Tasks: Effect of Different Locations on the Lower Limb. *Front. Bioeng. Biotechnol.* **2021**, *9*, 721900. [CrossRef] [PubMed]
8. Nazarahari, M.; Rouhani, H. 40 Years of Sensor Fusion for Orientation Tracking via Magnetic and Inertial Measurement Units: Methods, Lessons Learned, and Future Challenges. *Inf. Fusion* **2020**, *68*, 67–84. [CrossRef]
9. Vitali, R.V.; Perkins, N.C. Determining anatomical frames via inertial motion capture: A survey of methods. *J. Biomech.* **2020**, *106*, 109832. [CrossRef]
10. Pacher, L.; Chatellier, C.; Vauzelle, R.; Fradet, L. Sensor-to-Segment Calibration Methodologies for Lower—Body Kinematic Analysis with Inertial Sensors: A Systematic Review. *Sensors* **2020**, *20*, 3322. [CrossRef]
11. Poitras, I.; Dupuis, F.; Bielmann, M.; Campeau-Lecours, A.; Mercier, C.; Bouyer, L.J.; Roy, J.S. Validity and reliability ofwearable sensors for joint angle estimation: A systematic review. *Sensors* **2019**, *19*, 1555. [CrossRef]
12. Mokkink, L.B.; Terwee, C.B.; Patrick, D.L.; Alonso, J.; Stratford, P.W.; Knol, D.L.; Bouter, L.M.; de Vet, H.C.W. The COSMIN study reached international consensus on taxonomy, terminology, and definitions of measurement properties for health-related patient-reported outcomes. *J. Clin. Epidemiol.* **2010**, *63*, 737–745. [CrossRef] [PubMed]
13. Cloete, T.; Scheffer, C. Benchmarking of a full-body inertial motion capture system for clinical gait analysis. In Proceedings of the 2008 30th Annual International Conference of the IEEE Engineering in Medicine and Biology Society, Vancouver, BC, Canada, 20–25 August 2008; pp. 4579–4582. [CrossRef]
14. Zhang, J.T.; Novak, A.C.; Brouwer, B.; Li, Q. Concurrent validation of Xsens MVN measurement of lower limb joint angular kinematics. *Physiol. Meas.* **2013**, *34*, N63–N69. [CrossRef]
15. Van Den Noort, J.C.; Ferrari, A.; Cutti, A.G.; Becher, J.G.; Harlaar, J. Gait analysis in children with cerebral palsy via inertial and magnetic sensors. *Med. Biol. Eng. Comput.* **2013**, *51*, 377–386. [CrossRef] [PubMed]
16. Leboeuf, F.; Baker, R.; Barré, A.; Reay, J.; Jones, R.; Sangeux, M. The conventional gait model, an open-source implementation that reproduces the past but prepares for the future. *Gait Posture* **2019**, *69*, 235–241. [CrossRef]
17. Gløersen, Ø.; Federolf, P. Predicting missing marker trajectories in human motion data using marker intercorrelations. *PLoS ONE* **2016**, *11*, e0152616. [CrossRef]
18. Zeni, J.A.; Richards, J.G.; Higginson, J.S. Two simple methods for determining gait events during treadmill and overground walking using kinematic data. *Gait Posture* **2008**, *27*, 710–714. [CrossRef] [PubMed]
19. Carcreff, L.; Payen, G.; Grouvel, G.; Massé, F.; Armand, S. Three-dimensional lower limb kinematics from accelerometers and gyroscopes with simple and minimal functional calibration tasks: Validation on asymptomatic participants. *Sensors* **2022**, *22*, 5657. [CrossRef]
20. Mariani, B.; Rochat, S.; Christophe, J.B.; Aminian, K. Heel and Toe Clearance Estimation for Gait Analysis Using Wireless Inertial Sensors. *IEEE Trans. Biomed. Eng.* **2012**, *59*, 3162–3168. [CrossRef]
21. Baker, R. *Measuring Walking: A Handbook of Clinical Gait Analysis*; Mac Keith Press: London, UK, 2013.
22. Baker, R.; McGinley, J.L.; Schwartz, M.H.; Beynon, S.; Rozumalski, A.; Graham, H.K.; Tirosh, O. The gait profile score and movement analysis profile. *Gait Posture* **2009**, *30*, 265–269. [CrossRef]
23. Nuesch, C.; Roos, E.; Pagenstert, G.; Mundermann, A.; Nüesch, C.; Roos, E.; Pagenstert, G.; Mündermann, A. Measuring joint kinematics of treadmill walking and running: Comparison between an inertial sensor based system and a camera-based system. *J. Biomech.* **2017**, *57*, 32–38. [CrossRef] [PubMed]
24. McGinley, J.L.; Baker, R.; Wolfe, R.; Morris, M.E. The reliability of three-dimensional kinematic gait measurements: A systematic review. *Gait Posture* **2009**, *29*, 360–369. [CrossRef] [PubMed]
25. Altman, D.G.; Altman, E. *Practical Statistics for Medical Research*; Chapman: Boca Raton, FL, USA, 1999.
26. Chia, K.; Sangeux, M. Quantifying sources of variability in gait analysis. *Gait Posture* **2017**, *56*, 68–75. [CrossRef] [PubMed]
27. McGraw, K.O.; Wong, S.P. Forming Inferences about Some Intraclass Correlation Coefficients. *Psychol. Methods* **1996**, *1*, 30–46. [CrossRef]
28. Bates, D.; Mächler, M.; Bolker, B.M.; Walker, S.C. Fitting linear mixed-effects models using lme4. *J. Stat. Softw.* **2015**, *67*, 1–48. [CrossRef]
29. Al-Amri, M.; Nicholas, K.; Button, K.; Sparkes, V.; Sheeran, L.; Davies, J.L. Inertial Measurement Units for Clinical Movement Analysis: Reliability and Concurrent Validity. *Sensors* **2018**, *18*, 719. [CrossRef]
30. Koo, T.K.; Li, M.Y. A Guideline of Selecting and Reporting Intraclass Correlation Coefficients for Reliability Research. *J. Chiropr. Med.* **2016**, *15*, 155–163. [CrossRef]
31. Rasmussen, H.M.; Nielsen, D.B.; Pedersen, N.W.; Overgaard, S.; Holsgaard-Larsen, A. Gait Deviation Index, Gait Profile Score and Gait Variable Score in children with spastic cerebral palsy: Intra-rater reliability and agreement across two repeated sessions. *Gait Posture* **2015**, *42*, 133–137. [CrossRef]

https://doi.org/10.3390/s26061746

32. Devetak, G.F.; Martello, S.K.; de Almeida, J.C.; Correa, K.P.; Iucksch, D.D.; Manffra, E.F. Reliability and minimum detectable change of the gait profile score for post-stroke patients. *Gait Posture* **2016**, *49*, 382–387. [CrossRef]
33. Santos Bueno, G.A.; Ribeiro, D.M.; Gervásio, F.M.; Correia, A.M.; de Menezes, R.L. Gait profile score identifies changes in gait kinematics in nonfaller, faller and recurrent faller older adults women. *Gait Posture* **2019**, *72*, 76–81. [CrossRef]
34. Carcreff, L. KINEMAGICS Dataset: 3D Gait Kinematics of Asymptomatic Individuals and Patients with Heterogenous Gait Disorders from IMU and Optoelectronic Systems. 2025. Available online: https://zenodo.org/records/18068953 (accessed on 3 March 2026). [CrossRef]
35. Kobsar, D.; Charlton, J.; Tse, C.; Esculier, J.F.; Graffos, A.; Krowchuk, N.M.; Thatcher, D.; Hunt, M.A. Validity and reliability of wearable inertial sensors in healthy adult walking: A systematic review and meta-analysis. *J. Neuroeng. Rehabil.* **2020**, *17*, 62.
36. Prisco, G.; Pirozzi, M.A.; Santone, A.; Esposito, F.; Cesarelli, M.; Amato, F.; Donisi, L. Validity of Wearable Inertial Sensors for Gait Analysis: A Systematic Review. *Diagnostics* **2025**, *15*, 36. [CrossRef]
37. Berner, K.; Cockcroft, J.; Morris, L.D.; Louw, Q. Concurrent validity and within-session reliability of gait kinematics measured using an inertial motion capture system with repeated calibration. *J. Bodyw. Mov. Ther.* **2020**, *24*, 251–260. [CrossRef] [PubMed]
38. Shuai, Z.; Dong, A.; Liu, H.; Cui, Y. Reliability and Validity of an Inertial Measurement System to Quantify Lower Extremity Joint Angle in Functional Movements. *Sensors* **2022**, *22*, 863. [CrossRef] [PubMed]
39. Buganè, F.; Benedetti, M.G.; D'Angeli, V.; Leardini, A. Estimation of pelvis kinematics in level walking based on a single inertial sensor positioned close to the sacrum: Validation on healthy subjects with stereophotogrammetric system. *Biomed. Eng. Online* **2014**, *13*, 146. [CrossRef] [PubMed]
40. Ortigas Vásquez, A.; Maas, A.; List, R.; Schütz, P.; Taylor, W.R.; Grupp, T.M. A Framework for Analytical Validation of Inertial-Sensor-Based Knee Kinematics Using a Six-Degrees-of-Freedom Joint Simulator. *Sensors* **2023**, *23*, 348. [CrossRef]
41. Teufl, W.; Miezal, M.; Taetz, B.; Fröhlich, M.; Bleser, G. Validity, test-retest reliability and long-term stability of magnetometer free inertial sensor based 3D joint kinematics. *Sensors* **2018**, *18*, 1980. [CrossRef]
42. Chau, T.; Young, S.; Redekop, S. Managing variability in the summary and comparison of gait data. *J. Neuroeng. Rehabil.* **2005**, *2*, 22. [CrossRef]
43. Hossain, S.B.; Dranetz, J.; Choi, H.; Guo, Z. DeepBBWAE-Net: A CNN-RNN Based Deep SuperLearner for Estimating Lower Extremity Sagittal Plane Joint Kinematics Using Shoe-Mounted IMU Sensors in Daily Living. *IEEE J. Biomed. Health Inform.* **2022**, *26*, 3906–3917. [CrossRef]
44. Sung, J.; Han, S.; Park, H.; Cho, H.; Hwang, S.; Park, J.W.; Youn, I. Prediction of Lower Extremity Multi-Joint Angles during Overground Walking by Using a Single IMU with a Low Frequency Based on an LSTM Recurrent Neural Network. *Sensors* **2022**, *22*, 53. [CrossRef]
45. Patel, G.; Mullerpatan, R.; Agarwal, B.; Shetty, T.; Ojha, R.; Shaikh-Mohammed, J.; Sujatha, S. Validation of wearable inertial sensor-based gait analysis system for measurement of spatiotemporal parameters and lower extremity joint kinematics in sagittal plane. *Proc. Inst. Mech. Eng. Part H J. Eng. Med.* **2022**, *236*, 686–696. [CrossRef]
46. Bonato, P.; Feipel, V.; Corniani, G.; Arin-Bal, G.; Leardini, A. Position paper on how technology for human motion analysis and relevant clinical applications have evolved over the past decades: Striking a balance between accuracy and convenience. *Gait Posture* **2024**, *113*, 191–203. [CrossRef] [PubMed]

https://doi.org/10.3390/s26061746

sensors

MDPI

Article

Depth Sensor-Based Instrumentation of the Fukuda Stepping Test: Reliability and Clinical Associations in Older Adults

Hasan Tolga Ünal [1], Mertcan Koçak [2], Sebahat Yaprak Çetin [1], Özgün Kaya Kara [1] and Mert Doğan [1,*]

[1] Department of Physiotherapy and Rehabilitation, Faculty of Health Sciences, Akdeniz University, Antalya 07070, Türkiye

[2] Department of Mechatronics Engineering, Faculty of Engineering and Architecture, İzmir Kâtip Çelebi University, İzmir 35620, Türkiye

* Correspondence: mertdogan@akdeniz.edu.tr

Highlights

What are the main findings?

- Depth sensor-based instrumentation of the Fukuda Stepping Test demonstrated moderate-to-good test–retest reliability for most segmental kinematic parameters in older adults.
- Trunk flexion and rotational kinematic parameters showed clinically meaningful associations with cognitive function, physical activity, balance performance, and quality of life.

What are the implications of the main findings?

- Markerless depth-sensing technology provides objective and clinically relevant information beyond conventional Fukuda Stepping Test outcomes.
- Segmental kinematic parameters, particularly trunk flexion, may serve as practical indicators for multidomain functional assessment and fall-risk screening in older adults.

Abstract

This study evaluated the test–retest reliability of a depth sensor-based Fukuda Stepping Test and examined associations between sensor-derived kinematic parameters and established clinical outcomes in older adults. Eighty-six community-dwelling older adults (mean age 70.3 ± 4.7 years) performed an eyes-closed stepping task monitored by a Microsoft Kinect v2 sensor. Clinical assessments included the Berg Balance Scale, Timed Up and Go test, Five Times Sit-to-Stand, Montreal Cognitive Assessment, International Physical Activity Questionnaire, and WHOQOL-OLD. Test–retest reliability was assessed using intraclass correlation coefficients in a randomly selected subgroup. Reliability estimates varied across parameters, with temporal and displacement-based measures demonstrating more consistent agreement across sessions, whereas selected angular variables showed greater variability. Correlation analyses identified statistically significant associations between trunk kinematic changes and clinical measures, with effect sizes generally ranging from weak to moderate magnitude. Upper trunk rotation was associated with functional mobility measures, while traditional displacement-based metrics demonstrated limited clinical relationships. These findings support the feasibility of markerless depth-sensing technology for objective quantification of movement during the Fukuda Stepping Test and highlight the potential contribution of segmental kinematic parameters to multidimensional functional assessment in older adults.

Keywords: Fukuda stepping test; depth sensor; markerless motion capture; postural control; balance assessment; older adults; reliability; kinematic analysis; Kinect v2

1. Introduction

Falls constitute a major public health concern among older adults, leading to increased morbidity, mortality, and healthcare costs. The annual prevalence of falls among community-dwelling older individuals has been reported to be 26.5%, and more than 14 million adults aged 65 years and older experience a fall each year in the United States alone [1,2]. Approximately 10–20% of falls result in fractures, and falls are estimated to have contributed to 684,000 deaths globally among older adults in 2017 [2,3]. Although exercise-based interventions are effective in reducing fall incidence [2], early and objective identification of individuals at increased fall risk remains a key challenge in clinical practice.

Age-related deterioration in vestibular, proprioceptive, and visual systems adversely affects postural control and increases susceptibility to falls [4,5]. Visual impairment has been associated with a 16% higher fall rate, rising to nearly 40% in individuals with blindness, while vestibular dysfunction has been shown to exert moderate negative effects on postural control parameters [5,6]. These sensory declines, often coexisting in older age, compromise multisensory integration and contribute to balance instability. Accordingly, clinical assessments that challenge sensory integration—especially by limiting visual feedback—play a critical role in evaluating postural control in older adults. The Fukuda Stepping Test (FST) is a simple and widely used assessment in which individuals step in place with their eyes closed while trunk rotation and linear displacement are observed relative to the starting position [7,8]. The test has been applied in vestibular and balance assessments, including studies involving older adults [7,8]. However, conventional implementations of the FST rely largely on subjective visual estimation or manual measurements, lack standardized protocols, and provide limited kinematic information. Reported test–retest reliability is moderate, and substantial variability exists across protocols [7,8]. Moreover, traditional FST outcomes are typically restricted to rotation and displacement measures, without capturing detailed three-dimensional kinematic behavior of the head and trunk [8].

The integration of motion-sensing technologies into traditional clinical balance tests reflects a broader transition from observational scoring toward quantitative movement analysis.

Depth-sensing technologies, such as Microsoft Kinect and Azure Kinect, enable markerless, relatively low-cost motion capture and provide automated access to three-dimensional kinematic data. These systems have been shown to be feasible and reliable for evaluating functional tasks, stepping, and balance in older adults in both laboratory and home-based settings [9,10]. Markerless depth sensors reduce preparation time, facilitate automation, and allow quantification of anterior–posterior, mediolateral, and vertical displacements, as well as head and trunk angular movements [9–11]. Previous studies have demonstrated meaningful associations between depth sensor-derived kinematic parameters and conventional clinical balance and mobility measures [9,10], and recent systematic reviews have reported good-to-excellent reliability of markerless camera-based systems compared with marker-based motion capture in gait and balance assessments [11].

Despite these advances, direct studies modernizing the FST using depth-sensing technology remain limited. While the reliability and protocol characteristics of the conventional FST in older adults have been explored [7,8], depth sensor-based implementations aligned specifically with the classical FST framework are scarce. Conversely, although Kinect-based stepping and balance tasks provide rich kinematic data, they have generally not been designed as standardized adaptations of the FST [9–11]. Importantly, recent inertial

https://doi.org/10.3390/s26051623

measurement unit-based analyses of the Fukuda–Unterberger test have demonstrated the feasibility of objective sensor-based quantification, reporting moderate test–retest reliability for stepping-on-the-spot paradigms [12]. However, these approaches primarily focused on rotational dynamics and did not examine multidomain clinical associations. This gap highlights the need for a standardized, sensor-based version of the FST with clearly defined protocols and objective outcome measures. Therefore, the aim of this study was to evaluate the test–retest reliability of a depth sensor-based, modernized FST in older adults and to examine the associations between sensor-derived kinematic parameters and established clinical measures of balance, mobility, fear of falling, and quality of life. The present study contributes to the literature by (1) proposing a standardized depth sensor-based adaptation of the Fukuda Stepping Test, (2) evaluating its test–retest reliability using a structured statistical framework, and (3) examining its associations with multidomain clinical outcomes in older adults.

2. Materials and Methods

2.1. Study Design and Participants

This cross-sectional methodological study was approved by the Akdeniz University Non-Interventional Clinical Research Ethics Committee (30.01.2025, No: TBAEK-168). All participants provided written informed consent according to the Declaration of Helsinki. The study was conducted between January and November 2025 at the Department of Physiotherapy and Rehabilitation, Faculty of Health Sciences, Akdeniz University, Antalya, Türkiye.

A total of 86 community-dwelling older adults (52 females, 34 males) were recruited using snowball sampling. Inclusion criteria were age $\geq$65 years, independent ambulation, ability to stand and perform stepping in place, and Montreal Cognitive Assessment (MoCA) score $\geq$26 to exclude cognitive impairment. Exclusion criteria included self-reported physician-diagnosed neurodegenerative disorders (e.g., Parkinson's disease, atypical parkinsonism, cerebellar ataxia, or dementia), as these conditions are characterized by disease-specific impairments in axial control, gait initiation, and rotational stability that could directly influence trunk and head kinematics during the Fukuda Stepping Test. Other focal neurological conditions such as prior stroke, peripheral neuropathy, or multiple sclerosis were excluded due to their potential to produce asymmetric stepping patterns, altered proprioception, or impaired motor coordination. Clinically diagnosed vestibular disorders (e.g., benign paroxysmal positional vertigo, unilateral vestibular hypofunction, Ménière's disease) were excluded because the FST is specifically sensitive to vestibular asymmetry and spatial orientation deficits, which would confound interpretation of age-related balance mechanisms. Severe visual impairment not adequately corrected with lenses (e.g., advanced macular degeneration or glaucoma) was excluded, as visual system deficits may alter baseline postural alignment and sensory reweighting even prior to eye closure. Recent lower-extremity fractures, joint replacement surgery, or major orthopedic procedures within six months were excluded to avoid compensatory stepping strategies, reduced weight-bearing symmetry, or incomplete biomechanical recovery affecting rotational displacement outcomes. Additionally, individuals with uncontrolled cardiovascular or systemic conditions that could compromise safe performance of repeated eyes-closed stepping were not included. Eligibility was determined through structured health screening interviews, participant medical history self-report, and cognitive testing prior to data collection. The retest subgroup (n = 23) was randomly selected from the full cohort using a simple random sampling procedure (computer-generated random allocation), thereby reducing selection bias and supporting representativeness of the reliability sample. The research workflow is shown in Figure 1.

https://doi.org/10.3390/s26051623

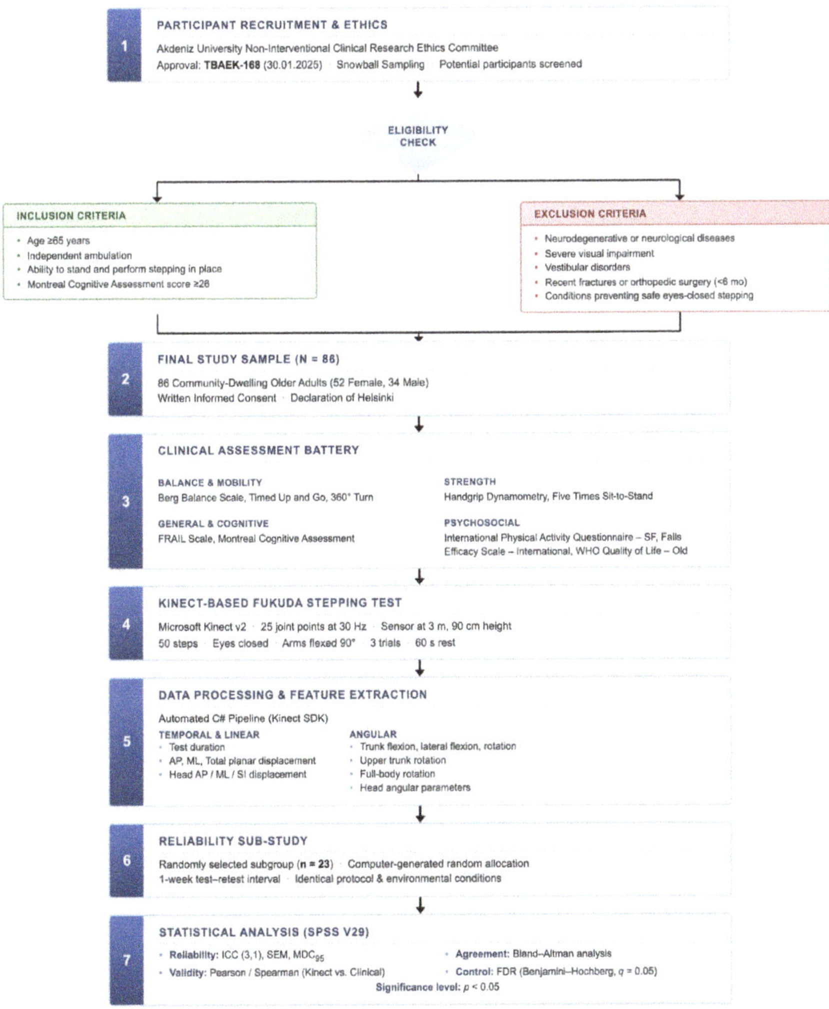

Figure 1. Research workflow showing participant selection, assessment procedures, testing protocol, data processing, and statistical analysis stages.

2.2. Depth Sensor-Based Fukuda Stepping Test

Three-dimensional kinematic data were collected using Microsoft Kinect v2 (Microsoft Corp., Redmond, WA, USA), a time-of-flight depth sensor tracking 25 anatomical joint points at 30 Hz. The sensor was positioned 3 m in front of participants at 90 cm height to ensure optimal skeletal tracking and minimize occlusion. Raw joint coordinate data were recorded in real time using custom-developed C# language utilizing the Kinect Software Development Kit version 2.0 (Microsoft Corp., Redmond, WA, USA), enabling synchronized recording, trial labeling, and structured data storage. Kinect provides acceptable validity and reliability for postural control and balance assessment [13,14].

2.2.1. Test Protocol

Participants stood in the center of the capture area with arms flexed to approximately 90° in front of the body. They were instructed to close their eyes and perform 50 consecutive steps in place at a self-selected comfortable pace without intentionally correcting body

https://doi.org/10.3390/s26051623

orientation or step direction. The selection of 50 steps was based on previous methodological evidence indicating that a 50-step protocol provides superior test–retest reliability compared with longer step counts while minimizing fatigue-related variability [8]. Verbal instructions were standardized, and no feedback was given during execution. Each participant completed three trials with 60 s rest intervals between trials; mean values were used for analysis. A physiotherapist supervised the test, which was terminated prematurely if balance was lost, eyes were opened, or external support was required (Figure 2).

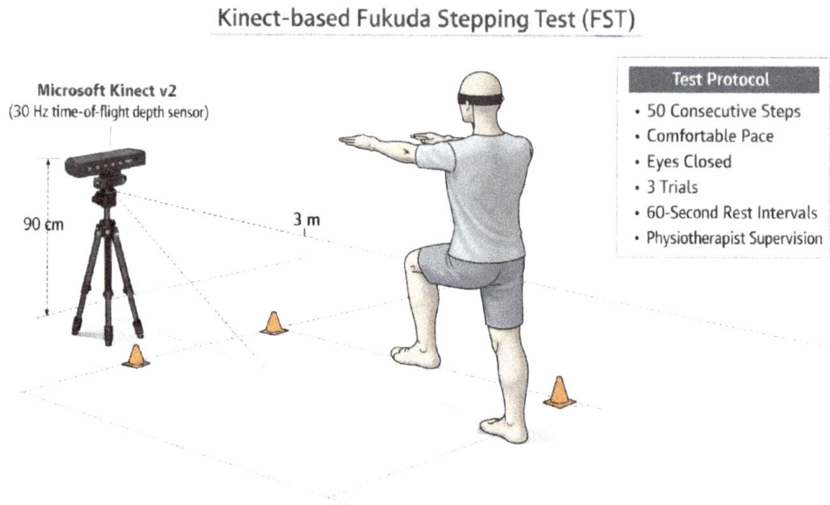

Figure 2. Kinect-based Fukuda Stepping Test.

For test–retest reliability assessment, the protocol was repeated one week later in a subgroup of 23 participants under identical conditions. Traditional visual estimation was replaced by continuous three-dimensional kinematic recording, enabling objective quantification of trunk and head movements throughout the test. The same processing pipeline was applied for initial assessment and retest session.

2.2.2. Data Extraction and Validation

Raw three-dimensional skeletal data were processed using custom-developed C# software that automatically extracted temporal, linear displacement, and angular kinematic parameters from joint coordinates defined within a global reference frame. The cut-off frequency was selected in accordance with the system's sampling rate of 30 Hz and the low-frequency characteristics of slow stepping and postural movements.

Skeletal tracking outputs were visually verified against synchronized video recordings to identify potential tracking artifacts, including transient deviations related to occlusion. Derived kinematic parameters were additionally examined against physiologically plausible anatomical range-of-motion limits to detect potential outliers. Verification procedures were limited to artifact identification and, where necessary, exclusion of affected segments; no manual editing of coordinate trajectories was performed.

Extracted parameters were categorized as follows:

Temporal parameter: Test duration defined as the time elapsed between the first and last detected step.

Linear displacement parameters: Trunk anterior–posterior and medio-lateral displacement relative to the initial position. Total planar displacement calculated as horizontal

https://doi.org/10.3390/s26051623

Euclidean distance. Head displacement in anterior–posterior, medio-lateral, and superior–inferior directions.

Angular parameters: Trunk flexion–extension, lateral flexion, and axial rotation relative to global coordinates. Upper trunk rotation and full-body rotation based on relative orientation between upper-body and lower-extremity segments. Head flexion–extension, lateral flexion, and axial rotation relative to the neck segment (Supplementary Materials File S2).

All parameters were calculated separately for each trial, and mean values across three trials were used for statistical analysis.

2.3. *Clinical Outcomes*

All clinical assessments were administered prior to the sensor-based FST in a standardized sequence by same physiotherapist. Tests were conducted in a controlled indoor environment under consistent lighting and safety conditions. Standardized verbal instructions were provided for each assessment according to validated administration guidelines, and no additional feedback was given during performance. Rest intervals were allowed when necessary to prevent fatigue-related performance bias. Assessments were selected to characterize balance performance, functional mobility, muscle strength, frailty status, cognitive function, physical activity level, fear of falling, and quality of life—domains commonly associated with postural control and fall risk in older adults.

2.3.1. Balance Assessment

Balance performance was evaluated using the Berg Balance Scale (BBS). The BBS consists of 14 items assessing static and dynamic balance during functional tasks (e.g., standing, transfers, turning, reaching). Each item is scored from 0 to 4, yielding a total score ranging from 0 to 56, with higher scores indicating better balance performance [15].

2.3.2. Functional Mobility

Functional mobility was assessed using the Timed Up and Go (TUG) test. Participants rose from a chair, walked 3 m at a comfortable pace, turned, returned to the chair, and sat down. Total time was recorded in seconds, with longer times indicating poorer functional mobility and dynamic balance [16].

2.3.3. Dynamic Turning Ability

Dynamic turning ability was evaluated using the 360° Turn Test. Participants completed a full 360-degree turn, and both turning time and number of steps were recorded. Turning performance is an important component of dynamic balance and has been associated with fall risk in older adults [17].

2.3.4. Upper Extremity Muscle Strength

Upper extremity muscle strength was assessed using handgrip strength, measured with a hand dynamometer (Fabrication Enterprises Inc., White Plains, NY, USA) and recorded in kilograms. Handgrip strength is commonly used as a proxy for overall muscle strength and has been associated with functional performance and adverse health outcomes in older adults [18].

2.3.5. Lower-Extremity Functional Strength

Lower-extremity functional strength was evaluated using the Five Times Sit-to-Stand Test (5xSTS). Participants stood up from and sat down on a standard chair five consecutive times as quickly as possible without using their arms. Total time was recorded in seconds, with longer times indicating reduced lower limb strength and functional capacity [19].

https://doi.org/10.3390/s26051623

2.3.6. Frailty

Frailty status was assessed using the FRAIL scale, a brief screening tool consisting of five components: fatigue, resistance, ambulation, illnesses, and weight loss. Each component is scored dichotomously, yielding a total score from 0 to 5, with higher scores indicating greater frailty [20].

2.3.7. Cognitive Performance

Global cognitive function was screened using the Montreal Cognitive Assessment (MoCA). The MoCA assesses multiple cognitive domains, including attention, executive functions, memory, language, visuospatial abilities, and orientation, with a maximum score of 30 [21].

2.3.8. Physical Activity

Physical activity level was assessed using the International Physical Activity Questionnaire–Short Form (IPAQ-SF), which captures self-reported physical activity over the previous seven days across different intensity levels. Results were expressed as metabolic equivalent task minutes per week (MET-min/week), providing an estimate of overall physical activity [22,23].

2.3.9. Fear of Falling

Fear of falling was evaluated using the Falls Efficacy Scale–International (FES-I), which assesses concern about falling during a range of daily activities. Higher scores indicate greater fear of falling [24].

2.3.10. Quality of Life

Quality of life was assessed using the World Health Organization Quality of Life Instrument–Older Adults Module (WHOQOL-OLD). This instrument evaluates domains particularly relevant to aging, including sensory abilities, autonomy, social participation, and intimacy. Higher scores reflect better perceived quality of life [25].

2.4. Statistical Analysis

All analyses were performed using IBM SPSS Statistics version 29.0 (IBM Corp., Armonk, NY, USA). Descriptive statistics were calculated for all variables; continuous variables were summarized as mean $\pm$ standard deviation, categorical variables as frequencies and percentages. Distribution was examined using histograms and Shapiro–Wilk test.

Test–retest reliability was assessed using intraclass correlation coefficient (ICC) based on a two-way mixed-effects model with absolute agreement (ICC [3,1]). ICC values were interpreted according to Koo and Li as poor (<0.50), moderate (0.50–0.75), good (0.75–0.90), or excellent (>0.90) [26]. Measurement precision was quantified by standard error of measurement (SEM = SD $\times$ $\sqrt{(1 - \text{ICC})}$) and minimal detectable change at 95% confidence level (MDC_{95} = SEM $\times$ 1.96 $\times$ $\sqrt{2}$). Also, 95% confidence intervals were calculated for ICC estimates to provide precision bounds consistent with established reliability reporting standards. Agreement was examined using Bland–Altman analysis with mean differences and 95% limits of agreement (Supplementary Materials File S1).

Associations between sensor-derived kinematic parameters and clinical outcomes were analyzed using Pearson's correlation coefficient for normally distributed variables and Spearman's rank correlation coefficient otherwise. To control for Type I error across multiple comparisons, clinical outcomes were categorized into five a priori hypothesized domains: (1) Balance and Mobility, (2) Strength, (3) Physical Activity, (4) Cognitive Function, and (5) Quality of Life/Frailty. Within each domain, we hypothesized that specific depth sensor–derived kinematic parameters would show significant clinical associations. The

https://doi.org/10.3390/s26051623

statistical significance of correlations within these hypothesized domains was tested using the Benjamini–Hochberg false discovery rate (FDR) procedure to adjust *p*-values. Any correlations identified outside these predefined hypothesized structures were treated as exploratory and interpreted with caution. Correlation magnitudes were interpreted according to Evans as very weak (<0.20), weak (0.20–0.39), moderate (0.40–0.59), strong (0.60–0.79), or very strong (≥0.80) [27]. Statistical significance was set at $p < 0.05$.

3. Results

A total of 86 community-dwelling older adults were included in the analysis. The cohort had a mean age of 70.3 ± 4.7 years and a mean body mass index of 27.52 ± 3.61 kg/m^2; 60.5% were female. Comprehensive demographic and clinical characteristics are summarized in Table 1.

Table 1. Demographic and clinical characteristics of participants.

Variable	Total (n = 86)
Age (years), mean ± SD	70.3 ± 4.7
Sex, n (%)	
Female	52 (60.5)
Male	34 (39.5)
Height (cm), mean ± SD	163.6 ± 10.2
Body weight (kg), mean ± SD	73.7 ± 11.7
Body mass index (kg/m^2), mean ± SD	27.52 ± 3.61
Dominant side, n (%)	
Right	80 (93.0)
Left	6 (7.0)
History of falls, n (%)	
Yes	71 (82.6)
No	15 (17.4)

Descriptive statistics for all sensor-derived kinematic parameters obtained during the test and retest sessions are presented in Table 2.

Test duration and linear displacement measures were lower at retest compared with the initial assessment. Anterior–posterior displacement represented the largest linear component in both sessions, whereas medio-lateral displacement remained smaller. Total planar displacement followed a similar pattern.

Table 2. Test–retest descriptive statistics of the depth sensor-based Fukuda Stepping Test parameters.

Parameter	Test (n = 86) Mean ± SD	Retest (n = 23) Mean ± SD
Test duration (s)	236.92 ± 39.28	216.60 ± 32.48
AP displacement (cm)	−72.88 ± 29.85	−66.69 ± 23.57
ML displacement (cm)	−6.06 ± 30.31	1.10 ± 24.77
Total planar displacement (cm)	80.52 ± 29.28	72.80 ± 23.53
Trunk flexion change (°)	4.10 ± 5.21	2.57 ± 3.42
Trunk lateral flexion change (°)	−9.79 ± 7.67	−13.60 ± 8.59
Upper trunk rotation change (°)	83.54 ± 7.40	88.03 ± 17.64
Full-body rotation change (°)	83.73 ± 14.12	83.71 ± 23.18

https://doi.org/10.3390/s26051623

Table 2. *Cont.*

Parameter	Test (n = 86) Mean ± SD	Retest (n = 23) Mean ± SD
Head flexion/extension change (°)	2.60 ± 10.22	1.83 ± 10.90
Head axial rotation change (°)	−45.09 ± 71.84	−31.49 ± 102.40
Head lateral flexion change (°)	3.25 ± 13.67	3.60 ± 15.96
Head ML displacement (cm)	−5.41 ± 29.50	1.90 ± 25.88
Head SI displacement (cm)	−5.47 ± 5.96	−2.64 ± 4.10
Head AP displacement (cm)	−74.77 ± 29.07	−68.28 ± 22.74

SD, standard deviation; AP, anterior–posterior; ML, mediolateral; SI, superior–inferior.

Among angular parameters, upper trunk rotation and full-body rotation exhibited the greatest angular deviations in both sessions. Head axial rotation showed the highest variability among head kinematic measures (Table 2). Test–retest reliability and post hoc power results for the depth sensor-based Fukuda Stepping Test parameters are summarized in Table 3.

Table 3. Test–retest reliability of the depth sensor-based Fukuda Stepping Test parameters (n = 23).

Parameter (n = 23)	r	*p*	ICC (3,1)	95% CI	*p* (ICC)	Power	SEM	MDC95
Test duration (s)	0.65	<0.001	0.65	0.33–0.83	<0.001	0.96	23.24	64.41
AP displacement (cm)	0.67	<0.001	0.63	0.30–0.82	<0.001	0.98	18.16	50.33
ML displacement (cm)	0.35	0.10	0.35	−0.07–0.66	0.04	0.37	24.44	67.74
Total planar displacement (cm)	0.62	<0.001	0.58	0.23–0.80	0.001	0.93	18.98	52.60
Trunk flexion change (°)	0.47	0.03	0.46	0.07–0.73	0.01	0.64	3.83	10.61
Trunk lateral flexion change (°)	0.03	0.89	−0.20	−0.56–0.23	0.82	0.05	8.40	23.29
Upper trunk rotation change (°)	−0.03	0.88	0.01	−0.40–0.41	0.48	0.05	7.36	20.41
Full-body rotation change (°)	0.30	0.17	0.47	0.08–0.73	0.01	0.28	10.28	28.49
Head flexion/extension change (°)	0.19	0.38	0.29	−0.13–0.62	0.09	0.14	8.61	23.87
Head axial rotation change (°)	0.31	0.15	0.35	−0.06–0.66	0.04	0.30	57.92	160.54
Head lateral flexion change (°)	0.70	<0.001	0.68	0.38–0.85	<0.001	0.99	7.73	21.43
Head ML displacement (cm)	0.39	0.07	0.29	−0.13–0.62	0.08	0.46	24.86	68.90
Head SI displacement (cm)	0.38	0.07	0.38	−0.03–0.68	0.04	0.44	4.69	13.01
Head AP displacement (cm)	0.63	<0.001	0.60	0.25–0.81	0.001	0.94	18.39	50.96

AP, anterior–posterior; ML, mediolateral; SI, superior–inferior; n, sample size; r, correlation coefficient; *p*, significance level; ICC (3,1), intraclass correlation coefficient based on a two-way mixed-effects model for single measurements; 95% CI, 95% confidence interval; *p* (ICC), significance level of the ICC; Power, statistical power; SEM (Standard Error of Measurement), standard error of measurement; MDC95 (Minimal Detectable Change at the 95% confidence level), the smallest change that exceeds measurement error with 95% confidence.

Test duration and anterior–posterior displacement demonstrated similar correlation and intraclass correlation coefficients, with confidence intervals remaining within positive ranges. Total planar displacement and head anterior–posterior displacement showed comparable levels of agreement. Medio-lateral trunk displacement exhibited lower correlation and wider confidence intervals. Among angular measures, trunk flexion change and full-body rotation change showed moderate ICC values, whereas trunk lateral flexion and upper trunk rotation demonstrated low agreement with confidence intervals crossing zero.

Head kinematic parameters displayed heterogeneous results. Head lateral flexion showed relatively higher agreement, while head flexion–extension, axial rotation, and selected displacement measures presented lower ICC values and broader confidence intervals.

https://doi.org/10.3390/s26051623

SEM and MDC values varied substantially across parameters, with greater measurement error observed in selected angular variables (Table 3).

Summary of clinical assessments, physical performance, and quality of life outcomes of the participants are presented in Table 4.

Table 4. Summary of clinical assessments, physical performance, and quality-of-life outcomes (n = 86).

Clinical Measure	Mean ± SD	Min–Max
Physical Performance and Strength		
Handgrip strength (kg)	28.76 ± 8.01	15.36–49.43
5xSTS (s)	11.72 ± 2.20	6.33–16.20
Mobility and Balance		
360° Turning time (s)	2.99 ± 0.55	1.97–4.67
TUG (s)	10.55 ± 1.93	6.43–15.40
BBS (score)	52.20 ± 2.60	45–56
Frailty and Cognitive Status		
Frailty score	0.30 ± 0.60	0.00–2.00
MoCA (score)	26.70 ± 0.70	26.00–28.00
Fear of Falling		
Falls Efficacy Scale–International (FES-I), score	23.20 ± 4.03	15–34
Low fear, n (%)	15 (17.4)	—
Moderate fear, n (%)	58 (67.4)	—
High fear, n (%)	13 (15.1)	—
Physical Activity (IPAQ)		
Moderate-intensity activity (days/week)	2.40 ± 2.00	0.00–7.00
Moderate-intensity activity (hours/week)	0.64 ± 0.50	0.00–2.00
Walking (days/week)	5.20 ± 1.80	0.00–7.00
Walking (hours/week)	0.71 ± 0.30	0.00–1.50
Sitting time (hours/day)	7.30 ± 1.40	3.00–10.00
Total MET value (MET-min/week)	1373.16 ± 756.01	198.00–3096.00
Low activity, n (%)	11 (12.8)	—
Moderate activity, n (%)	72 (83.7)	—
High activity, n (%)	3 (3.5)	—
Quality of Life (WHOQOL-OLD)		
Sensory Abilities (SAB)	65.70 ± 25.03	6.25–100.00
Past, Present and Future Activities (PPF)	75.15 ± 10.89	43.75–100.00
Autonomy (AUT)	77.54 ± 17.19	6.25–100.00
Social Participation (SOP)	72.17 ± 15.33	18.75–100.00
Death and Dying (DAD)	86.05 ± 18.88	25.00–100.00
Intimacy (INT)	75.29 ± 16.66	31.25–100.00
WHOQOL-OLD total score	75.31 ± 11.04	34.38–94.7

5xSTS, Five Times Sit-to-Stand Test; BBS, Berg Balance Scale; FES-I, Falls Efficacy Scale–International; IPAQ, International Physical Activity Questionnaire; MET, metabolic equivalent of task; MoCA, Montreal Cognitive Assessment; SAB, Sensory Abilities; PPF, Past, Present and Future Activities; AUT, Autonomy; SOP, Social Participation; DAD, Death and Dying; INT, Intimacy; SD, standard deviation.

https://doi.org/10.3390/s26051623

Handgrip strength and Five Times Sit-to-Stand performance averaged at 28.76 ± 8.01 kg and 11.72 ± 2.20 s, respectively. Mobility and balance assessments yielded mean values of 10.55 ± 1.93 s for TUG, 2.99 ± 0.55 s for 360° turning time, and 52.20 ± 2.60 for the Berg Balance Scale. Frailty and cognitive assessments showed mean scores of 0.30 ± 0.60 and 26.70 ± 0.70, respectively. The mean Falls Efficacy Scale–International score was 23.20 ± 4.03. Total physical activity averaged 1373.16 ± 756.01 MET-min/week. The WHOQOL-OLD total score was 75.31 ± 11.04, with domain-specific values detailed in Table 4.

Correlation analyses between depth sensor-based Fukuda Stepping Test parameters and muscle strength, frailty status, cognitive function, physical activity level are presented in Table 5.

Table 5. Correlations between depth sensor-based Fukuda Stepping Test parameters and muscle strength, frailty status, cognitive function, physical activity level outcomes (n = 86).

FST Parameter	Handgrip (r)	5xSTS (r)	MoCA (r)	Frailty (r)	IPAQ (MET) (r)
Test duration (s)	−0.14	0.04	−0.01	−0.17	0.08
AP displacement (cm)	−0.03	0.06	−0.16	−0.09	0.07
ML displacement (cm)	0.14	0.07	0.001	0.01	0.01
Total planar displacement (cm)	−0.04	−0.06	0.10	−0.01	−0.04
Trunk flexion change (°)	−0.19	−0.09	−0.38 ** q	−0.10	−0.39 ** q
Trunk lateral flexion change (°)	0.22	−0.01	0.02	−0.15	−0.19
Upper trunk rotation change (°)	−0.08	−0.30 ** q	0.09	−0.20	−0.02
Full-body rotation change (°)	−0.11	0.12	0.01	0.01	0.24 * q
Head flexion change (°)	−0.04	−0.02	−0.06	−0.20	−0.15
Head axial rotation change (°)	−0.08	−0.17	0.16	−0.18	0.02
Head lateral flexion change (°)	−0.03	0.14	−0.02	0.13	0.01
Head ML displacement (cm)	0.14	0.22	−0.15	0.22 * q	0.01
Head SI displacement (cm)	0.11	0.04	−0.01	0.00	0.29 **
Head AP displacement (cm)	0.04	0.07	−0.10	0.01	0.07

Frailty, Frail Scale; 5xSTS, Five Times Sit-to-Stand Test; MoCA, Montreal Cognitive Assessment; IPAQ, International Physical Activity Questionnaire; MET, metabolic equivalent of task, * $p < 0.05$, ** $p < 0.01$ (two-tailed for exploratory analysis), q indicates Benjamini–Hochberg FDR-adjusted p-values within each clinical domain (q < 0.05 considered significant).

Correlation analyses between depth sensor-based Fukuda Stepping Test parameters and balance, functional mobility and quality of life outcomes are presented in Table 6.

Correlation analyses revealed several statistically significant associations between depth sensor-based FST parameters and clinical assessment outcomes. Trunk flexion change demonstrated moderate negative correlations with cognitive performance and total physical activity. Weak-to-moderate negative associations were observed between trunk flexion change and quality of life, while a weak negative correlation was identified with balance performance.

Upper trunk rotation change showed moderate negative correlations with functional lower-extremity performance and TUG duration. Full-body rotation change exhibited a weak positive correlation with total physical activity.

Head kinematic parameters were also associated with selected clinical measures. Head mediolateral displacement showed weak positive correlations with frailty, lower-extremity performance, and fear of falling. Head superior–inferior displacement demonstrated weak-to-moderate positive associations with physical activity and weak positive correlations with quality of life. Head flexion change showed a weak negative correlation with daily

https://doi.org/10.3390/s26051623

walking duration. No statistically significant correlations were observed between test duration, anterior–posterior displacement, or total planar displacement and most clinical outcomes ($p < 0.05$) (Tables 5 and 6).

Table 6. Correlations between depth sensor-based Fukuda Stepping Test parameters and balance, functional mobility, and quality of life outcomes (n = 86).

FST Parameter	QoL	360° (r)	TUG (r)	BBS (r)	FES-I (r)
Test duration (s)	0.03	−0.08	0.12	0.21	−0.02
AP displacement (cm)	0.06	0.04	0.06	−0.12	0.03
ML displacement (cm)	0.06	0.14	**0.23 * q**	−0.16	**0.28 * q**
Total planar displacement (cm)	−0.05	−0.04	−0.06	0.10	−0.01
Trunk flexion change (°)	−0.29 ** q	−0.07	−0.04	**−0.24 * q**	0.05
Trunk lateral flexion change (°)	−0.001	**0.22 ***	−0.01	0.02	−0.15
Upper trunk rotation change (°)	0.05	−0.08	**−0.30 ** q**	0.09	−0.20
Full-body rotation change (°)	0.13	−0.11	0.12	0.01	0.01
Head flexion change (°)	−0.02	−0.04	−0.02	−0.06	−0.20
Head axial rotation change (°)	0.001	−0.08	−0.17	0.16	−0.18
Head lateral flexion change (°)	−0.01	−0.03	0.14	−0.02	0.13
Head ML displacement (cm)	0.04	0.14	**0.22 * q**	−0.15	**0.22 ***
Head SI displacement (cm)	0.26 *	0.11	0.04	−0.01	0.001
Head AP displacement (cm)	0.07	0.04	0.07	−0.10	0.01

QoL, WHOQOL-OLD; 360°, 360 Turn Test; TUG, Timed up and go test; BBS, Berg Balance Scale; FES-I, Falls Efficacy Scale–International; * $p < 0.05$, ** $p < 0.01$ (two-tailed for exploratory analysis), q indicates Benjamini–Hochberg FDR-adjusted p-values within each clinical domain (q < 0.05 considered significant).

4. Discussion

This study evaluated the test–retest reliability of a depth sensor-based FST and examined correlations between sensor-derived kinematic parameters and clinical outcomes in community-dwelling older adults. The Microsoft Kinect v2-based protocol yielded poor-to-moderate reliability for temporal and several kinematic parameters, with angular parameters exhibiting superior repeatability compared to linear displacement measures. Trunk flexion emerged as a multidimensional indicator, showing weak-to-moderate negative associations with cognitive function, physical activity, balance performance, and quality of life. Upper trunk rotation correlated negatively with functional mobility, while head superior–inferior displacement showed positive associations with physical activity and quality of life. Notably, traditional global displacement metrics showed weak or non-significant associations with clinical outcomes, suggesting that markerless depth sensor technology captures clinically meaningful segmental kinematic patterns that substantially exceed conventional FST metrics.

Recent FST modernization efforts have primarily employed wearable inertial measurement units (IMUs). Belluscio et al. developed the instrumented FST using five IMU sensors, demonstrating that upper-body stability indices discriminated between motor deficit levels in stroke patients despite non-significant differences in conventional displacement measures [28]. Hemm et al. employed nine Xsens IMU sensors, reporting excellent correlation between IMU-derived and manual angular measurements with moderate test–retest reliability for global displacement but emphasized that dynamic parameters provided additional discriminatory value [12]. Miwa et al. (2023) used three motion sensors at head, thorax, and lumbar spine, demonstrating that segmental motion increased substantially under eyes-closed conditions [29].

https://doi.org/10.3390/s26051623

The present depth sensor approach offers distinct advantages: markerless capture eliminates donning/doffing procedures and sensor placement variability, avoids cumulative IMU drift error, and provides direct skeletal tracking without complex sensor fusion algorithms. However, single-camera configurations are vulnerable to occlusion artifacts, whereas body-worn IMUs maintain contact throughout motion sequences. Critically, both IMU-based and depth sensor studies converge on a key finding: segmental kinematic parameters provide superior clinical information compared to traditional global displacement metrics.

The poor-to-moderate reliability observed for angular kinematic parameters aligns with recent findings supporting markerless motion capture. Guess et al. reported that Azure Kinect-based gait analyses demonstrated high agreement with gold-standard optical motion capture systems, with Pearson correlation coefficients exceeding 0.87 for spatiotemporal parameters such as step length and step width [30]. Similarly, Albert et al. showed that Azure Kinect provides significantly higher accuracy in spatial gait parameters compared with previous-generation sensors, with notable improvements in foot tracking performance [31]. In addition, Bertrand et al. reported excellent agreement (ICC > 0.90) between tablet-based markerless motion capture applications and expert clinician ratings during the 30-Second Sit-to-Stand and Timed Up and Go tests, supporting the clinical validity of this technology [32]. However, as noted by Lim et al. while flexion–extension movements of lower extremity joints exhibit high reliability, movements in the abduction–adduction plane demonstrate lower reliability [33]. The lower reliability observed in selected parameters may reflect both technical and biological sources of variability, including single-camera occlusion, skeletal tracking limitations, and intra-individual motor variability during repeated stepping. Differences across studies may also arise from variations in measurement methodology and sample characteristics. As previous reports do not consistently provide detailed clinical profiles of study populations, direct comparison of reliability estimates remains limited. The present study contributes by combining transparent kinematic processing with clearly described clinical characteristics, thereby supporting more interpretable integration of technological and clinical assessments in future research. Also, the relatively small retest subsample may have influenced the stability of certain reliability estimates. It is well established that intraclass correlation coefficients can be sensitive to sample size and data imbalance, particularly in movement analysis research. As discussed by Trabassi et al., small-sample designs may limit the robustness and generalizability of classification and reliability outcomes in sensor-based assessments [34]. Therefore, while the present findings provide preliminary reliability evidence, future multicenter studies including larger and more heterogeneous samples are warranted to further strengthen the stability and external validity of depth sensor-based FST implementations.

The moderate negative correlation between trunk flexion and cognitive performance indicates that individuals with lower cognitive capacity exhibit greater trunk flexion during eyes-closed stepping, reflecting impaired cognitive–motor integration or reduced executive control over posture. In line with the cognitive penetrability hypothesis proposed by Boisgontier et al., postural control in aging shifts away from automatic regulation toward increased reliance on cortical and voluntary control mechanisms, thereby heightening dependence on general cognitive resources to maintain balance [35]. Marusic et al. further emphasized that increased activation of the dorsolateral prefrontal cortex, particularly during challenging tasks, reflects this elevated cognitive load and highlights the critical role of executive function capacity in shaping postural strategies [36]. Within this framework, when executive resources become insufficient, the postural control system may revert to more primitive or seemingly safer postural configurations. Excessive trunk flexion may therefore represent a compensatory strategy adopted by individuals with reduced

 https://doi.org/10.3390/s26051623

processing speed, whereby stability is enhanced through lowering the center of mass at the expense of postural efficiency. The weaker association observed between trunk flexion and balance performance compared with cognitive performance suggests that cognitive–motor integration may be a stronger determinant of trunk kinematic behavior than balance capacity alone.

Trunk flexion showed consistent negative associations with physical activity, balance performance, and quality of life in the present study, suggesting a potential relationship between sagittal-plane trunk kinematics and multidomain functional status. Previous biomechanical research has reported associations between sagittal-plane trunk kinematics and balance recovery outcomes following destabilizing perturbations, indicating that trunk motion may be relevant in fall-related contexts [37]. In addition, sensor-based fall-risk assessment studies have reported that trunk-related kinematic features are associated with frailty and fall-risk indicators beyond global performance measures alone [38]. In contrast, the absence of significant associations for trunk lateral flexion suggests that, within the present eyes-closed stepping paradigm, sagittal-plane trunk motion was more closely related to the selected clinical measures than frontal-plane trunk motion. Given the cross-sectional design, these findings should be interpreted as associative rather than mechanistic.

The negative associations observed between upper trunk rotation and functional mobility measures (5xSTS and TUG) suggest that excessive axial rotation during eyes-closed stepping may reflect underlying limitations in transfer- and turning-related mobility. This interpretation is consistent with evidence from instrumented TUG studies demonstrating that trunk kinematic features, particularly during transitional and turning phases, provide clinically relevant information beyond total task duration alone [39]. In addition, the positive associations between head mediolateral displacement and frailty, 5xSTS performance, and fear of falling suggest that increased lateral head motion reflects compromised mediolateral stability and reduced balance confidence. Supporting this interpretation, recent evidence indicates that fear of falling in older adults is associated with impaired mediolateral postural control and altered balance strategies [40]. Furthermore, the positive relationships observed between head superior–inferior displacement and both physical activity level and quality of life may reflect greater stepping amplitude and more dynamic movement execution in physically active individuals, whereas reduced vertical head motion may indicate more constrained and cautious movement strategies in less confident or less active individuals.

Traditional FSTs typically rely on subjective tools such as protractors and tape measures to quantify performance. However, research indicates that these global metrics often show weak associations with specific underlying impairments, whereas segmental kinematics provide robust insights into functional control. Global measures (e.g., distance or duration) reflect cumulative endpoints that can be achieved through diverse compensatory strategies, thereby obscuring specific neuromuscular control deficits [41,42]. In contrast, kinematic analysis reveals the subtle mechanisms of movement invisible to traditional scoring. This limitation underscores the need for modernizing clinical balance assessments to incorporate objective, instrumented measurement, a perspective strongly advocated in recent translational research [43]. In this perspective, depth sensor-based FST offers substantial clinical applicability. Trunk flexion's associations with cognitive function, physical activity, balance, and quality of life suggest this parameter may serve as a pragmatic screening indicator for multidomain functional capacity and cognitive–motor integration deficits. Integration into digital health platforms could support remote fall-risk stratification, telerehabilitation monitoring, and early functional decline detection. Multiplanar kinematic data may facilitate individualized intervention targeting by identifying specific movement deficits warranting therapeutic attention.

https://doi.org/10.3390/s26051623

Strengths include automated objective kinematic analysis, comprehensive clinical assessment battery, rigorous reliability evaluation, and markerless approach advantages (no donning/doffing, no drift error, simplified processing). However, single-sensor setup may introduce occlusion artifacts and precludes ambulatory assessment possible with IMU devices. The cross-sectional design precludes causal inference regarding whether trunk flexion causes or results from reduced cognitive/physical function. Sample characteristics (relatively high functional capacity, no diagnosed cognitive/neurological conditions) may limit generalizability to frailer populations. In addition, the high prevalence of self-reported fall history in the sample may have influenced postural strategies and movement variability, potentially affecting the magnitude of observed associations. Although this may limit generalizability to non-faller populations, it increases the ecological relevance of the findings for individuals at elevated fall risk. Also, given the cross-sectional design, the observed associations should be interpreted as relational findings rather than evidence of causal or mechanistic pathways. Weak-to-moderate correlation magnitudes ($r = 0.22–0.39$) suggest individual parameters have limited standalone predictive value, emphasizing need for multivariate modeling. The present findings should be interpreted within the context of a relatively high-functioning community-dwelling sample; therefore, clinically meaningful thresholds or predictive cut-off values cannot be derived from the current data. Future research should focus on validating these kinematic markers in frailer and clinically vulnerable populations and on establishing clinically interpretable reference values.

Longitudinal studies should determine whether trunk flexion patterns prospectively predict cognitive decline, falls, or functional deterioration, establishing directionality and potential causality. Machine learning integration could identify kinematic patterns associated with specific clinical phenotypes (sarcopenia, mild cognitive impairment, vestibular dysfunction). Validation of unsupervised home-based protocols using consumer-grade depth sensors could democratize objective balance screening. Domain-specific cognitive assessments could clarify which subdomains (executive function, visuospatial processing) most strongly drive trunk flexion associations. Comparative studies evaluating depth sensor versus IMU-based protocols regarding clinical utility, cost-efficiency, and user acceptability would inform technology selection. Intervention studies examining whether targeted rehabilitation modifies FST kinematics and associated outcomes would establish parameter responsiveness and therapeutic relevance.

5. Conclusions

This study demonstrates that depth sensor-based modernization of the Fukuda Stepping Test provides reliable, objective assessment of postural control in community-dwelling older adults, with segmental kinematic parameters, particularly trunk flexion, exhibiting clinically meaningful associations with cognitive function, physical activity, balance performance, and quality of life that substantially exceed traditional displacement-only metrics. The convergence of findings across multiple sensor modalities (depth sensors and IMUs) establishes that conventional FST scoring inadequately captures the biomechanical complexity of postural control, supporting the integration of affordable, markerless motion capture technology into geriatric assessment protocols. Trunk flexion emerges as a promising multidimensional marker warranting investigation as an early indicator of cognitive–motor integration deficits and a potential target for individualized rehabilitation interventions in aging populations.

https://doi.org/10.3390/s26051623

Supplementary Materials: The following supporting information can be downloaded at https://www.mdpi.com/article/10.3390/s26051623/s1.

Author Contributions: Conceptualization, M.D.; methodology, M.D. and H.T.Ü.; software, M.K.; validation, H.T.Ü., S.Y.Ç. and Ö.K.K.; formal analysis, M.D.; investigation, H.T.Ü. and M.K.; resources, M.D.; data curation, H.T.Ü., S.Y.Ç., Ö.K.K. and M.D.; writing—original draft preparation, M.D.; writing—review and editing, M.D., H.T.Ü., M.K., S.Y.Ç. and Ö.K.K.; visualization, M.D. and M.K.; supervision, M.D.; project administration, M.D.; funding acquisition, M.D. All authors have read and agreed to the published version of the manuscript.

Funding: This research received no external funding.

Institutional Review Board Statement: This cross-sectional methodological study was approved by the Akdeniz University Non-Interventional Clinical Research Ethics Committee (30.01.2025, No: TBAEK-168).

Informed Consent Statement: Informed consent was obtained from all subjects involved in the study.

Data Availability Statement: The data presented in this study are not publicly available due to ethical and privacy restrictions, as they contain information that could compromise the confidentiality of research participants. Data supporting the findings of this study are available from the corresponding author upon reasonable request and with permission of the relevant institutional ethics committee.

Conflicts of Interest: The authors declare no conflicts of interest.

Abbreviations

The following abbreviations are used in this manuscript:

FST	Fukuda Stepping Test
AP	Anterior–posterior
ML	Mediolateral
SI	Superior–inferior
ICC	Intraclass correlation coefficient
SEM	Standard error of measurement
MDC	Minimal detectable change
MDC95	Minimal detectable change at the 95% confidence level
SD	Standard deviation
MoCA	Montreal Cognitive Assessment
BBS	Berg Balance Scale
TUG	Timed Up and Go test
5xSTS	Five Times Sit-to-Stand Test
FES-I	Falls Efficacy Scale–International
IPAQ	International Physical Activity Questionnaire
IPAQ-SF	International Physical Activity Questionnaire–Short Form
MET	Metabolic equivalent of task
WHOQOL-OLD	World Health Organization Quality of Life Instrument–Older Adults Module
IMU	Inertial measurement unit
QoL	Quality of life

References

1. Salari, N.; Darvishi, N.; Ahmadipanah, M.; Shohaimi, S.; Mohammadi, M. Global prevalence of falls in the older adults: A comprehensive systematic review and meta-analysis. *J. Orthop. Surg. Res.* **2022**, *17*, 334. [CrossRef]
2. Colón-Emeric, C.S.; McDermott, C.L.; Lee, D.S.; Berry, S.D. Risk Assessment and Prevention of Falls in Older Community-Dwelling Adults: A Review. *JAMA* **2024**, *331*, 1397–1406. [CrossRef]
3. Lewis, S.R.; Griffin, X.L. Preventing falls in older people: The evidence for environmental interventions and why history matters. *Cochrane Database Syst. Rev.* **2023**, *2023*, ED000162. [CrossRef] [PubMed]

4. Quijoux, F.; Vienne-Jumeau, A.; Bertin-Hugault, F.; Zawieja, P.; Lefèvre, M.; Vidal, P.-P.; Ricard, D. Center of pressure displacement characteristics differentiate fall risk in older people: A systematic review with meta-analysis. *Ageing Res. Rev.* **2020**, *62*, 101117. [CrossRef] [PubMed]
5. Singh, R.R.; Maurya, P. Visual impairment and falls among older adults and elderly: Evidence from longitudinal study of ageing in India. *BMC Public Health* **2022**, *22*, 2324. [CrossRef] [PubMed]
6. McLaren, R.; Smith, P.F.; Taylor, R.L.; Ravindran, S.; Rashid, U.; Taylor, D. Efficacy of nGVS to improve postural stability in people with bilateral vestibulopathy: A systematic review and meta-analysis. *Front. Neurosci.* **2022**, *16*, 1010239. [CrossRef]
7. Paquet, N.; Jehu, D.A.; Lajoie, Y. Age-related differences in Fukuda stepping and Babinski-Weil tests, within-day variability and test–retest reliability. *Aging Clin. Exp. Res.* **2017**, *29*, 223–230. [CrossRef]
8. Paquet, N.; Jehu, D.A.; Lajoie, Y. Impact of the Number of Steps on the Fukuda Stepping Test in Older Adults. *Phys. Occup. Ther. Geriatr.* **2016**, *34*, 104–111. [CrossRef]
9. Liu, X.-T.; Nikkhoo, M.; Wang, L.; Chen, C.P.C.; Chen, H.-B.; Chen, C.-J.; Cheng, C.-H. Feasibility of a kinect-based system in assessing physical function of the elderly for home-based care. *BMC Geriatr.* **2023**, *23*, 495. [CrossRef]
10. Takeshima, N.; Kohama, T.; Kusunoki, M.; Okada, S.; Fujita, E.; Oba, Y.; Brechue, W.F. A 20-sec Stepping Test and KINECTTM Sensor Provides Objective Quantification of Movement/Balance Dysfunction in Older Individuals. *Exp. Aging Res.* **2020**, *46*, 244–256. [CrossRef]
11. Scataglini, S.; Abts, E.; Van Bocxlaer, C.; Van den Bussche, M.; Meletani, S.; Truijen, S. Accuracy, Validity, and Reliability of Markerless Camera-Based 3D Motion Capture Systems versus Marker-Based 3D Motion Capture Systems in Gait Analysis: A Systematic Review and Meta-Analysis. *Sensors* **2024**, *24*, 3686. [CrossRef]
12. Hemm, S.; Baumann, D.; Duarte da Costa, V.; Tarnutzer, A.A. Test-re-test reliability and dynamics of the Fukuda–Unterberger stepping test. *Front. Neurol.* **2023**, *14*, 1128760. [CrossRef]
13. Clark, R.A.; Pua, Y.-H.; Fortin, K.; Ritchie, C.; Webster, K.E.; Denehy, L.; Bryant, A.L. Validity of the Microsoft Kinect for assessment of postural control. *Gait Posture* **2012**, *36*, 372–377. [CrossRef]
14. Guffanti, D.; Brunete, A.; Hernando, M.; Rueda, J.; Navarro Cabello, E. The Accuracy of the Microsoft Kinect V2 Sensor for Human Gait Analysis. A Different Approach for Comparison with the Ground Truth. *Sensors* **2020**, *20*, 4405. [CrossRef]
15. Sahin, F.; Yilmaz, F.; Ozmaden, A.; Kotevoglu, N.; Sahin, T.; Kuran, B. Reliability and validity of the Turkish version of the Berg Balance Scale. *J. Geriatr. Phys. Ther.* **2008**, *31*, 32–37. [CrossRef]
16. Podsiadlo, D.; Richardson, S. The timed "Up & Go": A test of basic functional mobility for frail elderly persons. *J. Am. Geriatr. Soc.* **1991**, *39*, 142–148. [CrossRef] [PubMed]
17. Gill, T.M.; Williams, C.S.; Tinetti, M.E. Assessing risk for the onset of functional dependence among older adults: The role of physical performance. *J. Am. Geriatr. Soc.* **1995**, *43*, 603–609. [CrossRef] [PubMed]
18. Roberts, H.C.; Denison, H.J.; Martin, H.J.; Patel, H.P.; Syddall, H.; Cooper, C.; Sayer, A.A. A review of the measurement of grip strength in clinical and epidemiological studies: Towards a standardised approach. *Age Ageing* **2011**, *40*, 423–429. [CrossRef]
19. Bohannon, R.W. Reference values for the five-repetition sit-to-stand test: A descriptive meta-analysis of data from elders. *Percept. Mot. Ski.* **2006**, *103*, 215–222. [CrossRef] [PubMed]
20. Thompson, M.Q.; Theou, O.; Tucker, G.R.; Adams, R.J.; Visvanathan, R. FRAIL scale: Predictive validity and diagnostic test accuracy. *Australas. J. Ageing* **2020**, *39*, e529–e536. [CrossRef]
21. Sweet, L.; Van Adel, M.; Metcalf, V.; Wright, L.; Harley, A.; Leiva, R.; Taler, V. The Montreal Cognitive Assessment (MoCA) in geriatric rehabilitation: Psychometric properties and association with rehabilitation outcomes. *Int. Psychogeriatr.* **2011**, *23*, 1582–1591. [CrossRef]
22. Saglam, M.; Arikan, H.; Savci, S.; Inal-Ince, D.; Bosnak-Guclu, M.; Karabulut, E.; Tokgozoglu, L. International physical activity questionnaire: Reliability and validity of the Turkish version. *Percept. Mot. Ski.* **2010**, *111*, 278–284. [CrossRef]
23. Craig, C.L.; Marshall, A.L.; Sjöström, M.; Bauman, A.E.; Booth, M.L.; Ainsworth, B.E.; Pratt, M.; Ekelund, U.; Yngve, A.; Sallis, J.F.; et al. International physical activity questionnaire: 12-country reliability and validity. *Med. Sci. Sports Exerc.* **2003**, *35*, 1381–1395. [CrossRef]
24. Yardley, L.; Beyer, N.; Hauer, K.; Kempen, G.; Piot-Ziegler, C.; Todd, C. Development and initial validation of the Falls Efficacy Scale-International (FES-I). *Age Ageing* **2005**, *34*, 614–619. [CrossRef] [PubMed]
25. Eser, S.; Saatli, G.; Eser, E.; Baydur, H.; Fidaner, C. The reliability and validity of the Turkish Version of the World Health Organization Quality of Life Instrument-Older Adults Module (WHOQOL-Old). *Turk. Psikiyatr. Derg.* **2010**, *21*, 37–48.
26. Koo, T.K.; Li, M.Y. A Guideline of Selecting and Reporting Intraclass Correlation Coefficients for Reliability Research. *J. Chiropr. Med.* **2016**, *15*, 155–163. [CrossRef] [PubMed]
27. Evans, J.D. *Straightforward Statistics for the Behavioral Sciences*; Thomson Brooks/Cole Publishing Co.: Pacific Grove, CA, USA, 1996.
28. Belluscio, V.; Bergamini, E.; Iosa, M.; Tramontano, M.; Morone, G.; Vannozzi, G. The iFST: An instrumented version of the Fukuda Stepping Test for balance assessment. *Gait Posture* **2018**, *60*, 203–208. [CrossRef]

 https://doi.org/10.3390/s26051623

29. Miwa, T.; Yasuda, T.; Kouga, T.; Sunami, K. Modified Fukuda stepping motion assessment of young healthy adults using portable inertial measurement units. *Heliyon* **2023**, *9*, e15018. [CrossRef]
30. Guess, T.M.; Bliss, R.; Hall, J.B.; Kiselica, A.M. Comparison of Azure Kinect overground gait spatiotemporal parameters to marker based optical motion capture. *Gait Posture* **2022**, *96*, 130–136. [CrossRef] [PubMed]
31. Albert, J.A.; Owolabi, V.; Gebel, A.; Brahms, C.M.; Granacher, U.; Arnrich, B. Evaluation of the Pose Tracking Performance of the Azure Kinect and Kinect v2 for Gait Analysis in Comparison with a Gold Standard: A Pilot Study. *Sensors* **2020**, *20*, 5104. [CrossRef]
32. Bertrand, J.K.; McNeely, M.L.; Bates, J.; Joy, J.; Kanani, J.; Ezeugwu, V.E.; Tandon, P. Validation of a markerless motion capture app for automated scoring of sit-to-stand, timed up and go, and short physical performance battery tests in adults with chronic disease. *PLoS Digit. Health* **2026**, *5*, e0001172. [CrossRef]
33. Lim, D.; Kim, C.; Jung, H.; Jung, D.; Chun, K.J. Use of the Microsoft Kinect system to characterize balance ability during balance training. *Clin. Interv. Aging* **2015**, *10*, 1077–1083. [CrossRef]
34. Trabassi, D.; Castiglia, S.F.; Bini, F.; Marinozzi, F.; Ajoudani, A.; Lorenzini, M.; Chini, G.; Varrecchia, T.; Ranavolo, A.; De Icco, R.; et al. Optimizing Rare Disease Gait Classification through Data Balancing and Generative AI: Insights from Hereditary Cerebellar Ataxia. *Sensors* **2024**, *24*, 3613. [CrossRef] [PubMed]
35. Boisgontier, M.P.; Beets, I.A.; Duysens, J.; Nieuwboer, A.; Krampe, R.T.; Swinnen, S.P. Age-related differences in attentional cost associated with postural dual tasks: Increased recruitment of generic cognitive resources in older adults. *Neurosci. Biobehav. Rev.* **2013**, *37*, 1824–1837. [CrossRef]
36. Marusic, U.; Taube, W.; Morrison, S.A.; Biasutti, L.; Grassi, B.; De Pauw, K.; Meeusen, R.; Pisot, R.; Ruffieux, J. Aging effects on prefrontal cortex oxygenation in a posture-cognition dual-task: An fNIRS pilot study. *Eur. Rev. Aging Phys. Act.* **2019**, *16*, 2. [CrossRef]
37. Grabiner, M.D.; Kaufman, K.R. Developing and Establishing Biomechanical Variables as Risk Biomarkers for Preventable Gait-Related Falls and Assessment of Intervention Effectiveness. *Front. Sports Act. Living* **2021**, *3*, 722363. [CrossRef]
38. Ruiz-Ruiz, L.; Jimenez, A.R.; Garcia-Villamil, G.; Seco, F. Detecting Fall Risk and Frailty in Elders with Inertial Motion Sensors: A Survey of Significant Gait Parameters. *Sensors* **2021**, *21*, 6918. [CrossRef]
39. Ortega-Bastidas, P.; Gómez, B.; Aqueveque, P.; Luarte-Martínez, S.; Cano-de-la-Cuerda, R. Instrumented Timed Up and Go Test (iTUG)—More Than Assessing Time to Predict Falls: A Systematic Review. *Sensors* **2023**, *23*, 3426. [CrossRef] [PubMed]
40. Sarvari, M.; Shanbehzadeh, S.; Shavehei, Y.; ShahAli, S. Postural control among older adults with fear of falling and chronic low back pain. *BMC Geriatr.* **2024**, *24*, 862. [CrossRef]
41. Dunne, H.; Strongman, C.; Tarini Clark, S.; Fawden, T.; Roberts, I.; Bance, M. Exploring clinicians' perspectives on falls, balance and gait assessments to inform wearable device development. "Adding to the bigger picture of the patient in falls assessments". *Front. Digit. Health* **2025**, *7*, 1659786. [CrossRef] [PubMed]
42. Kristoffersson, A.; Lindén, M. A Systematic Review of Wearable Sensors for Monitoring Physical Activity. *Sensors* **2022**, *22*, 573. [CrossRef] [PubMed]
43. Horak, F.; King, L.; Mancini, M. Role of body-worn movement monitor technology for balance and gait rehabilitation. *Phys. Ther.* **2015**, *95*, 461–470. [CrossRef] [PubMed]

 https://doi.org/10.3390/s26051623

sensors

MDPI

Article

Impact of Lower-Limb Muscle Fatigue on Dynamic Postural Control During Stair Descent: A Study Using Stair-Embedded Force Plates

Liangsen Wang, Wenyue Ma, Wenfei Zhu, Qian Xie * and Yuliang Sun *

School of Physical Education, Shaanxi Normal University, Xi'an 710119, China; wlsen13@snnu.edu.cn (L.W.); mawnyue@snnu.edu.cn (W.M.); wzhu@snnu.edu.cn (W.Z.)

* Correspondence: qq77@snnu.edu.cn (Q.X.); ysun@snnu.edu.cn (Y.S.); Tel.: +86-186-3882-8666 (Q.X.); +86-135-7200-0729 (Y.S.)

Abstract: This study used stair-embedded force plates to investigate the effects of lower-limb muscle fatigue on dynamic postural control during stair descent in young adults. Twenty-five healthy male adults (age = 19.2 $\pm$ 1.5 years) were tested for stair descent gait in pre-fatigue and post-fatigue conditions. To induce fatigue, participants performed a sit-to-stand task. The kinematic and kinetic data were collected synchronously, and gait parameters were analyzed. Data were analyzed using one-dimensional statistical parametric mapping (SPM1d) and paired t-tests in SPSS. After fatigue, the right knee flexion angle increased significantly across all phases (0–14%, $p < 0.001$; 14–19%, $p = 0.032$; 42–50%, $p = 0.023$; 60–65%, $p = 0.022$; 80–100%, $p = 0.012$). Additionally, the step width widened notably ($p < 0.001$), while the proportion of the swing phase decreased ($p = 0.030$). During the event of right-foot release, the left knee flexion ($p = 0.005$) and ankle dorsiflexion ($p = 0.001$) angle increased significantly, along with a larger left ankle plantarflexion moment ($p = 0.032$). After fatigue, the margin of stability in the anterior–posterior direction (MoS-AP) ($p = 0.002$, $p = 0.014$) and required coefficient of friction (RCOF) ($p = 0.031$, $p = 0.021$) significantly increased at the left-foot release and right-foot release moments. This study demonstrates that lower-limb muscle fatigue increases dynamic instability during stair descent. Participants adopted compensatory strategies, including widening step width, reducing single-support duration, and enhancing ankle plantarflexion to offset knee strength deficits. These adaptations likely reflect central nervous system mechanisms prioritizing stability, highlighting the ankle's compensatory role as a potential target for joint-specific interventions in fall prevention and rehabilitation. Future studies should investigate diverse populations, varying fatigue levels, and comprehensive neuromuscular indicators.

Keywords: force plate; stair descent; muscle fatigue; lower-limb biomechanics

1. Introduction

Stair descent, a highly demanding locomotor task, is common in daily life [1]. Staircases are among the most hazardous locations, both in the workplace and at home, with approximately one out of ten fall accidents occurring on them [2]. These incidents are associated with a significantly higher risk of severe injury or death compared to falls on level ground [3]. This is because stair negotiation, unlike level-ground walking, requires greater lower-limb muscle strength to propel the body upward or downward simultaneously [4]. Studies have shown that, compared with stair ascent, stair descent demands greater balance and postural control and is associated with a higher risk of falls [2]. This requires a coordinated interplay of joint moments and muscle activation to manage impact forces and maintain balance, especially during phases of single-limb support [5]. Therefore, sufficient muscle strength is crucial for preventing falls during stair descent [6,7]. However, muscle fatigue is a frequent occurrence in daily life, which reduces muscle strength and potentially affects motor performance [8]. For example, fatigue occurs after prolonged walking, stair climbing, or other repetitive occupational tasks. Indeed, evidence indicates that approximately one-third of the U.S. workforce engages in substantial physical exertion to meet job demands, often resulting in work-related fatigue [9]. Muscle fatigue is defined as a reduction in the force-generating capacity of the entire neuromuscular system, resulting in a significant decline in muscle strength, impaired neuromuscular control, and reduced motor control performance [8,10]. Extended periods of physical activity can lead to a deterioration of physical capabilities and performance, exacerbating the effects of fatigue, reducing stability, and increasing the risk of accidents, injuries, and falls [8]. Muscle fatigue is a primary task-related risk factor influencing gait [11,12]. Given the heightened biomechanical demands of stair descent compared to level walking, understanding fatigue-induced adaptations in this context is critical for fall prevention. Therefore, investigating the effects of muscle fatigue on stair descent gait is of significant importance.

Although the impact of muscle fatigue on gait has been widely studied in previous research, most studies have focused on level-ground walking or have only analyzed kinematics while ignoring the use of force platforms [13]. Previous studies have demonstrated that fatigue can disrupt joint coordination, delay reflex responses, and reduce dynamic balance during walking [14]. A study found that muscle fatigue may reduce ground dynamic balance, affecting lower-limb kinematics [15]. Conversely, some research has also concluded that after a fatigue intervention, an adult with greater lower-limb strength can still maintain sufficient strength reserves for submaximal-intensity activities, such as walking [16]. From a kinematic perspective, a study found that lower-limb muscular fatigue compromised stair gait during descent [17]. However, few studies have investigated stair gait biomechanics before and after fatigue, particularly utilizing force platforms to quantify kinetic parameters. In addition, most existing studies have primarily used the center of mass (CoM) or center of pressure (CoP) to evaluate balance control. However, the margin of stability in the anterior–posterior direction (MoS-AP) incorporates both CoM velocity and CoP position, providing a more comprehensive measure of dynamic stability [18], while required coefficient of friction (RCOF) reflects the frictional requirements at the foot–ground interface and the associated risk of falling [19,20].

Therefore, this study used stair-embedded force plates to collect kinematic and kinetic data simultaneously and to calculate MoS-AP and RCOF, aiming to investigate the effects of lower-limb muscle fatigue on dynamic postural control during stair descent in young adults. Similar to previous fatigue protocols, we selected multi-joint muscular fatigue of the lower limbs to simulate more realistic scenarios of real-life muscle fatigue [17]. Unlike traditional zero-dimensional (0D) analyses that may overlook time-dependent changes, this study used SPM to analyze stair descent gait, preserving the entire continuous waveform and providing a more sensitive assessment of biomechanical differences [21]. We hypothesize that muscle fatigue will impair dynamic stability during stair descent, altering lower-limb joint kinematics and kinetics. Specifically, fatigue will lead to (1) decreased dynamic stability, as evidenced by increased MoS-AP and RCOF values measured by force platforms, and (2) increased knee flexion and greater ankle plantarflexion as compensatory strategies to counteract reduced quadriceps control.

2. Materials and Methods

2.1. Participants

This study recruited 25 male participants (age = 19.2 ± 1.5 years; height = 1.79 ± 0.06 m; mass = 74.3 ± 6.8 kg). Inclusion criteria included healthy young adults without neurological or musculoskeletal disorders, capable of independently performing stair ascent and descent. Exclusion criteria included (1) neurological or musculoskeletal injuries affecting gait or balance and (2) history of lower-limb fractures. Foot dominance was assessed based on each participant's preferred foot for kicking a ball, and all participants were found to be right-foot dominant [22]. Participation was voluntary, and informed consent was obtained in accordance with the principles outlined in the Declaration of Helsinki. This study was approved by the Ethics Committee of the School of Physical Education at Shaanxi Normal University (202516041).

2.2. Apparatus and Procedures

Our stair-embedded force plate system comprised five steps, with force plates embedded in the first four steps to ensure the collection of complete gait cycles during stair descent. The height between steps is 19 cm, and the width is 30 cm, according to the Chinese national stair standards [23]. Four force plates (Model 9260AA6, Kistler Instrument, Winterthur, Switzerland, 500 × 600 mm, 1000 Hz) were embedded in the staircase's first, second, third, and fourth steps to collect ground reaction force (GRF) data. Kinematics were obtained using a Qualisys capture system (Oqus 700+, Qualisys AB, Gothenburg, Sweden, 200 Hz) with 10 infrared cameras. Kinematic and kinetic data were collected simultaneously (Figure 1).

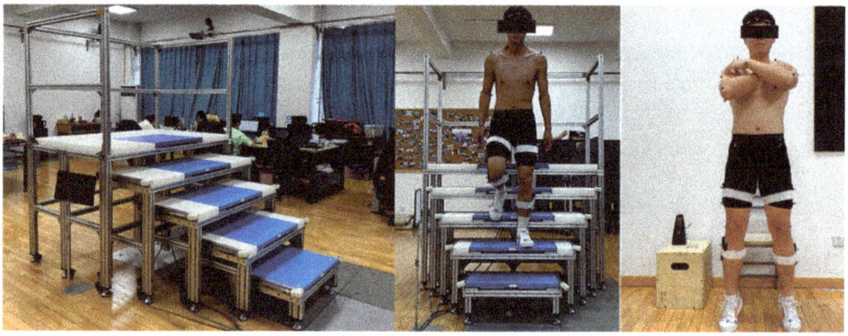

Figure 1. The experiment environment and fatigue protocol.

All participants wore standardized apparel and footwear throughout the study. Participants first completed a baseline stair gait assessment under non-fatigued conditions, followed by a standardized fatigue intervention 72 h later, with immediate post-fatigue stair gait testing. After warming up with stair-stepping exercises (step-ups on a 20 cm step, 2 min, self-selected cadence) [24], each participant had 58 reflective markers placed at specific anatomical locations [25]. All investigations were conducted under constant conditions, with the temperature controlled at 27 °C and the environment kept quiet during the tests. Each subject completed three valid trials, ensuring the left foot stepped on the fourth step during descent.

2.3. Fatigue Protocol

The fatigue protocol targeted the primary muscle groups involved in stair descent, including the quadriceps, gluteal muscles, hamstrings, and gastrocnemius–soleus complex [26]. The fatigue protocol and evaluation methods were referenced from previous studies [27,28]. To induce fatigue, participants performed a sit-to-stand task. A standard chair (42 cm in height, 40 cm in width, and 40 cm in depth) without armrests was used for all participants. The fatigue protocol was terminated when participants reported being unable to continue, failed to maintain the required movement frequency, the Rating of Perceived Exertion (RPE) scale reached 17 or above, or after 30 min had elapsed [27,28]. In addition, conventional sit-to-stand tasks do not adequately fatigue the calf muscles; thus, participants performed the task with simultaneous heel raises to effectively target this muscle group, simulating the fatigue typically experienced in the calf muscles during daily walking (Figure 1).

2.4. Data Analysis

The data were extracted as C3D files into Visual 3D (version 6.0, C-Motion, Germantown, MD, USA) for further processing. Kinematic data were filtered using a fourth-order Butterworth low-pass digital filter with a cutoff frequency of 14 Hz, whereas kinetic data were filtered with a cutoff frequency of 50 Hz [29]. An X-Y-Z Cardan sequence defined the order of rotations according to the right-hand rule about the segment coordinate axes. Inverse dynamics were employed to compute sagittal joint torques in the lower limb, with subsequent normalization to the entire gait cycle set at 100% [23,30]. The position data of the CoM were calculated using a 13-segment model with the weighted sum method, where segmental mass proportions and anthropometric parameters were based on Winter's standardized dataset [30,31]. During the single stance phase, the CoP position in the anterior–posterior (AP) and medial–lateral (ML) directions was determined using the GRFs and torques measured by the force plates at a sampling rate of 1000 Hz [30].

Gait events of the foot contact and release were identified with the Visual 3D Software (version 6.0, C-Motion, Germantown, MD, USA) using a threshold of 10N on the vertical ground reaction force [23]. The stair gait cycle of this research begins when the right foot contacts the third step (0% of the gait cycle) and ends when the right foot contacts the first step (100% of the gait cycle). The 5 steps are numbered from bottom to top, from step 1 to step 5. These contact moments define the start and end of the cycle (Figure 2). Gait events include right-foot contact (third step), left-foot release (fourth step), left-foot contact (second step), right-foot release (third step), and right-foot contact (first step). The first four gait events mark the start of the first double-support phase, the single-support phase, the second double-support phase, and the swing phase.

To quantify dynamic stability while descending stairs, we applied the inverted pendulum model defined by Hof et al. [18]. In this model, the MoS-AP (Figure 3) is calculated as the difference between the anterior–posterior boundary of the highest base of support (BoS) and the extrapolated center of mass (CM). Given that foot placement and step length, and consequently, the position of the CoP under the foot, are constrained by the staircase geometry, the anterior boundary of the BoS was defined by the edge of the stair [32].

$$CM = pCoM + \frac{vCoM}{\sqrt{g/l}}$$

$$\text{MoS} = \text{BoS} - \text{CM}.$$

The BoS represents the maximum support boundary, denoted by CoP in this study. CM is the projection of the center of mass position on the horizontal plane, influenced by velocity. pCoM and vCoM represent the position and velocity of the CoM in the AP direction, respectively; g is gravitational acceleration, and l is the perpendicular distance from the CoM to the ankle joint midpoint. In the sagittal plane, positive values indicate backwards CoM velocity; in the coronal plane, positive values indicate outward CoM velocity. As MoS-AP increases, the distance between CM and CoP increases, reducing stability [33]. The heel position, representing the posterior boundary of the BoS, was determined from reflective markers placed on the posterior aspect of the calcaneus using the 3D motion capture system.

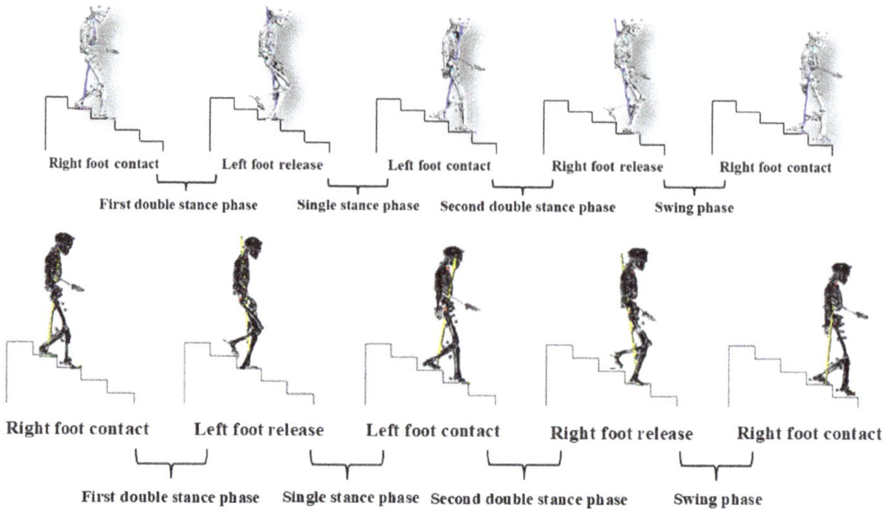

Figure 2. Gait cycle division of the dominant leg (right leg).

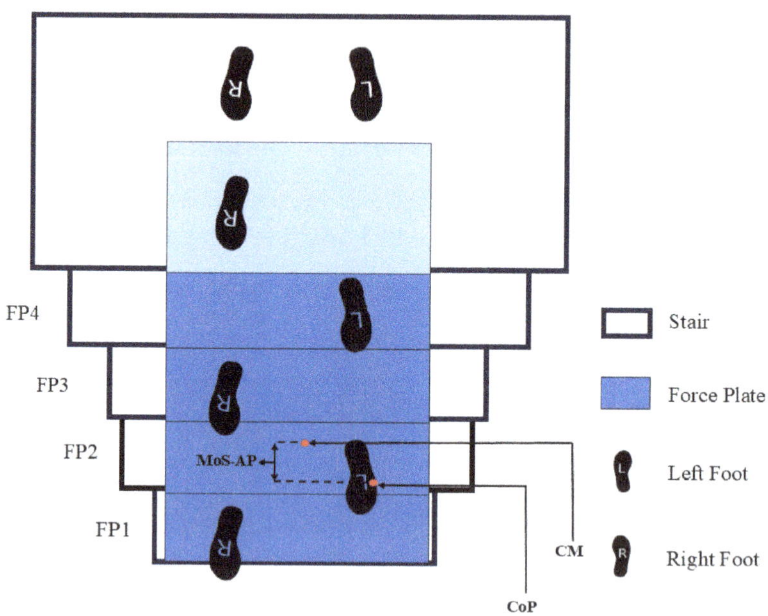

Figure 3. Dynamic stability diagram during stair descent. MoS-AP (margin of stability in the anterior–posterior direction), CoP (center of pressure), and CM (extrapolated center of mass).

The RCOF is the peak ratio of the horizontal ground reaction force to the vertical ground reaction force. It serves as an indicator of slip risk, which is the leading external factor contributing to postural instability and falls. The RCOF is computed separately during the single-support and swing phases [20]. F_{AP} = anterior–posterior GRF, F_{ML} = medial–lateral GRF, and F_V = vertical GRF.

$$RCOF = \frac{\sqrt{F_{AP}^2 + F_{ML}^2}}{F_V}.$$

We extracted discrete values of MoS-AP and RCOF at two critical gait events for analysis: left-foot release and right-foot release.

2.5. Statistical Analysis

Statistical analyses were conducted using SPSS (SPSS Statistics v25, IBM Corp., Armonk, NY, USA) and MATLAB (MATLAB R2018b, The MathWorks Inc., Natick, MA, USA). The analysis compared conditions before and after fatigue. As key variables, our study included joint angles, moments, MoS-AP, and RCOF. All variables passed the Shapiro–Wilk normality test. Using the spm1d toolbox [34], we conducted paired *t*-tests to compare the right lower-limb joint angles/moments during the gait cycle, where effect size measures were computed for each time point and averaged across significant clusters [35]. Additionally, we conducted paired-samples *t*-tests in SPSS to compare both the left lower-limb joint angles/moments during gait events and the MoS-AP and RCOF between pre-fatigue and post-fatigue conditions. Statistical significance was assumed when $p < 0.05$.

3. Results

All 25 participants completed both the fatigue and non-fatigue conditions, with an average fatigue duration of 12.15 ± 2.43 min. Fatigue resulted in a more extended double-support phase (p = 0.005) and a shorter swing phase (p = 0.030). Step width increased significantly ($p < 0.001$), with no significant changes in step length, walking speed, and cadence (Figure 4) (Table 1).

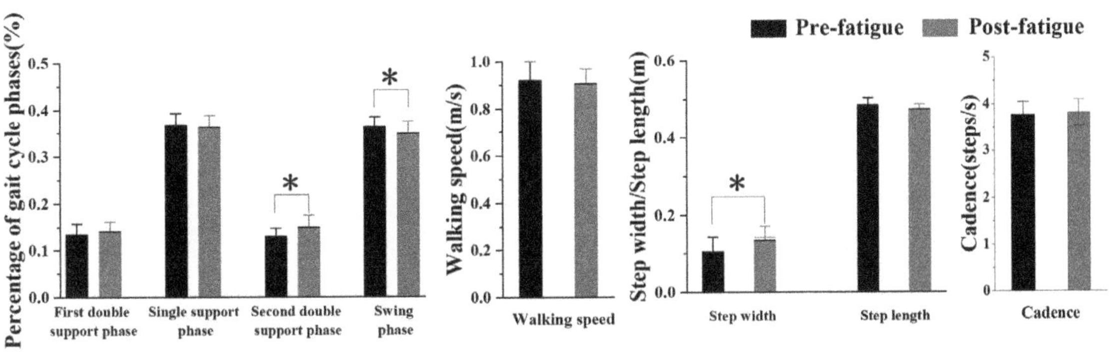

Figure 4. Comparison of spatiotemporal parameters pre- and post-fatigue. *, indicating a statistically significant difference ($p < 0.05$).

SPM1D analysis revealed statistically significant differences in lower-limb joint kinematics and kinetics during stair descent. We found that the right knee flexion angle increased significantly during the first double-support phase (0–14%, $p < 0.001$), single-support phase (14–19%, p = 0.032; 42–50%, p = 0.023), second double-support phase (60–65%, p = 0.022), and swing phase (80–100%, p = 0.012). The ankle plantarflexion moment increased significantly during the first double-support phase (1–14%, p = 0.006) (Figure 5).

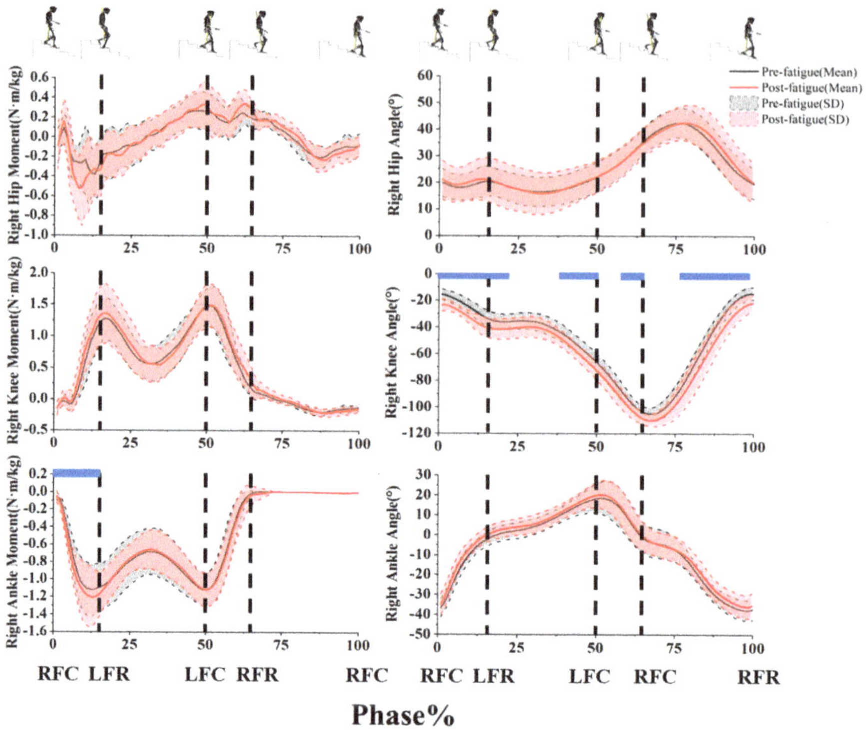

Figure 5. Right lower-limb joint kinetics and kinematics during stair descent derived from SPM1d analysis. The black dashed lines represent gait events: RFC (right-foot contact), LFR (left-foot release), LFC (left-foot contact), and RFR (right-foot release) Blue indicates a statistically significant difference ($p < 0.05$) between pre- and post-fatigue conditions during this phase of the gait cycle. Positive values on the vertical axis represent dorsiflexion, knee extension, and hip flexion, whereas negative values represent plantarflexion, knee flexion, and hip extension.

Additionally, we analyzed the left lower-limb joint angles and moments at the gait event of right-foot release. After fatigue, an increased knee flexion angle ($p = 0.005$) and an increased ankle dorsiflexion angle ($p = 0.001$) were observed in the fatigue condition versus the no-fatigue condition, and the ankle plantarflexion moment ($p = 0.032$) significantly increased. After fatigue, the MoS-AP ($p = 0.002$, $p = 0.014$) and RCOF ($p = 0.031$, $p = 0.021$) increased dramatically at the left-foot release and right-foot release moments (Table 1).

Table 1. Mean (SD) of joint angle and moment measures during stair descent.

Dependent Measures		Pre-Fatigue	Post-Fatigue	*p*
Spatiotemporal parameters	First double-support phase (% of gait cycle)	0.14 (0.03)	0.14 (0.02)	0.571
	Single-support phase (% of gait cycle)	0.36 (0.03)	0.37 (0.03)	0.251
	Second double-support phase (% of gait cycle)	0.12 (0.02)	0.13 (0.03)	0.005 *
	Swing phase (% of gait cycle)	0.37 (0.02)	0.36 (0.03)	0.030 *
	Cadence (steps/s)	3.77 (0.27)	3.81 (0.28)	0.645
	Walking speed (m/s)	0.92 (0.08)	0.92 (0.10)	0.960
	Step width (m)	0.11 (0.04)	0.12 (0.03)	$p < 0.001$ *
	Step length (m)	0.49 (0.03)	0.48 (0.02)	0.414

Table 1. *Cont.*

Dependent Measures		Pre-Fatigue	Post-Fatigue	p
Left Hip joint angle (°)	Right-foot release	21.32 (4.36)	22.78 (8.96)	0.312
Left Knee joint angle (°)	Right-foot release	−36.36 (5.24)	−41.11 (6.15)	0.005 *
Left Ankle joint angle (°)	Right-foot release	88.42 (4.55)	91.39 (3.44)	0.001 *
Left Hip joint moment (N·m/kg)	Right-foot release	−0.25 (0.12)	−0.30 (0.20)	0.386
Left Knee joint moment (N·m/kg)	Right-foot release	1.27 (0.33)	1.25 (0.26)	0.796
Left Ankle joint moment (N·m/kg)	Right-foot release	−1.12 (0.21)	−1.24 (0.26)	0.032 *
MoS-AP(m)	Left-foot release	−0.13 (0.04)	−0.23 (0.17)	0.002 *
RCOF	Right-foot release	−0.13 (0.03)	−0.16 (0.02)	0.014 *
	Left-foot release	0.20 (0.05)	0.23 (0.04)	0.031 *
	Right-foot release	0.15 (0.02)	0.17 (0.02)	0.021 *

'*' indicates statistical significance. Positive values on the vertical axis represent dorsiflexion, knee extension, and hip flexion, whereas negative values represent plantarflexion, knee flexion, and hip extension.

4. Discussion

In this study, we investigate the effects of lower-limb muscle fatigue on dynamic postural control during stair descent in young adults, focusing on lower-limb kinematics and kinetics. Muscle fatigue significantly impaired dynamic stability during stair descent, as reflected by increased MoS-AP and RCOF. Kinematically, fatigue increased knee flexion angles across all gait phases and ankle dorsiflexion angles, widened step width, and reduced the swing phase proportion. Regarding joint kinetics, ankle plantarflexion moments increased in both limbs after fatigue. These findings align with our hypothesis that muscle fatigue impairs dynamic stability during stair descent, while the human body employs a series of compensatory neuromuscular adaptations to maintain postural control.

Descending stairs is more challenging than walking on level ground or ascending stairs [36]. Therefore, fatigued participants may have difficulty controlling their body movements while descending stairs [17]. This phenomenon is validated by the changes in the MoS-AP and the RCOF during stair descent after fatigue observed in our study. Dynamic stability refers to the body's ability to resist external disturbances and maintain stability during movement [18]. During stair descent, dynamic stability manifests as a series of balance losses and recoveries, particularly in the three-dimensional dynamic process where the CoM constantly changes [32]. Studies have shown that the MoS-AP is typically positive when walking on level ground, indicating a low risk of falling [37]. However, during stair descent, our study found that all participants showed negative MoS-AP values, regardless of fatigue. This is consistent with previous studies [32]. From a mechanical perspective, a negative MoS-AP is necessary and energy-efficient during stair descent, facilitating a smooth transition to the next step [32,38]. Although a negative MoS-AP is typically associated with a higher risk of falling, it does not necessarily indicate an immediate fall. Negative MoS-AP suggests that the individual must take immediate corrective actions to prevent falls [17,39]. The heart center position used in this study is consistent with the dynamic stability concept proposed by Hof [18]. After fatigue, the moment of right-foot release showed a higher RCOF and sagittal plane MoS-AP, indicating increased fall risk and reduced balance control, reflecting compromised postural stability. This change was attributed to muscle fatigue leading to decreased muscle strength, impairing neuromuscular control [12].

Stair descent imposes fundamentally different biomechanical demands than level-ground walking, requiring greater muscle strength to propel the body forward and resist gravitational acceleration [4,36]. This explains why significant kinematic alterations emerge at lower-limb joints when comparing pre-fatigue and post-fatigue conditions during descent [17]. Our study revealed substantial post-fatigue adaptations during stair descent. We found that knee flexion angles increased during all phases, and ankle plantarflexion angles and moments increased during the double-support phase, particularly at single stance. Simultaneously, increased ankle plantarflexion moments in both limbs may compensate for diminished triceps surae strength by requiring greater distal output to sustain propulsion. This suggests an increased reliance on the gastrocnemius and other plantar flexors to enhance push-off power, offsetting diminished proximal joint efficiency [40,41]. In parallel, the widened step width and reduced proportion of swing phase suggest adjustments in spatiotemporal parameters to enhance stability. Studies have demonstrated that an increased step width can expand the BOS, a critical factor in strengthening dynamic stability [42–44]. These adaptations align with previous studies reporting greater knee flexion after quadriceps fatigue [45], and that ankle kinematics may adjust through enhanced plantarflexion following fatigue [11,45]. They highlight a joint-specific redistribution of mechanical workload across the lower limb, whereby different joints compensate for reduced function in other segments. Such alterations likely represent coordinated neuromuscular strategies by the central nervous system to preserve postural control and ensure safe forward progression under fatigue [46–49].

While these findings contrast with reports of reduced joint flexion under limb muscle fatigue protocols [17,50], these discrepancies may stem from differences in experimental design. Multi-joint fatigue may provoke increased leg stiffness [50], resulting in smaller plantarflexion, knee flexion, and hip flexion [17]. This may be due to differences in the fatigue protocols. In contrast, the increased knee flexion observed in our study may reflect a shock absorption strategy. Specifically, the increased knee flexion observed here could reflect an attempt to reduce impact forces by lengthening the time for energy dissipation during landing [51,52]. This strategy could help maintain step-to-step stability even in decreased muscular stiffness or delayed reflexes.

Although some of the statistically significant differences identified were small in magnitude, this could be due to the fatigue level not being sufficiently severe. Participants may not have been required to drastically modify their gait strategies to cope with the demands of stair descent. Similar phenomena have been observed in level-ground walking, where modest fatigue levels can still induce subtle compensatory gait changes. Furthermore, although stair descent requires greater muscular strength than level walking, it may not reach the threshold at which dynamic balance is critically impaired or injury risk substantially increases [15].

Limitations of this study include the following: due to difficulties in recruiting participants and the potential risks of stair descent for older adults, we chose to study healthy young male adults. Future studies should consider including females and older adults with pathologies. Furthermore, this study did not examine the role of upper limb joints, though the upper limbs are crucial for balance. Lastly, future studies should employ varying fatigue protocols to explore the fatigue threshold that disrupts stair descent balance and the compensatory mechanisms of lower-limb joints after fatigue. Moreover, we did not include objective indicators of fatigue intensity, such as electromyographic median frequency or torque decline, which could provide a more precise quantification of fatigue depth. Future studies should employ varying fatigue protocols to explore the threshold at which compensatory mechanisms fail, and to characterize potential nonlinearities in the response of joint kinematics and kinetics to progressive fatigue. Lastly, although the chair and stair heights were fixed, participants had different statures, which may have influenced joint kinematics and kinetics during stair ascent.

5. Conclusions

In summary, this study demonstrates that lower-limb muscle fatigue increases dynamic instability during stair descent, which is partially compensated by adaptive postural control mechanisms, including widening step width and reducing single-support duration, while increasing ankle plantarflexion output to compensate for strength deficits in the knee joint. These postural controls likely reflect central nervous system compensatory strategies prioritizing stability to mitigate balance impairments caused by diminished muscular capacity. Although it may appear common sense that fatigue compromises stair descent stability, our study provides quantitative biomechanical evidence to specify how joint kinematics, kinetics, and neuromuscular coordination adapt under localized fatigue. This mechanistic insight is critical because clinical fall prevention and rehabilitation programs require precise targets for intervention rather than general assumptions. For example, identifying the ankle's compensatory role under knee fatigue may inform joint-specific strengthening or fatigue-resistance training protocols. While the underlying neuromuscular mechanisms remain unclear, future studies should investigate diverse populations, varying fatigue levels, and comprehensive neuromuscular indicators.

Author Contributions: Data curation: L.W. and W.M.; formal analysis: L.W.; investigation: W.Z.; methodology: L.W.; writing—original draft: L.W.; writing—review and editing: Q.X. and Y.S. All authors have read and agreed to the published version of the manuscript.

Funding: This study was supported by the General Project of Education, National Social Science Foundation of China (Grant No. BLA230108).

Institutional Review Board Statement: This study was conducted by the Declaration of Helsinki and approved by the Ethics Committee of Shaanxi Normal University (No. 202516041, 19 June 2025).

Informed Consent Statement: Informed consent was obtained from all subjects involved in the study.

Data Availability Statement: The data presented in this study are available on request from the corresponding author. The data are not publicly available due to confidentiality reasons.

Acknowledgments: The authors thank the other investigators, the staff, and the participants of this study for their valuable contributions.

Conflicts of Interest: The authors declare no conflicts of interest.

References

1. Cayless, S.M. Slip, Trip and Fall Accidents: Relationship to Building Features and Use of Coroners' Reports in Ascribing Cause. *Appl. Ergon.* **2001**, *32*, 155–162. [CrossRef] [PubMed]
2. Startzell, J.K.; Owens, D.A.; Mulfinger, L.M.; Cavanagh, P.R. Stair Negotiation in Older People: A Review. *J. Am. Geriatr. Soc.* **2000**, *48*, 567–580. [CrossRef]
3. Jacobs, J.V. A Review of Stairway Falls and Stair Negotiation: Lessons Learned and Future Needs to Reduce Injury. *Gait Posture* **2016**, *49*, 159–167. [CrossRef]
4. McFadyen, B.J.; Winter, D.A. An Integrated Biomechanical Analysis of Normal Stair Ascent and Descent. *J. Biomech.* **1988**, *21*, 733–744. [CrossRef]
5. Winter, D. Human Balance and Posture Control during Standing and Walking. *Gait Posture* **1995**, *3*, 193–214. [CrossRef]
6. Karamanidis, K.; Arampatzis, A. Altered Control Strategy between Leading and Trailing Leg Increases Knee Adduction Moment in the Elderly While Descending Stairs. *J. Biomech.* **2011**, *44*, 706–711. [CrossRef]
7. Reeves, N.D.; Spanjaard, M.; Mohagheghi, A.A.; Baltzopoulos, V.; Maganaris, C.N. The Demands of Stair Descent Relative to Maximum Capacities in Elderly and Young Adults. *J. Electromyogr. Kinesiol.* **2008**, *18*, 218–227. [CrossRef]
8. Hsiao, H.; Simeonov, P. Preventing Falls from Roofs: A Critical Review. *Ergonomics* **2001**, *44*, 537–561. [CrossRef]

9. Swaen, G.M.H.; Van Amelsvoort, L.G.P.M.; Bültmann, U.; Kant, I.J. Fatigue as a Risk Factor for Being Injured in an Occupational Accident: Results from the Maastricht Cohort Study. *Occup. Env. Med.* **2003**, *60* (Suppl. S1), i88–i92. [CrossRef]
10. Yu, C.; Zhan, J.; Xu, L.; Zhou, J.; Fu, W. Motor Control Performance-Related Modulation of Beta-Band EEG–sEMG Coherence Differs between General and Local Muscular Exercise-Induced Fatigue. *Eur. J. Appl. Physiol.* **2025**, *125*, 1869–1879. [CrossRef] [PubMed]
11. Lew, F.L.; Qu, X. Effects of Multi-Joint Muscular Fatigue on Biomechanics of Slips. *J. Biomech.* **2014**, *47*, 59–64. [CrossRef] [PubMed]
12. Rapp Van Roden, E.A.; George, J.; Milan, L.T.; Bove, R.T. Evaluation of Injury Patterns and Accident Modality in Step Ladder-Related Injuries. *Appl. Ergon.* **2021**, *96*, 103492. [CrossRef]
13. Hunt, M.A.; Hatfield, G.L. Ankle and Knee Biomechanics during Normal Walking Following Ankle Plantarflexor Fatigue. *J. Electromyogr. Kinesiol.* **2017**, *35*, 24–29. [CrossRef]
14. Mizrahi, J.; Verbitsky, O.; Isakov, E.; Daily, D. Effect of Fatigue on Leg Kinematics and Impact Acceleration in Long Distance Running. *Hum. Mov. Sci.* **2000**, *19*, 139–151. [CrossRef]
15. Wang, L.; Ma, W.; Zhu, W.; Zhai, L.; Sun, Y. Effects of Experimentally Induced Lower Limb Muscle Fatigue on Healthy Adults' Gait: A Systematic Review. *Bioengineering* **2025**, *12*, 225. [CrossRef]
16. Longpré, H.S.; Potvin, J.R.; Maly, M.R. Biomechanical Changes at the Knee after Lower Limb Fatigue in Healthy Young Women. *Clin. Biomech.* **2013**, *28*, 441–447. [CrossRef] [PubMed]
17. Qu, X. Effects of Lower-Limb Muscular Fatigue on Stair Gait. *J. Biomech.* **2015**, *48*, 4059–4064. [CrossRef]
18. Hof, A.L.; Gazendam, M.G.J.; Sinke, W.E. The Condition for Dynamic Stability. *J. Biomech.* **2005**, *38*, 25. [CrossRef]
19. Predicting Slips and Falls Considering Required and Available Friction (Topic)–1–All Databases. Available online: https://webofscience.clarivate.cn/wos/alldb/summary/ba979230-3bab-4963-8c41-dac46fe83edd-0140a32994/relevance/1 (accessed on 7 January 2025).
20. Liu, J.; Qu, X. Postural Stability and Risk of Slips in Lifting Tasks: Effects of Load Weight and Load Knowledge. *Int. J. Ind. Ergon.* **2025**, *105*, 103675. [CrossRef]
21. Huang, R.-F.; Yick, K.-L.; Shi, Q.-Q.; Liu, L.; Li, C.-H. Acute Effects of Different Types of Compression Legwear on Biomechanics of Countermovement Jump: A Statistical Parametric Mapping Analysis. *J. Funct. Morphol. Kinesiol.* **2025**, *10*, 257. [CrossRef] [PubMed]
22. van Melick, N.; Meddeler, B.M.; Hoogeboom, T.J.; Nijhuis-van der Sanden, M.W.; van Cingel, R.E. How to Determine Leg Dominance: The Agreement between Self-Reported and Observed Performance in Healthy Adults. *PLoS ONE* **2017**, *12*, e0189876. [CrossRef]
23. Lu, Z.; Mao, C.; Tan, Y.; Liu, T.; Li, X.; Li, Z.; Zhu, W.; Sun, Y. The Impact of Backpack Load on Adolescent's Stair Descent Gait. *J. Biomech.* **2024**, *166*, 112029. [CrossRef]
24. Barkley, J.E.; Lepp, A. Cellular Telephone Use during Free-Living Walking Significantly Reduces Average Walking Speed. *BMC Res. Notes* **2016**, *9*, 195. [CrossRef]
25. Cappozzo, A.; Catani, F.; Croce, U.D.; Leardini, A. Position and Orientation in Space of Bones during Movement: Anatomical Frame Definition and Determination. *Clin. Biomech.* **1995**, *10*, 171–178. [CrossRef]
26. Kowalski, E.; Pelegrinelli, A.R.M.; Ryan, N.; Dervin, G.; Lamontagne, M. Muscle Activity and Biomechanics While Descending a Staircase after Total Knee Arthroplasty: A Study Comparing Different Posterior Stabilized and Medial Ball-and-Socket Designs. *J. Arthroplast.* **2024**, *39*, 3076–3083.e2. [CrossRef]
27. Barbieri, F.A.; Gobbi, L.T.B.; Lee, Y.J.; Pijnappels, M.; Van Dieën, J.H. Effect of Triceps Surae and Quadriceps Muscle Fatigue on the Mechanics of Landing in Stepping down in Ongoing Gait. *Ergonomics* **2014**, *57*, 934–942. [CrossRef] [PubMed]
28. Barbieri, F.A.; Dos Santos, P.C.R.; Vitório, R.; Van Dieën, J.H.; Gobbi, L.T.B. Effect of Muscle Fatigue and Physical Activity Level in Motor Control of the Gait of Young Adults. *Gait Posture* **2013**, *38*, 702–707. [CrossRef] [PubMed]
29. Sun, Y.; Wei, S.; Zhong, Y.; Fu, W.; Li, L.; Liu, Y. How Joint Torques Affect Hamstring Injury Risk in Sprinting Swing-Stance Transition. *Med. Sci. Sports Exerc.* **2015**, *47*, 373–380. [CrossRef] [PubMed]
30. Winter, D.A.; Prince, F.; Stergiou, P.; Powell, C. Medial-Lateral and Anterior-Posterior Motor Responses Associated with Centre of Pressure Changes in Quiet Standing. *Neurosci. Res. Commun.* **1993**, *12*, 141–148.
31. Biomechanics and Motor Control of Human Movement | Wiley Online Books. Available online: https://onlinelibrary.wiley.com/doi/book/10.1002/9780470549148 (accessed on 4 September 2024).
32. Bosse, I.; Oberländer, K.D.; Savelberg, H.H.; Meijer, K.; Brüggemann, G.-P.; Karamanidis, K. Dynamic Stability Control in Younger and Older Adults during Stair Descent. *Hum. Mov. Sci.* **2012**, *31*, 1560–1570. [CrossRef]
33. Zheng, H.; Song, Q.; Zhang, C.; Sun, W.; Mao, M.; Zhang, X.; Zhu, X.; Ma, G.; Mao, D. The Effect of Text-Based Math Task on Dynamic Stability Control during Stair Descent (ID: BM-D-20-00079R3). *J. Biomech.* **2020**, *113*, 110088. [CrossRef] [PubMed]
34. Pataky, T.C. One-Dimensional Statistical Parametric Mapping in Python. *Comput. Methods Biomech. Biomed. Engin.* **2012**, *15*, 295–301. [CrossRef] [PubMed]

35. Yang, C.; Bouffard, J.; Srinivasan, D.; Ghayourmanesh, S.; Cantú, H.; Begon, M.; Côté, J.N. Changes in Movement Variability and Task Performance during a Fatiguing Repetitive Pointing Task. *J. Biomech.* **2018**, *76*, 212–219. [CrossRef]
36. Zachazewski, J.E.; Riley, P.O.; Krebs, D.E. Biomechanical Analysis of Body Mass Transfer during Stair Ascent and Descent of Healthy Subjects. *J. Rehabil. Res. Dev.* **1993**, *30*, 412–422.
37. Bohm, S.; Mersmann, F.; Bierbaum, S.; Dietrich, R.; Arampatzis, A. Cognitive Demand and Predictive Adaptational Responses in Dynamic Stability Control. *J. Biomech.* **2012**, *45*, 2330–2336. [CrossRef]
38. Kuo, A.D.; Donelan, J.M.; Ruina, A. Energetic Consequences of Walking like an Inverted Pendulum: Step-to-Step Transitions. *Exerc. Sport. Sci. Rev.* **2005**, *33*, 88–97. [CrossRef] [PubMed]
39. Young, P.M.M.; Wilken, J.M.; Dingwell, J.B. Dynamic Margins of Stability during Human Walking in Destabilizing Environments. *J. Biomech.* **2012**, *45*, 1053–1059. [CrossRef]
40. Gallinger, T.L.; Fletcher, J.R.; MacIntosh, B.R. Mechanisms of Reduced Plantarflexor Function in Cerebral Palsy: Smaller Triceps Surae Moment Arm and Reduced Muscle Force. *J. Biomech.* **2020**, *110*, 109959. [CrossRef]
41. Shi, F.; Rymer, W.Z.; Son, J. Mechanomyogram Amplitude vs. Isometric Ankle Plantarflexion Torque of Human Medial Gastrocnemius Muscle at Different Ankle Joint Angles. *J. Electromyogr. Kinesiol.* **2021**, *61*, 102609. [CrossRef]
42. Young, P.M.M.; Dingwell, J.B. Voluntary Changes in Step Width and Step Length during Human Walking Affect Dynamic Margins of Stability. *Gait Posture* **2012**, *36*, 219–224. [CrossRef]
43. Bauby, C.E.; Kuo, A.D. Active Control of Lateral Balance in Human Walking. *J. Biomech.* **2000**, *33*, 1433–1440. [CrossRef] [PubMed]
44. Yocum, D.; Weinhandl, J.T.; Fairbrother, J.T.; Zhang, S. Wide Step Width Reduces Knee Abduction Moment of Obese Adults during Stair Negotiation. *J. Biomech.* **2018**, *75*, 138–146. [CrossRef]
45. Parijat, P.; Lockhart, T.E. Effects of Quadriceps Fatigue on the Biomechanics of Gait and Slip Propensity. *Gait Posture* **2008**, *28*, 568–573. [CrossRef]
46. Tresch, M.C.; Jarc, A. The Case for and against Muscle Synergies. *Curr. Opin. Neurobiol.* **2009**, *19*, 601–607. [CrossRef]
47. Latash, M.L. The Bliss (Not the Problem) of Motor Abundance (Not Redundancy). *Exp. Brain Res.* **2012**, *217*, 1–5. [CrossRef] [PubMed]
48. Kellis, E.; Liassou, C. The Effect of Selective Muscle Fatigue on Sagittal Lower Limb Kinematics and Muscle Activity during Level Running. *J. Orthop. Sports Phys. Ther.* **2009**, *39*, 210–220. [CrossRef] [PubMed]
49. Paillard, T. Effects of General and Local Fatigue on Postural Control: A Review. *Neurosci. Biobehav. Rev.* **2012**, *36*, 162–176. [CrossRef]
50. Hortobágyi, T.; DeVita, P. Altered Movement Strategy Increases Lower Extremity Stiffness during Stepping down in the Aged. *J. Gerontol. A Biol. Sci. Med. Sci.* **1999**, *54*, B63–B70. [CrossRef]
51. DeVita, P.; Helseth, J.; Hortobagyi, T. Muscles Do More Positive than Negative Work in Human Locomotion. *J. Exp. Biol.* **2007**, *210*, 3361–3373. [CrossRef]
52. Boyer, K.A.; Nigg, B.M. Muscle Activity in the Leg Is Tuned in Response to Impact Force Characteristics. *J. Biomech.* **2004**, *37*, 1583–1588. [CrossRef]

sensors

MDPI

Article

QAMT: An LLM-Based Framework for Quality-Assured Medical Time-Series Data Generation

Yi Luo [1,2], Yong Zhang [2,*], Chunxiao Xing [2], Peng Ren [2] and Xinhao Liu [1]

1 School of Computer Science and Technology, Beijing Institute of Technology, Beijing 100081, China; luoyi@bit.edu.cn (Y.L.); liuxinhao@bit.edu.cn (X.L.)
2 BNRist, DCST, RIIT, Tsinghua University, Beijing 100084, China; xingcx@tsinghua.edu.cn (C.X.); renpeng@tsinghua.edu.cn (P.R.)
* Correspondence: zhangyong05@tsinghua.edu.cn

Abstract: The extensive deployment of diverse sensors in hospitals has resulted in the collection of various medical time-series data. However, these real-world medical time-series data suffer from limited volume, poor data quality, and privacy concerns, resulting in performance degradation in downstream tasks, such as medical research and clinical decision-making. Existing studies provide generated medical data as a supplement or alternative to real-world data. However, medical time-series data are inherently complex, including temporal data such as laboratory measurements and static event data such as demographics and clinical outcomes, with each patient's temporal data being influenced by their static event data. This intrinsic complexity makes the generation of high-quality medical time-series data particularly challenging. Traditional methods typically employ Generative Adversarial Networks (GANs) or Variational Autoencoders (VAEs), but these methods struggle to generate high-quality static event data of medical time-series data and often lack interpretability. Currently, large language models (LLMs) introduce new opportunities for medical data generation, but they face difficulties in generating temporal data and have challenges in specific domain generation tasks. In this study, we are the first to propose an LLM-based framework for modularly generating medical time-series data, QAMT, which generates quality-assured data and ensures the interpretability of the generation process. QAMT constructs a reliable health knowledge graph to provide medical expertise to the LLMs and designs dual modules to simultaneously generate static event data and temporal data, constituting high-quality medical time-series data. Moreover, QAMT introduces a quality assurance module to evaluate the generated data. Unlike existing methods, QAMT preserves the interpretability of the data generation process. Experimental results show that QAMT can generate higher-quality time-series medical data compared with existing methods.

Keywords: medical time-series data generation; large language models; health knowledge graph; data quality assurance

1. Introduction

With the development of Internet-of-Things (IoT) technologies, hospitals are increasingly equipped with a wide array of sensors that monitor and capture various medical data. Since these sensors record data continuously (e.g., vital signs), most of the medical data exist in the form of time-series data. These data support medical research through effective exploration and mining [1] and offer valuable assistance in tasks such as health monitoring [2], clinical decision-making [3], disease diagnosis [4,5], and treatment recommendation [6].

Although medical time-series data have demonstrated substantial value in various downstream tasks, their acquisition remains a significant bottleneck. On one hand, although sensors collect data in real time, issues such as sensor malfunctions and the inherent complexity of medical time-series data often lead to challenges in data quality. As a result, medical time-series datasets commonly used, such as eICU [7] and MIMIC-III [8], suffer from missing values, inaccuracies, incompleteness, etc., ultimately resulting in poor data quality [9]. On the other hand, because medical time-series data contain highly sensitive information (e.g., patients' age, gender), their access is typically subject to strict privacy regulations and governance controls [10], resulting in limited data volume and privacy concerns. Some studies attempt to mitigate privacy risks by removing personal information from medical data. However, even de-identified data remain vulnerable to re-identification [11,12]. Other studies also explore the use of federated learning to address this issue. However, these works still face challenges related to system deployment and security attacks [13]. Motivated by these limitations, generated medical time-series data offer a supplement and alternative to real-world data.

However, due to the complexity of medical time-series data, which encompass a wide variety of variables that interact in complex and nonlinear ways, generating such data is challenging. A critical aspect of medical time-series data is the presence of static event data (e.g., demographics, clinical outcomes), which can significantly influence temporal data (e.g., vital signs, laboratory measurements). For instance, an in-hospital patient's temporal data, which would include hundreds of variables, such as heart rate and blood pressure, can be influenced by several static event variables, including age, comorbidities, etc. We define an effective medical time-series generation model as one that jointly models both static event data and temporal data. Consequently, generating medical time-series data presents several significant challenges:

- **Challenge 1: Joint Generation of Temporal Data and Static Event Data.** Medical time-series data consist of diverse data types, each with unique characteristics. Among them, static event data are generally high-dimensional and discrete (e.g., demographics, clinical outcomes), whereas temporal data tend to be lower-dimensional and continuous (e.g., vital signs, laboratory measurements). Therefore, jointly generating medical time-series data that include both static and temporal components is essential for producing comprehensive and realistic datasets.
- **Challenge 2: Clinical Constraints and Variable Dependencies.** Many variables in medical time-series data are governed by clinical constraints, and their values often exhibit strong interdependencies. For example, a patient's systolic blood pressure value of zero is clinically impossible. Additionally, if a patient consistently exhibits systolic blood pressure readings above 150 mmHg, a final diagnosis would not be hypotension. Accurately modeling these constraints and dependencies is essential to ensure the clinical plausibility of generated data.
- **Challenge 3: Need for Interpretability.** The generation process of medical time-series data should be interpretable. In clinical research and practice, it is crucial for stakeholders to understand how the data are produced in order to evaluate their quality, support downstream applications, and maintain trust in data-driven healthcare systems.

Table 1 shows that traditional medical time-series data generation methods based on deep learning techniques primarily rely on generative models, such as Variational Autoencoders (VAEs) [14] and Generative Adversarial Networks (GANs) [15]. However, although GANs and VAEs have demonstrated strong performance in generating long-sequence temporal data, they face significant challenges in simultaneously producing high-quality static event data. A few studies have attempted to address this by incorporating statistical summaries of the generated temporal data as inputs to guide the generation of static event data [16]. Nevertheless, GANs and VAEs still struggle to generate sparse one-hot encoded representations [17] with satisfactory fidelity. In addition, it is challenging to incorporate clinical constraints and variable dependencies into these models, and the black-box nature of deep learning models further limits the interpretability of the generation process. The emergence of large language models (LLMs) presents a promising new direction for medical time-series data generation [9,10]. LLMs have demonstrated strong capabilities in modeling complex distributions over discrete data [18]. However, since LLMs are primarily designed for text-based tasks, they struggle to jointly generate high-quality continuous temporal data [10]. Moreover, due to their limited exposure to domain-specific medical knowledge, generating clinically meaningful and high-quality medical time-series data remains a challenge for LLMs [19]. A recent study [20] proposes a pipeline framework for medical time-series data generation. However, this framework lacks clinical constraints and variable dependencies, as well as interpretability in the data generation process. Therefore, jointly generating high-quality medical time-series data that include temporal data and static event data with an interpretable generation process remains a significant challenge in current research.

Table 1. Comparison of medical time-series data generation methods.

Method	Type	Domain	Joint Generation	Quality Assurance [1]	Interpretability
SynTEG [20]	GAN-based	Medical	✗	-	✗
BGAN [21]	GAN-based	Medical	✗	-	✗
MTGAN [22]	GAN-based	Medical	✗	-	✗
VGAE [23]	VAE-based	Medical	✗	-	✗
PromptEHR [19]	LLM-based	Medical	✗	CC, VD	✗
CEHR-GPT [24]	LLM-based	Medical	✗	CC	✗
RTSGAN [25]	GAN-based	Medical	✗	-	✗
TimEHR [26]	GAN-based	Medical	✗	-	✗
EHR-Safe [27]	GAN-based	Medical	✗	-	✗
CodeAR [28]	VAE-based	Medical	✗	-	✗
GenG [29]	LLM-based	Energy	✗	-	✗
HGAN [16]	GAN-based	Medical	✓	CC, VD	✗
HALO [9]	LLM-based	Medical	✓	CC	✗
SynEHRgy [10]	LLM-based	Medical	✓	-	✗
QAMT (ours)	LLM-based	medical	✓	CC, VD	✓

[1] "CC" refers to clinical constraints, and "VD" refers to variable dependencies. "✓" indicates that the method has solved the corresponding challenge, while "✗" indicates that it has not been solved.

In this study, we propose a modularized LLM-based framework for **Q**ality-**A**ssured **M**edical **T**ime-series data generation, QAMT. To incorporate domain-specific medical knowledge into LLMs and guide both the generation and quality assurance of medical time-series data, we leverage a reliable Health Knowledge Graph Builder, HKGB [30], to build a health knowledge graph (HKG). During the medical time-series generating process, QAMT adopts dual modules, combining GANs for generating temporal data and health knowledge graph-based retrieval augmented generation for generating static event data. This design leverages the advantages of existing methods and jointly generates medical time-series data. To assess the quality of the generated medical time-series data, QAMT integrates a quality assurance module that evaluates the data through clinical constraints and inter-variable dependencies, using the capabilities of the HKGB, LLMs, and a chain-of-thought prompting over HKG. In contrast to existing methods, QAMT maintains interpretability throughout the medical time-series data generation process. The main contributions of this work are as follows:

- QAMT jointly generates medical time-series data, which include both continuous temporal data and discrete static event data.
- QAMT ensures the quality assurance of the generated data by accounting for real-world clinical constraints and variable dependencies.
- QAMT enables interpretability in the medical time-series data generation process.
- QAMT is evaluated on the eICU and MIMIC-III datasets, and demonstrates superior performance compared to state-of-the-art models in terms of fidelity, utility, and privacy.

In this paper, we outline the problem definition and review related work in Section 2. Section 3 introduces the overall architecture and workflow of the proposed QAMT framework. In Section 4, we delve into the design of the health knowledge graph (HKG) module and explain the motivation for incorporating HKG. Section 5 details the dual modules responsible for jointly generating medical time-series data, including the static event data generation module and temporal data generation module. We present our data quality assurance module, including how clinical constraints and variable dependencies are enforced in Section 6. In Section 7, we provide an explanation of the interpretability of QAMT. In Section 8, we provide experimental results to evaluate our proposed QAMT. Finally, Section 9 summarizes the key contributions of this work and discusses potential directions for future research.

2. Preliminaries and Related Work

In this chapter, Section 2.1 provides the definition of our medical time-series data generation task. Sections 2.2–2.4 discuss the limitations of existing work in jointly generating high-quality and interpretable medical time-series data.

2.1. Definition

A medical time-series dataset is denoted as $D = \{P_i\}_{i=1}^{N}$, where N represents the number of patients, and P_i denotes the medical time-series data of the ith patient. Each patient record P_i consists of a sequence of clinical visits, denoted as $P_i = \{V_{ij}, T_{ij}\}_{j=1}^{|V_i|}$, and each visit V_{ij} at time T_{ij} is defined as a tuple $V_{ij} = \{c_{ij}, y_{ij}, E_{ij}, t_{ij}\}$, where: c_{ij} is a set of covariates (e.g., demographics) of this patient, y_{ij} is a set of clinical outcomes (e.g., expire), $E_{ij} = \{e_m\}_{m=1}^{|E_j|}$ is a sequence of clinical events (e.g., ICD diagnosis codes), and t_{ij} is a collection of temporal data recorded in in-patient settings, including vital signs and laboratory measurements [10]. We refer to $m_{ij} = \{c_{ij}, y_{ij}, E_{ij}\}$ collectively as **static event data** (high-dimensional and discrete data), and t_{ij} as **temporal data** (low-dimensional and continuous data). Each temporal data element t_{ij} is further represented as $t_{ij} = \left\{time_k, \left\{(n_{kp}, x_{kp})\right\}_{p=1}^{|L_k|}\right\}_{k=1}^{|t_{ij}|}$, where $|t_{ij}|$ denotes the total number of measurement time points, $|L_k|$ denotes the number of variables measured at time point $time_k$, n_{kp} is the name of the pth variable measured at time $time_k$, and x_{kp} is the corresponding observed value. In this paper, our task is to jointly generate high-quality and interpretable medical time-series data, including both static event data and temporal data.

2.2. Medical Time-Series Data Generation

Due to the complexity of medical time-series data, as shown in Table 1, few studies have addressed the joint generation of medical time-series data.

A considerable amount of research has focused on the generation of a sequence of static event data. Traditional methods, such as GAN-based methods [20,21] and VAE-based methods [23], have been widely used. However, the discrete and high-dimensional nature of static event data makes it challenging for them to guarantee the quality of the generated data. Recently, LLMs have been explored for static event data generation [19,24]. Empirical studies have demonstrated that LLMs outperform GAN- and VAE-based methods in static event data generation [31,32], largely due to their superior capabilities in semantic reasoning and contextual understanding.

Other research on temporal data generation has primarily focused on producing data with meaningful patterns and trends. Traditional generative techniques, such as GANs [25–27] and VAEs [28], have demonstrated strong potential for generating temporal data. Existing studies have shown that GAN-based methods generally outperform VAE-based methods in generating temporal data [33]. The application of LLMs to temporal data generation remains relatively underexplored [10]. This limitation is due to LLMs being inherently designed for text-based tasks, and they must treat numerical temporal data as sequences of tokens.

Few studies can jointly generate both static event data and temporal data. Traditional GAN-based methods [16] generate medical time-series data by a hybrid multi-generator framework. LLM-based methods, such as HALO [9] and SynEHRgy [10], employ novel tokenization strategies. Although these methods achieve joint generation of medical time-series data, they face notable limitations: GAN-based methods often produce unreliable static event data, while LLM-based methods struggle to generate high-quality temporal data and face high computational cost. Therefore, modularizing the medical time-series data generation task using two modules to generate static event and temporal data may yield better performance than relying on a single model.

2.3. Medical Time-Series Data Quality Assurance

Data quality assurance plays a crucial role in guaranteeing the quality of generated medical time-series data. This assurance must consider not only clinical constraints (value range limits on temporal variables and logical limits among sequential diagnoses in static event variables) but also variable dependencies (logical inferences exist between different variables). In the following, we provide an overview of existing data assurance methods.

A few existing studies have addressed clinical constraints in medical time-series data generation. As shown in Table 1, some methods perform pre-generation quality assurance by incorporating quality assurance mechanisms directly within the generative model architecture [16,24] or leveraging the reasoning capabilities of LLMs to ensure logical consistency [24]. Other methods [34] perform post-generation quality assurance by employing sampling and data transformation techniques to maintain value ranges and data integrity. Although these methods can enforce numerical value ranges for temporal data, they fall short in validating static event data, which require logical constraints. For example, a patient cannot have a diagnosis of poliomyelitis immediately followed by Alzheimer's.

Due to the complex logical relationships among variables in medical time-series data, very few studies have addressed the assurance of variable dependencies in generated data. Existing work, such as PromptEHR [19], leverages the reasoning capabilities of LLMs to infer dependencies from one variable to another during the generation process. However, the generation process of LLMs is prone to hallucination [35], which limits the reliability of such inferred dependencies. As a result, though LLMs may capture dependencies among variables when generating data, it is also essential to validate the variable dependencies of the generated medical time-series data again by post-generation assurance.

2.4. Interpretability of Data Generation

Traditional medical time-series data generation methods, such as GANs [20–22,25,26] and VAEs [23,28], are inherently black-box models, making it difficult to provide meaningful explanations for the data generation process. Moreover, existing LLM-based methods [9,10,19,24] also lack the interpretability of data generation. However, the interpretability of medical time-series data generation is essential for researchers, making the interpretability of the data generation process crucial.

3. QAMT Overview

QAMT enables the joint generation of high-quality medical time-series data while preserving the interpretability throughout the generation process. As illustrated in Figure 1, this framework consists of four main modules. The health knowledge graph module provides domain-specific knowledge to LLMs. The GAN-based temporal data generation module and the LLM-based static event data generation module are responsible for jointly generating medical time-series data. The medical time-series data quality assurance module evaluates the generated data to ensure the overall quality and clinical plausibility of the final generated data. The entire medical time-series data generation and quality assurance process can be divided into the following steps:

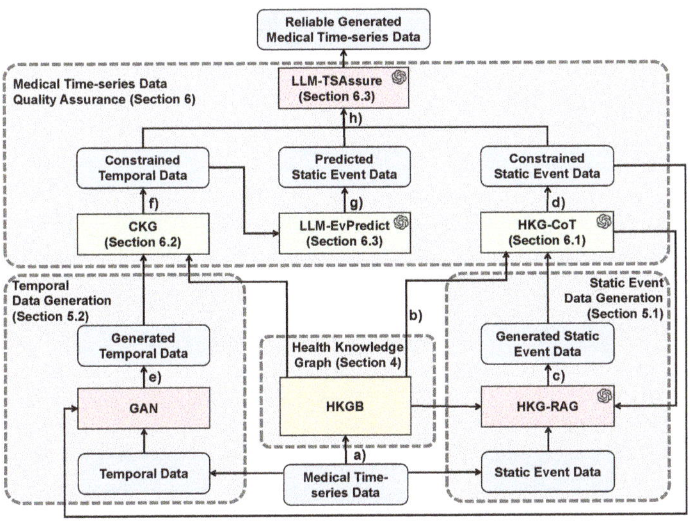

Figure 1. The framework of QAMT, where the blue rounded rectangles represent data, red rectangles represent generative models, green rectangles represent medical time-series data quality assurance models, and yellow rectangles represent the health knowledge graphs. The parts marked with icons indicate the use of LLMs.

(a) Based on real-world medical time-series datasets and open knowledge bases, a health knowledge graph (HKG) is constructed using an existing Health Knowledge Graph Builder (HKGB). This knowledge graph serves as a domain-specific knowledge resource for downstream LLMs (Section 4.1).

(b) Construct HKG-CoT, a chain-of-thought (CoT) reasoning process enriched with clinical knowledge from the HKG, which provides healthcare-specific inference capabilities (Section 4.2).

(c) The static event data generation module uses Retrieval-Augmented Generation (RAG) guided by the HKG to generate static event data m' (Section 5.1), referred to as Health Knowledge Graph-based Retrieval Augmented Generation (HKG-RAG):

$$m' = \text{HKG-RAG}(m, HKG) \tag{1}$$

(d) The medical time-series data quality assurance module then evaluates the generated static event data using HKG-CoT (Section 6.1), obtaining constrained static event data

m'' with clinical constraints and logical consistency. Then, the evaluation results are fed back to the static event data generation module:

$$\text{Constraint 1: } CC_1(m') = \text{HKG-CoT}(m', HKG) \in \{0,1\} \tag{2}$$

$$m'' = CC_1(m') \cdot m' \tag{3}$$

(e) The temporal data generation module uses a GAN to generate temporal data t' (Section 5.2), leveraging constrained static event data as conditional guidance:

$$t' = \text{GAN}(t, m'') \tag{4}$$

(f) The generated temporal data are further validated by the medical time-series data quality assurance module, using the Concept Knowledge Graph (CKG), which is a subgraph within the HKG (Section 6.2), to check for clinical value range constraints and plausibility and obtain constrained temporal data:

$$\text{Constraint 2: } CC_2(t') = \text{CKG}(t', HKG) \in \{0,1\} \tag{5}$$

(g) The constrained temporal data are then input to an LLM-based diagnostic model, LLM-EvPredict, which predicts its corresponding static event data. Then, the medical time-series data quality assurance module compares the predicted static event data with the previously constrained static event data using the LLM-TSAssure (Section 6.3):

$$\text{Constraint 3: } VD(m', t') = \text{LLM-TSAssure}(m', \text{LLM-EvPredict}(t')) \in \{0,1\} \tag{6}$$

(h) Finally, if the predicted static event data and constrained static event data are deemed consistent, the static event data and temporal data are considered to satisfy variable dependencies and are jointly assembled into final, reliable synthetic medical time-series data $V^{'}$ (Section 6.3):

$$V' = VD(m', t')(CC_1(m') \cdot m' \oplus CC_2(t') \cdot t') \tag{7}$$

The following sections provide a detailed description of each module and explain how they contribute to the joint generation and quality assurance of medical time-series data.

4. Health Knowledge Graph Module

The emergence of LLMs has introduced new potential for data generation. However, existing LLMs lack sufficient domain-specific knowledge in the medical time-series data generation task, leading them to generate hallucinated information. One solution involves pretraining LLMs on domain-specific corpora to enhance their medical knowledge [36,37]. However, it is computationally expensive. Another strategy involves prompt tuning to improve LLM performance [37,38], which requires relevant domain knowledge. Therefore, an increasingly explored method is to integrate knowledge graphs (KGs) with LLMs [39–43]. By leveraging the structured domain knowledge encoded in KGs, LLMs can be guided toward more accurate generation and reasoning. In this paper, we construct a health knowledge graph module to provide domain-specific expertise to the LLMs. In Section 4.1, we describe how we construct the HKG. In Section 4.2, we explain how HKG is used to guide reliable CoT reasoning. In Section 4.3, we demonstrate how HKG assists in generating trustworthy medical information.

4.1. HKG

To construct a reliable HKG, we leverage an existing HKGB [30]. As shown in Figure 2, the HKGB is an end-to-end platform designed to construct disease-specific and scalable health knowledge graphs (HKGs) from diverse sources as well as evaluate its generated HKG. In this study, we built a reliable HKG by leveraging external knowledge bases together with medical time-series data (e.g., MIMIC-III and eICU). The HKG constructed by the HKGB consists of three types of nodes: concept nodes, entity nodes, and event nodes. Among them, concept nodes are extracted from open knowledge bases and include medical information such as diseases, symptoms, drugs, treatments, clinical indicators, and clinical constraints. We refer to the subgraph containing only concept nodes as the Concept Knowledge Graph (CKG). Since the construction of the CKG relies on external knowledge bases, its update is relatively slow.

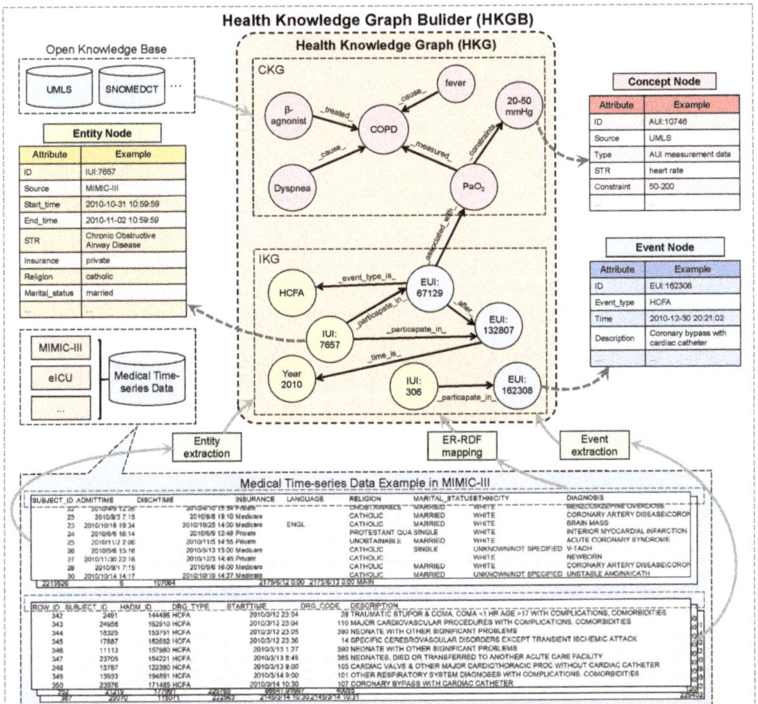

Figure 2. The detailed workflow of the HKGB to construct an HKG.

In contrast, entity nodes and event nodes are derived from real medical time-series datasets. Entity nodes, which represent patient information (e.g., demographic) or contextual information, are obtained through event extraction, and event nodes, which capture static clinical events in the medical time-series data, are extracted through entity extraction. Furthermore, we use ER-RDF to extract column-level information from the datasets to enrich the node information. We refer to the subgraph containing entity and event nodes as the Instance Knowledge Graph (IKG). Since the construction of the IKG depends on real medical time-series datasets, it is frequently updated and continuously expanded as new medical time-series data become available.

In this work, the HKG is defined as the integration of the CKG and IKG, thereby capturing both general medical knowledge and instance-level clinical information. Given the complexity and heterogeneity of these nodes, the HKG defines edges to represent relationships between them, with several illustrative examples provided in Table 2.

Table 2. Examples of edge relationships in the HKG.

Relation	Source Node [1,2]	Target Node [1,2]	Example
Concept → Concept			
_has_symptom_	Dis	Sym	COPD -_has_symptom_→ fever
_is_treated_by_	Dis/Sym	Med/Tre	COPD -_is_treated_by_→ β-agonist
_indicated_by_	Dis	CI	Diabetes -_indicated_by_→ HbA1c
_progresses_to_	Dis	Dis	Prediabetes -_progresses_to_→ Type 2 Diabetes
treated	Med/Tre	Dis/Sym	β-agonist -_treated_→ COPD
cause	Sym	Dis	fever -_cause_→ COPD
measured	CI	Dis	PaO_2 -_measured_→ COPD
constraints	CI	CC	PaO_2 -_constraints_→ 20–50 mmHg
		...	
Entity/Event → Concept			
_conforms_to_	Entity/Event	CC	EUI:67178.Lab Result -_conforms_to_→ HbA1c > 7%
diagnoses	Event	Dis	EUI:43613.Description -_diagnoses_→ Hypertension
prescribes	Event	Med/Tre	EUI:54241.Prescription -_prescribes_→ Amoxicillin
_associated_with_	Event	CI	EUI:67129.Indicators -_associated_with_→ PaO_2
		...	
Entity/Event → Entity/Event			
_participate_in_	Entity	Event	IUI:7657 -_participate_in_→ EUI:67129
_associate_with_	Entity (A/Eth)	Event	Age 65 -_associate_with_→ EUI:412546
_occurs_at_	Event	Entity (L)	EUI:63415 -_occurs_at_→ Beijing
_time_is_	Event	Entity (T)	EUI:132807 -_time_is_→ Year 2010
_event_type_is_	Event	Entity (ET)	EUI:67129 -_event_type_is_→ Surgery
after/_before_	Event	Event	EUI:67129 -_after_→ EUI:162308
		...	

[1] For concept nodes, "Dis" refers to diseases, "Sym" refers to symptoms, "Med" refers to medications, "Tre" refers to treatments, "CI" refers to clinical indicators, and "CC" refers to clinical constraints. [2] For entity nodes, "A" refers to age, "Eth" refers to ethnicity, "L" refers to location, "T" refers to time, "ET" refers to static event type.

4.2. HKG-CoT

Handling reasoning tasks such as arithmetic, commonsense, and symbolic reasoning has always been a challenge [44]. Previous studies have explored the use of KGs for logical reasoning [45]. However, selecting nodes based solely on similarity does not necessarily lead to correct or complete reasoning outcomes. With the rapid development of LLMs, recent work has explored LLMs' potential in reasoning by constructing a CoT [38], which aims to solve complex reasoning problems by generating a sequence of intermediate reasoning steps through manually designed prompts. Although CoT prompting has demonstrated promising performance in reasoning, it often suffers from hallucinations when applied to knowledge-intensive tasks, primarily due to the lack of related knowledge. To address this limitation, recent work proposed the concept of KG-CoT [41], which combines the explicit relational structure of KGs with the step-by-step reasoning capabilities of CoT prompting. This method decomposes complex problem-solving into manageable steps, enhancing the reasoning capabilities of LLMs while providing an observable reasoning process that improves interpretability [35]. However, its multi-turn question mechanism inevitably leads to increased computational cost [46]. Since the generated static event data consist of various events for multiple patients, and there should be certain logical constraints among the events of each individual, it is reasonable to use a CoT to infer the logic between a patient's multiple events. Therefore, in this paper, we construct an HKG-CoT to fulfill static event data quality assurance.

4.3. HKG-RAG

To address the limitations of LLMs in accessing external domain-specific knowledge, many studies have adopted Retrieval-Augmented Generation (RAG) [47], which enhances LLMs by incorporating few-shot prompts retrieved from external sources (e.g., real-world medical time-series data or open knowledge bases). However, information retrieved directly from data-rich databases often lacks reliability and interpretability [48]. Recent research has therefore focused on integrating KGs into retrieval strategies to strengthen LLMs' generation ability [49,50]. Compared to databases, KGs provide structured and inferable knowledge, making them more suitable for enhancing RAG. Recent work proposed KG-RAG [40], demonstrating that KGs could effectively enhance the performance of LLMs. In the generation tasks, RAG has the advantage of encoding sensitive data in a secure manner, thereby reducing the risk of privacy leakage. Furthermore, compared to a CoT, RAG can eliminate certain intermediate generation steps, thus lowering computational costs. Given that static event data generation requires the rapid synthesis of large volumes of data while minimizing privacy risks during the generation process, we propose a method, HKG-RAG, to support the generation of static event data.

5. Medical Time-Series Data Generation Module

As analyzed in Section 2.2, since GANs and LLMs each have their advantages in static event data generation and temporal data generation, respectively, QAMT adopts dual modules to accomplish joint medical time-series data generation: the LLM-based static event data generation module (Section 5.1) and the GAN-based temporal data generation module (Section 5.2).

5.1. Static Event Data Generation Module

In the static event data generation module, we employ the HKG-RAG introduced in Section 4.3. As illustrated in Figure 3, the process of static event data generation consists of the following steps:

Step 1: Privacy-insensitive demographic sampling and prompt customization. Demographic information c, such as patient ID, age, diagnosis time, religion, and marital status, is randomly sampled from real-world medical time-series data. This information is first checked by an LLM, obtaining reliable demographic information c' as shown in Equation (8), and then used to construct a customized prompt q, which is fed into another LLM:

$$c' = \text{LLM-judge}(Sample(c)) \quad (8)$$

Step 2: Entity recognition. Entities $e = \{e_{n1}, e_{n2} \dots e_{nm}\} \subseteq q$ are extracted from the input prompt $q = NL(c')$ and matched to corresponding nodes in the HKG:

$$N_q = \{n \mid sim(n, e_{ni}) > \tau\} \subseteq \text{HKG} \quad (9)$$

where τ is the threshold value. During the entity extraction, QAMT employs zero-shot prompting on an LLM distinct from the previous one to ensure accuracy. Next, entity linking is performed to obtain the corresponding entities in the HKG. We utilize the sentence embedding model "all-MiniLM-L6-v2" to encode entity nodes into dense vector representations [51]. The similarity between the extracted entities and HKG nodes is then calculated in this embedding space to determine the best match.

Step 3: Contextual retrieval of clinical outcomes and events based on the HKG. In the HKG, entity nodes are connected to event nodes through contextual triples (Subject, Predicate, Object), following a defined schema as shown in Table 2. Since multiple entities may be extracted from one prompt and may correspond to several entity nodes in the HKG, we select the top-K_{rag} most frequently occurring objects across inferred triples as the final context targets. These are then structured into n-ary tuples: (Subject_1, Subject_2, …, Subject_n, Predicate, Object). These tuples can be directly transformed into a natural language sentence using the following pattern: (Subject_1, Subject_2, …, Subject_n, Predicate, Object) $\rightarrow$ Subjects predicateName Object.

Step 4: Generation of clinical outcomes and events. The prompt-aware content is used as few-shot prompts for the LLM to generate the final output. The generated clinical outcomes and events are then matched with the corresponding demographic information to form the final generated static event data.

$$m' = \tilde{m_S} \oplus \text{HKG-RAG}(\tilde{m_S}, fewshot). \tag{10}$$

Step 5: Feedback mechanism. In the initial generation process, we set the size of top-$Krag$ to 5. Subsequently, the medical time-series data quality assurance module evaluates the generated static event data (Section 6.1). If the data generated by HKG-RAG fail to satisfy the clinical constraints, we expand top-$Krag$ to 10, enabling the generation of more diverse results and correcting previously erroneous outputs. By feeding back the evaluation results from the time-series data quality assurance module to the static event data generation module, the efficiency of data generation can be improved.

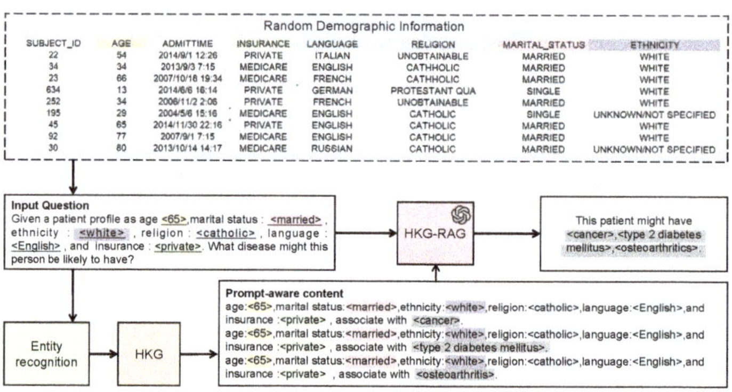

Figure 3. HKG-RAG workflow. The demographic information is randomly generated. An LLM is then used to assess its validity to avoid logically inconsistent cases, such as an age of "6" with a marital status of "MARRIED". Only demographic data that pass this logical check are used as input for generating complete static event data.

5.2. Temporal Data Generation Module

In the temporal data generation module, we adopt a GAN, which has demonstrated strong performance in continuous data generation. To establish a connection between it and the static event data generation module, we incorporate the statistical information of constrained static event data m' after validation (Section 6.1) as input to assist the generation of temporal data [52]:

$$\min_{G} \max_{D} \mathbb{E}_{t \sim P_{\text{real}}} [\log D(t \mid \theta_D)] + \mathbb{E}_{z \sim P_z} [\log(1 - D(G(z, m' \mid \theta_G)))] \tag{11}$$

In this GAN model, we introduce an attention mechanism to mitigate the impact of noise. Additionally, we incorporate learned positional encoding to integrate position information into the attention computation process, preserving relative distance information within the sequence.

The generator accepts input random noise vectors z and statistical information derived from the static event data c, generating time-series data $D_T = G_t(z, c; \theta_g)$, which captures variable dependencies between the static event data and temporal data.

6. Medical Time-Series Data Quality Assurance Module

Real-world medical time-series data often exhibit multiple constraints, including logical constraints within a single patient's static event data (e.g., a patient diagnosed with poliomyelitis is unlikely to be diagnosed with Alzheimer's disease in a short time) and value constraints on temporal data (e.g., heart rate values cannot be zero). Moreover, there are typically variable dependencies among variables (e.g., patients with high systolic blood pressure are more likely to have hypertension).

To ensure higher quality in the generated medical time-series data, we apply the HKG-CoT (Section 6.1) and CKG (Section 6.2) to enforce constraints on the outputs from the static event data and the temporal data generation module, respectively. In addition, we use LLMs to predict the static event data corresponding to the constrained temporal data. By comparing the predicted static event data with the constrained static event data, we validate the variable dependencies in the generated medical time-series data (Section 6.3).

6.1. Clinical Constraint Assurance in Static Event Data

Since static event data often exhibit inherent logical relationships, we apply the HKG-CoT introduced in Section 4.2 to enforce logical constraints on the generated static event data. Figure 4 illustrates the detailed process of the HKG-CoT:

Step 1: Step-by-step graph reasoning model. Let n denote the number of entities in the HKG. We first initialize an entity state $e^0 \in [0,1]^n$. If the ith entity is mentioned in the question, $e_i^0 \in e^0$ is initialized to 1; otherwise, it is set to 0. The question is transformed into a one-dimensional vector q through embedding. Divide the graph reasoning process into T steps, and through the design of an attention-related function $f^t()$, obtain the question representation $q^t = f^t(q)$ at step t, $t = 1, 2, \ldots, T$. The question representations focus on different parts of the question context at different steps. Then, we can obtain the scores of all relations in the HKG at step t by using a multi-layer perception (MLP). We define a transition matrix W^t and the update formula for e is $e^t = e^{t-1}W^t$. After T steps of reasoning, we can obtain the confidence score of each entity based on e^k.

Step 2: Reasoning path generation method. Based on the results of the step-by-step graph reasoning model, obtain the k entities with the highest confidence, denoted as E^k. Perform T steps of reasoning, and for each reasoning path, calculate its score based on rules. Take the entity in question as the initial entity. At step t, select the top-k intermediate reasoning paths based on the scores and add these paths to the candidate paths. After T steps of reasoning, select the top-k candidate reasoning paths with the highest scores to form the final reasoning paths: ($\text{Path}_1, \text{Path}_2, \ldots, \text{Path}_k$).

Step 3: Reasoning. Serialize the selected k final reasoning paths and use detailed instructions to prompt the LLM to generate answers using these reasoning paths. Since generated static event data consist of multiple events of a single patient, we form a question by combining the first event (the event data are sorted by timestamp) with a pre-defined prompt. If the answer produced by the HKG-CoT does not contain the entity of the second event, we consider the patient's event data to be incorrect. If this is the first time the data undergo validation, the validation results are fed back to the static event generation module for correction (Section 5.1). If the data remain incorrect even after feedback-based revision, they are discarded. Conversely, if the event data are deemed correct, we consider the first event as passed and continue to ask questions about the second event until the set of event data is discarded or all events of this patient are passed. After the data quality assurance conducted by the HKG-CoT, we consider the generated event data to be constrained.

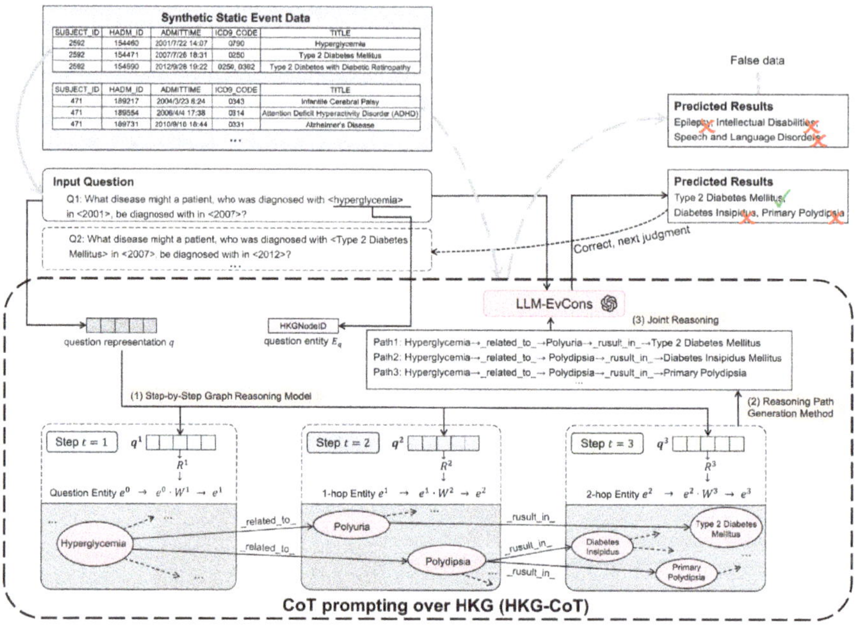

Figure 4. HKG-CoT workflow. It presents an example of how the HKG-CoT is used for static event data quality assurance. The solid lines indicate that the first event in the generated data is extracted and inserted into a customized prompt, which is then submitted to the LLM for reasoning. The HKG-CoT generates a response Type 2 Diabetes Mellitus, Diabetes Insipidus, Primary Polydipsia for the question. Since Type 2 Diabetes Mellitus matches the second event in the data sequence, showing as a check mark, the first event passes the validation, and the process continues by questioning the next event, and so on. If all events pass this validation, the entire dataset is considered trustworthy. In contrast, as shown by the dashed lines, if the LLM's response does not match the event, showing as a cross mark, and does not contain the next event, the event sequence is deemed invalid and discarded.

6.2. Clinical Constraint Assurance in Temporal Data

Temporal data are often subject to value constraints. However, due to the structural complexity of medical time-series data, different datasets may record different variables, making it difficult to manually define value constraints for each variable. The CKG within the HKG introduced in Section 4.1 captures clinical constraints associated with various clinical indicators and serves as a reliable and comprehensive knowledge base. Therefore, QAMT leverages the CKG to apply value constraints to the temporal data, thereby enabling data quality assurance.

For the generated temporal data, we extract each variable and perform similarity matching with the clinical indicator nodes (concept nodes) in the CKG. By leveraging the relationships between clinical indicators and clinical constraints, we obtain the corresponding value constraints for each variable. If a variable in the temporal data exceeds its specified value range, the data are considered unreliable. Otherwise, if all values fall within the range, the data are constrained.

6.3. Assurance of Variable Dependencies

Due to the complex relationships among variables in medical time-series data, it is necessary to verify the dependencies between temporal data and static event data. For example, suppose a patient's temporal data show elevated systolic blood pressure (within the acceptable value range), but the corresponding static event data indicate a diagnosis of hypotension. This inconsistency suggests that the generated medical time-series data are unreliable. Given the diagnostic relationship between temporal data and static event data, we utilize an LLM to predict diagnostic outcomes in the static event data based on the constrained temporal data. If the predicted static event data are similar to the originally constrained static event data, the generated medical time-series data are considered reliable.

Previous studies have shown that with carefully designed prompts, LLMs can successfully predict static event data from temporal inputs [53,54]. Therefore, we designed LLM-EvPredict, a static event prediction model based on temporal data. This LLM formats a patient's temporal data into queries by organizing the variables into tuples (variable: values)$_n$, and construct the corresponding prompt as: Prompt = Instruction$_{start}$ + Context + Instruction$_{end}$.

The results obtained by LLM-EvPredict are written in the form of event tuples (PreEvent$_1$, PreEvent$_2$, ..., PreEvent$_n$). Another model, LLM-TSAssure, is then used to compare the predicted static event data (in tuple form) with the constrained static event data of the corresponding patient (also in tuple form). If LLM-TSAssure determines that the two sets of static events are likely to come from the same patient, we consider that variable dependencies exist between the generated temporal data and static event data and retain the data. Otherwise, the data are discarded.

7. The Interpretability of Medical Time-Series Data Generation

Due to the high modularity of QAMT and its multiple usage of LLMs throughout the time-series data generation and quality assurance, the framework offers an interpretable generation pipeline.

(1) Clearly defined modularization. As shown in Figure 1, QAMT consists of four modules, each functioning independently. Specifically, steps such as (c) static event data generation, (d) static event data quality assurance, (g) static event data prediction, and (h) assurance of variable dependencies involve the formulation of question prompts and the use of LLMs to produce outputs. These interactions with the LLMs, comprising formulated questions and corresponding answers, reflect the logical flow of medical time-series data generation, thereby providing QAMT's interpretability.

(2) Clear collaboration between modules. Figure 5 illustrates the process of generating and validating static event data based on randomly sampled demographic information. The output of each step serves as input for the next, and the question–answer pairs at each step make the generation process interpretable, demonstrating the collaboration of the data generation task. Moreover, some steps, such as the clinical constraint assurance in static event data based on the HKG-CoT, further enhance interpretability by providing insight into the reasoning steps within the step itself.

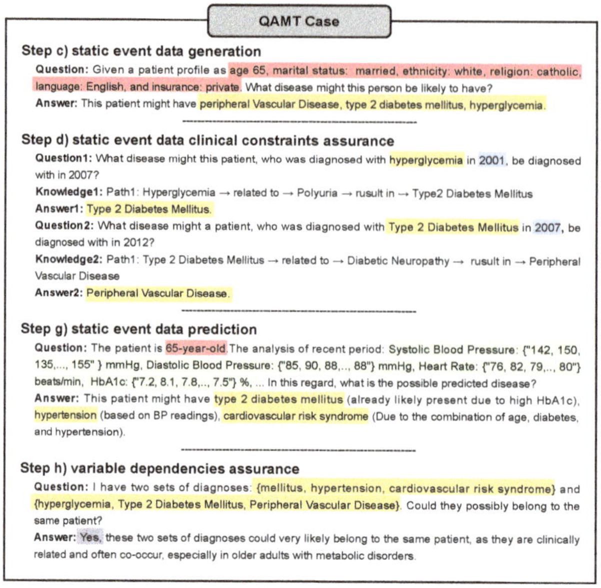

Figure 5. Interpretable prompt example, where (c), (d), (g) and (h) corresponds to the step introduced in Section 3.

8. Experimental Results

8.1. Experimental Setup

8.1.1. Datasets

We conducted experiments using the MIMIC-III [8] and eICU [7] datasets. MIMIC-III is a large, publicly available database that contains a wide range of medical time-series data. In this study, we extracted 40 thousand data samples following the methodology adopted in previous work [10]. Specifically, we employed the preprocessing pipeline in [55] to extract relevant data. Except for patient demographic information (age and gender), we selected 25 phenotype labels as static event data for each visit. Furthermore, we included 41 continuous temporal variables derived from vital signs and laboratory measurements. The eICU Collaborative Research Database is a comprehensive and publicly available resource that contains data on approximately 200,000 hospitalized patients. In this study, similar to previous work [52], we extracted 13 thousand data samples. In addition to patient demographic information (age, gender, and ethnicity), we selected seven phenotype labels as static event variables and 40 continuous vital signs and laboratory measurements as temporal variables.

8.1.2. Evaluation Metrics

We evaluated the generated time-series data based on fidelity, utility, and privacy [10,52]. **Fidelity.** For static event data, we assessed fidelity using the probabilities of unigram, bigram, and trigram within each visit, as well as the probabilities of sequential bigram between continuous visits. For example, the probability of a continuous visit as $[icd_{13}, icd_{920}]$ was computed by dividing its frequency by the total number of patients. We then calculated the Pearson Correlation between the top 1000 n-gram probabilities in the real and generated datasets to evaluate the similarity in their distributions.

For temporal data, we first constructed embeddings for each patient by calculating the statistical features of their first 48 h of temporal data (minimum, maximum, mean, and standard deviation). Using these embeddings, we evaluated the fidelity by calculating

the precision, recall, density, and coverage (PRDC) between the embeddings of the generated data and real data. Furthermore, we compared the correlation matrices of the real and generated data and report the mean squared error (MSE_{corr}) as overall correlation fidelity.
Utility. We assessed the utility of the generated data by evaluating their performance across four downstream tasks involving two disease types: sepsis clustering [56], sepsis treatment strategy modeling [57], ARDS (Acute Respiratory Distress Syndrome) prediction [58], and ARDS treatment strategy modeling [59]. A smaller difference between the generated data and real data in downstream tasks indicates more similarity. For the sepsis clustering task, the study evaluated the results before and after applying its proposed method using the Sum of Squares Error (SSE) metric. Accordingly, we adopted the difference in SSE improvement between real and generated data as our evaluation metric. For the sepsis treatment strategy task, we measured the difference in patient condition improvement ΔQ between models trained on generated data and real data. For ARDS prediction, we compared the AUROC scores of classifiers trained on real data and generated data within a 12 h window before onset. For ARDS treatment strategy modeling, we compared the average reduction in mortality achieved by reinforcement learning algorithms when trained on real or generated data.
Privacy. We adopted the Membership Inference Attack (MIA) as the evaluation metric of privacy to determine whether specific data points were included in the data [27]. We fit a K-Nearest Neighbors (KNN) model on the generated data and the real dataset and calculated their nearest distances for each patient. A significant disparity between the distance distributions in the generated and real sets indicates lower privacy. We used the Hamming distance for static event sequences and the Euclidean distance for temporal embeddings. We then fit Gaussian distributions to these distances and assess the differences between the two distributions using the Wasserstein Distance (WD), Jensen–Shannon Divergence (JSD), and Area Under the Receiver Operating Characteristic (AUROC) metrics.

8.1.3. Baselines

Since QAMT is capable of jointly generating medical time-series data, we compared our method, based on Table 1, with HGAN [16], HALO [9], SynEHRgy [10], and SynTEG [20].

8.1.4. Experimental Details

Since QAMT utilizes LLMs across multiple steps, including static event data generation, clinical constraint assurance in static event data, static event data prediction, and assurance of variable dependencies, we employed different models for different steps, including LLaMA 2 [60], GPT-3.5, Gemini [61], and GPT-4 [62]. The experiments were conducted in an environment equipped with an NVIDIA RTX A5000 GPU (Santa Clara, CA, USA).

8.2. Medical Time-Series Data Fidelity Evaluation

Table 3 presents the correlation values of n-gram probabilities between real and generated static event data. On the MIMIC-III dataset, although SynEHRgy achieved the best performance on trigram probabilities, our method consistently outperformed all baselines, particularly in unigram, bigram, and sequential bigram probabilities. Similarly, on the eICU dataset, our method demonstrated superior performance in unigram, trigram, and sequential bigram probabilities. These results confirm that our method is capable of generating high-quality static event data.

Table 4 reports the PRDC metrics, along with the correlation difference (MSE_{corr}) for evaluating the fidelity of temporal data. We found that our proposed framework achieved either the best or second-best performance in the PRDC metrics compared to the baselines. This is attributed to our modular design, which incorporates a GAN model better suited for temporal data generation. Moreover, our model achieved the lowest MSE_{corr}, indicating that QAMT excels in capturing variable-level correlations, thanks to the variable dependencies described in Sections 5.2 and 6.

Table 3. Fidelity metrics for static event data [1].

	MIMIC-III dataset			
	Unigram	Bigram	Trigram	Sequential Bigram
HGAN	0.832	0.445	0.513	0.487
SynEHRgy	<u>0.907</u>	<u>0.717</u>	**0.738**	<u>0.571</u>
HALO	0.872	0.287	0.313	0.521
SynTEG	0.858	0.501	0.647	0.562
QAMT	**0.928**	**0.721**	<u>0.718</u>	**0.631**
	eICU dataset			
	Unigram	Bigram	Trigram	Sequential Bigram
HGAN	0.799	0.319	0.483	0.366
SynEHRgy	<u>0.848</u>	**0.763**	<u>0.711</u>	<u>0.500</u>
HALO	0.769	0.293	0.281	0.474
SynTEG	0.787	0.579	0.663	0.492
QAMT	**0.897**	<u>0.755</u>	**0.736**	**0.592**

[1] **Bold** and <u>underline</u> values indicate the best and second-best results, respectively.

Table 4. Fidelity metrics for temporal data [1].

	MIMIC-III dataset				
	Precision	Recall	Density	Coverage	MSE_{corr}
SynEHRgy	<u>0.781 (0.011)</u>	<u>0.853 (0.003)</u>	0.711 (0.016)	**0.852 (0.008)**	<u>0.036</u>
HGAN	0.731 (0.023)	0.617 (0.028)	**0.745 (0.045)**	0.315 (0.004)	0.083
HALO	0.503 (0.038)	0.461 (0.002)	0.372 (0.029)	0.215 (0.009)	0.075
SynTEG	0.610 (0.013)	0.721 (0.006)	0.672 (0.031)	0.507 (0.005)	0.045
QAMT	**0.811 (0.009)**	**0.859 (0.011)**	<u>0.739 (0.036)</u>	<u>0.638 (0.003)</u>	**0.024**
	eICU dataset				
	Precision	Recall	Density	Coverage	MSE_{corr}
SynEHRgy	<u>0.814 (0.018)</u>	**0.822 (0.005)**	<u>0.701 (0.011)</u>	<u>0.714 (0.007)</u>	<u>0.042</u>
HGAN	0.669 (0.037)	0.691 (0.042)	**0.728 (0.032)**	0.594 (0.004)	0.045
HALO	0.400 (0.061)	0.417 (0.024)	0.296 (0.018)	0.395 (0.013)	0.062
SynTEG	0.523 (0.020)	0.806 (0.012)	0.633 (0.022)	0.678 (0.006)	0.051
QAMT	**0.863 (0.017)**	<u>0.817 (0.009)</u>	0.698 (0.025)	**0.743 (0.005)**	**0.033**

[1] **Bold** and <u>underline</u> values indicate the best and second-best results, respectively.

8.3. Medical Time-Series Data Utility Evaluation

Table 5 presents the performance of different models across various downstream tasks. We found that our proposed method achieved the best performance on all tasks in the MIMIC-III dataset. On the eICU dataset, our model performed best on the SepsisClustering and SepsisTreatment tasks. Compared with the MIMIC-III dataset, we included three demographic variables (static event variables) in the eICU dataset. Although these variables were also subject to quality assurance, they were randomly generated and thus carried uncertainty. As a result, our model showed some instability in the ARDSPrediction and ARDSTreatment tasks of eICU.

Table 5. Utility metrics for medical time-series data [1].

	SpesisClustering (ΔSSE %)	SpesisTreatment (ΔQ)	ARDSPrediction (AUROC)	ARDSTreatment ($\Delta MortalityRate$ %)
MIMIC-III dataset				
Real data	−32.37	0.217	0.809	−2.33
HGAN	−28.67	0.189	0.764	−1.73
SynEHRgy	−28.11	0.203	0.818	−2.17
HALO	−27.72	0.195	0.793	−2.11
SynTEG	−28.02	0.191	0.799	−2.13
QAMT	**−29.91**	**0.214**	**0.801**	**−2.28**
eICU dataset				
Real data	−40.82	0.172	0.813	−2.49
HGAN	−38.82	0.161	0.825	−2.01
SynEHRgy	−37.41	0.167	0.816	**−2.31**
HALO	−43.58	0.165	**0.814**	−2.16
SynTEG	−40.07	0.162	0.820	−2.20
QAMT	**−42.17**	**0.173**	0.807	−2.24

[1] **Bold** and underline values indicate the best and second-best results, respectively.

In addition, we applied QAMT to generate sepsis time-series data based on the real data provided by the Emergency Department of Peking University People's Hospital. First, the generated data improved the sepsis prediction model's accuracy as the training data [57], which assisted doctors in clinical decisions. Second, the generated data supported clinical research on sepsis, particularly in sepsis subphenotypes, revealing significant heterogeneity in inflammatory biomarkers, treatments, and consistency across cohorts [63]. Due to privacy considerations, the related medical data cannot be publicly released.

8.4. Medical Time-Series Data Privacy Evaluation

MIA metrics are reported in Table 6. Here, we computed privacy metrics by using the Hamming distance of static event data and the Euclidean distance of temporal data. We found that none of the methods showed a privacy risk.

Table 6. Privacy metrics for medical time-series data.

Static Event Data				Temporal Data			
MIMIC-III Dataset							
Method	JSD	WD	AUROC	Method	JSD	WD	AUROC
HGAN	0.015	0.001	0.482	HGAN	0.001	0.003	0.482
SynEHRgy	0.014	0.001	0.461	SynEHRgy	0.002	0.002	0.492
HALO	0.013	0.000	0.477	HALO	0.003	0.001	0.493
SynTEG	0.014	0.000	0.469	SynTEG	0.002	0.001	0.488
QAMT	0.013	0.000	0.456	QAMT	0.001	0.002	0.477
eICU Dataset							
Method	JSD	WD	AUROC	Method	JSD	WD	AUROC
HGAN	0.015	0.002	0.496	HGAN	0.001	0.001	0.497
SynEHRgy	0.015	0.002	0.479	SynEHRgy	0.001	0.002	0.508
HALO	0.015	0.001	0.486	HALO	0.003	0.002	0.509
SynTEG	0.015	0.001	0.481	SynTEG	0.002	0.002	0.506
QAMT	0.015	0.001	0.456	QAMT	0.001	0.002	0.504

8.5. Robustness Analysis

8.5.1. Statistical Significance Test

To ensure the statistical reliability of the experimental conclusions, we repeated several experiments using the MIMIC-III dataset. For the static event data, we selected the probability of unigram and bigram in each visit as the evaluation metrics. For temporal data, we chose precision and recall as evaluation metrics. For time-series data, we selected the performance in the downstream tasks of sepsis clustering and sepsis treatment as the evaluation index. We measured the mean of the results and their 95% confidence intervals. The results are shown in Table 7. We found that the experimental results fell in a certain interval with high probability, and the worst value of that interval was still better than the vast majority of baseline methods. Therefore, our experimental conclusions are statistically significant.

Table 7. Statistical significance test.

	Static event data	
	Unigram	Bigram
Value	0.928	0.721
95% CI	[0.905, 0.951]	[0.712, 0.730]
p	$p < 0.05$	$p < 0.05$
	Temporal data	
	Precision	Recall
Value	0.811	0.859
95% CI	[0.802, 0.820]	[0.852, 0.866]
p	$p < 0.05$	$p < 0.05$
	Time-series data	
	SpesisClustering (ΔSSE %)	SpesisTreatment (ΔQ)
Value	−29.91	0.214
95% CI	[−30.27, −29.55]	[0.211, 0.217]
p	$p < 0.05$	$p < 0.01$

8.5.2. Noise Robustness Analysis

We injected 0% to 20% Gaussian noise into MIMIC-III time-series data and used the fidelity of temporal data as the evaluation metric to evaluate the noise robustness of QAMT. The results are shown in Table 8. We found that QAMT was less affected by noise, as its quality assurance module performed quality validations on the generated data. As a result, QAMT demonstrates robustness against noise.

Table 8. Noise robustness analysis.

Noise Intensity	Precision	Recall	Density	Coverage	MSE_{corr}
0%	0.811	0.859	0.739	0.638	0.024
10%	0.806	0.848	0.735	0.636	0.025
20%	0.803	0.841	0.728	0.633	0.025
Influence	0.6%	2.1%	1.5%	0.8%	4.2%

8.6. Parameter Sensitivity Analysis

We also tested the sensitivity of QAMT by varying k, which is the number of inference paths selected in the KG-CoT (Section 5.1). The results are shown in Table 9. We found that the improvement of the fidelity of temporal data became smaller as k increased.

Moreover, when $k > 7$, the fidelity of temporal data was no longer significantly improved by increasing k.

Table 9. Parameter sensitivity analysis.

k	Precision	Recall	Density	Coverage	MSE_{corr}
3	0.790	0.837	0.728	0.615	0.022
5	0.811	0.859	0.739	0.638	0.024
7	0.814	0.871	0.743	0.644	0.025
10	0.815	0.873	0.744	0.646	0.025

8.7. Ablation Experiments

To validate the importance of each module in the proposed QAMT framework, we conducted an ablation study. Specifically, $QAMT_0$ only used a GAN to generate all static event data and temporal data, $QAMT_1$ only used an LLM to generate all static event data and temporal data, and $QAMT_2$ used a GAN to generate temporal data and an LLM to generate static event data. On the basis of $QAMT_2$, $QAMT_3$ added CoT prompting to ensure clinical constraints to the generated static event data. $QAMT_4$ introduced the external knowledge graph HKG based on $QAMT_3$. We compared the full QAMT against $QAMT_0$, $QAMT_1$, $QAMT_2$, $QAMT_3$, and $QAMT_4$ across fidelity, utility, and privacy metrics on the MIMIC-III dataset to demonstrate the contribution and necessity of each module in the framework.

8.7.1. Fidelity Evaluation

As shown in Figure 6, from QAMT to $QAMT_0$, the fidelity of the data generated with fewer modules in QAMT shows a step-like decreasing trend. The results of the fidelity evaluation experiments clearly demonstrate that omitting the variable dependencies assurance ($QAMT_4$) leads to a decline in data quality. Similarly, comparing $QAMT_4$ and $QAMT_3$, the lack of HKG leads to a significant performance decrease, with a more noticeable drop compared to the exclusion of variable dependencies. It shows that the external knowledge provided by the HKG helps with higher-quality medical time-series data generation. Moreover, the data generated without clinical constraints ($QAMT_2$) also result in decreased fidelity performance. This is because during the data generation process, both GANs and LLMs may produce incorrect outputs. Therefore, comparing $QAMT_2$ and $QAMT_4$, we find that the clinical constraints applied after generation using the HKG-CoT and CKG can effectively eliminate these errors. We find that the medical time-series data generated with the simultaneous usage of LLM and GAN ($QAMT_2$) have higher fidelity compared with the single usage of LLM or GAN for data generation ($QAMT_1$ and $QAMT_0$), showing that the modularization of the generation process in QAMT is important.

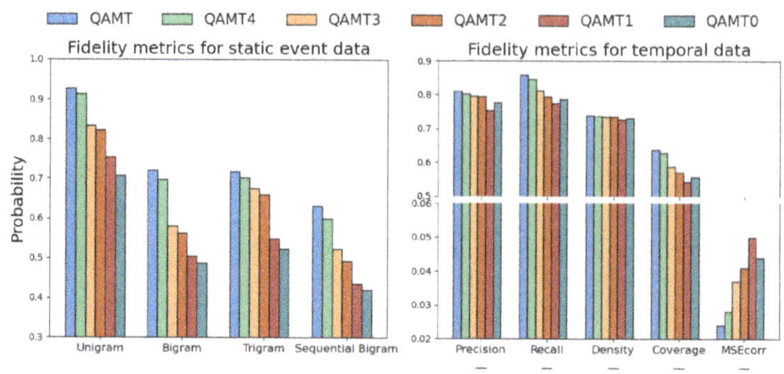

Figure 6. Fidelity metrics for medical time-series data in ablation experiments. Static event data results are on the left, and temporal data results are on the right.

8.7.2. Utility Evaluation

Figure 7 demonstrates that, in downstream tasks, medical time-series data generated without applying the simultaneous usage of the LLM and GAN, HKG, clinical constraints, and variable dependencies perform worse than the data generated by QAMT. This further demonstrates the importance of our modular design in QAMT.

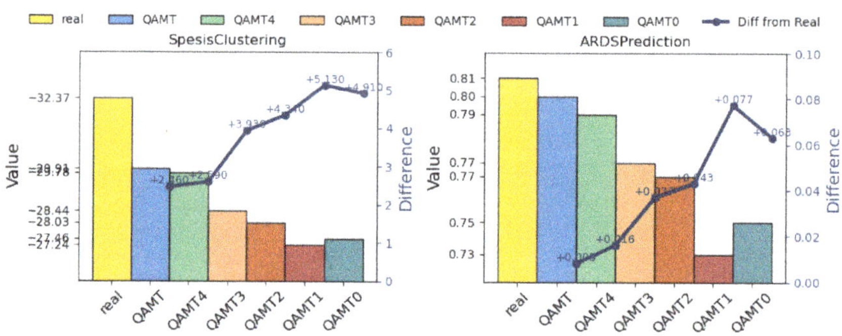

Figure 7. Utility metrics for medical time-series data in ablation experiments. Each point in the line chart represents the difference between $QAMT_n$ and the real data.

8.7.3. Privacy Evaluation

Table 10 reports that the privacy risk of the data generated by QAMT, $QAMT_4$, $QAMT_3$, $QAMT_2$, $QAMT_1$, and $QAMT_0$ show a step-like increasing trend, indicating the significance of our modular design in QAMT.

Table 10. Privacy metrics for medical time-series data in ablation experiments.

Static Event Data				Temporal Data			
Method	**JSD**	**WD**	**AUROC**	**Method**	**JSD**	**WD**	**AUROC**
QAMT	0.013	0.000	0.456	QAMT	0.001	0.002	0.477
$QAMT_4$	0.016	0.004	0.457	$QAMT_4$	0.002	0.002	0.478
$QAMT_3$	0.044	0.013	0.462	$QAMT_3$	0.005	0.006	0.484
$QAMT_2$	0.050	0.014	0.465	$QAMT_2$	0.005	0.007	0.486
$QAMT_1$	0.071	0.019	0.470	$QAMT_1$	0.006	0.008	0.491
$QAMT_0$	0.068	0.018	0.469	$QAMT_0$	0.006	0.008	0.492

9. Conclusions

In this study, we proposed QAMT, an LLM-based framework for quality-assured medical time-series data generation. The framework constructs a reliable HKG to inject medical expertise into LLMs and uses a dual-module method to jointly generate medical time-series data, including static event data and temporal data. In addition, QAMT incorporates a quality assurance module to evaluate the generated data. It provides clinical constraint assurance in static event data based on an HKG-CoT and in temporal data based on a CKG and employs LLM-based prediction to ensure variable dependencies. Unlike existing methods, QAMT maintains the modularity and high-level pipeline structure of the generation process, preserving interpretability.

Currently, the proposed QAMT is only applicable to the medical domain. Other domain-specific areas, such as energy [29], also involve time-series data generation tasks. Therefore, in the future, we plan to extend QAMT to other domains to support other time-series data generation tasks.

Author Contributions: Conceptualization, Y.L., Y.Z., C.X. and P.R.; Methodology, Y.L. and Y.Z.; Investigation, Y.L., P.R. and X.L.; Writing—original draft, Y.L. and X.L.; Writing—review & editing, Y.L., Y.Z., C.X. and P.R.; Visualization, Y.L. and X.L.; Supervision, Y.Z. and C.X. All authors have read and agreed to the published version of the manuscript.

Funding: This research was funded by the National Natural Science Foundation of China under Grant 42371480.

Institutional Review Board Statement: Not applicable.

Informed Consent Statement: Not applicable.

Data Availability Statement: We use the open source data MIMIC-III and eICU. The data provided by the Emergency Department of Peking University People's Hospital is unavailable due to privacy concerns.

Conflicts of Interest: The authors declare no conflicts of interest.

References

1. Cruz-Vega, I.B.; Ávila Vanzzini, N.; González-Gómez, G.H.; Springall, R.; Echeverría, J.C.; Lerma, C. Dynamic Response of Heart Rate Variability to Active Standing in Aortic Valve Disease: Insights from Recurrence Quantification Analysis. *Sensors* **2025**, *25*, 1535. [CrossRef]
2. Fan, Y.; Dang, Y.; Guo, Y. Fault Identification Model Using Convolutional Neural Networks with Transformer Architecture. *Sensors* **2025**, *25*, 3897. [CrossRef]
3. Rupprechter, S.; Morinan, G.; Peng, Y.; Foltynie, T.; Sibley, K.; Weil, R.S.; Leyland, L.A.; Baig, F.; Morgante, F.; Gilron, R.; et al. A Clinically Interpretable Computer-Vision Based Method for Quantifying Gait in Parkinson's Disease. *Sensors* **2021**, *21*, 5437. [CrossRef]
4. Kalahasty, R.; Yerrapragada, G.; Lee, J.; Gopalakrishnan, K.; Kaur, A.; Muddaloor, P.; Sood, D.; Parikh, C.; Gohri, J.; Panjwani, G.A.R.; et al. A Novel You Only Listen Once (YOLO) Deep Learning Model for Automatic Prominent Bowel Sounds Detection: Feasibility Study in Healthy Subjects. *Sensors* **2025**, *25*, 4735. [CrossRef]
5. Dang, T.H.; Kim, S.m.; Choi, M.s.; Hwan, S.n.; Min, H.k.; Bien, F. An Automated Algorithm for Obstructive Sleep Apnea Detection Using a Wireless Abdomen-Worn Sensor. *Sensors* **2025**, *25*, 2412. [CrossRef] [PubMed]
6. Randazzo, V.; Caligari, S.; Pasero, E.; Giustetto, C.; Saglietto, A.; Bertarello, W.; Averbuch, A.; Marcus-Kalish, M.; Zheludev, V.; Gaita, F. A Vision Transformer Model for the Prediction of Fatal Arrhythmic Events in Patients with Brugada Syndrome. *Sensors* **2025**, *25*, 824. [CrossRef] [PubMed]
7. Pollard, T.J.; Johnson, A.E.; Raffa, J.D.; Celi, L.A.; Mark, R.G.; Badawi, O. The eICU Collaborative Research Database, a freely available multi-center database for critical care research. *Sci. Data* **2018**, *5*, 180178. [CrossRef]
8. Johnson, A.E.; Pollard, T.J.; Shen, L.; Lehman, L.w.H.; Feng, M.; Ghassemi, M.; Moody, B.; Szolovits, P.; Anthony Celi, L.; Mark, R.G. MIMIC-III, a freely accessible critical care database. *Sci. Data* **2016**, *3*, 160035. [CrossRef]
9. Theodorou, B.; Xiao, C.; Sun, J. Synthesize high-dimensional longitudinal electronic health records via hierarchical autoregressive language model. *Nat. Commun.* **2023**, *14*, 5305. [CrossRef] [PubMed]

10. Karami, H.; Atienza Alonso, D.; Ionescu, A. SynEHRgy: Synthesizing Mixed-Type Structured Electronic Health Records using Decoder-Only Transformers. In Proceedings of the 38th Annual Conference on Neural Information Processing Systems, Vancouver, BC, Canada, 10–15 December 2024.
11. El Emam, K.; Buckeridge, D.; Tamblyn, R.; Neisa, A.; Jonker, E.; Verma, A. The re-identification risk of Canadians from longitudinal demographics. *BMC Med. Inform. Decis. Mak.* **2011**, *11*, 46. [CrossRef]
12. Benitez, K.; Malin, B. Evaluating re-identification risks with respect to the HIPAA privacy rule. *J. Am. Med. Inform. Assoc.* **2010**, *17*, 169–177. [CrossRef]
13. Abbas, S.R.; Abbas, Z.; Zahir, A.; Lee, S.W. Federated learning in smart healthcare: A comprehensive review on privacy, security, and predictive analytics with IoT integration. *Healthcare* **2024**, *12*, 2587. [CrossRef]
14. Kingma, D.P.; Welling, M. An introduction to variational autoencoders. *Found. Trends® Mach. Learn.* **2019**, *12*, 307–392. [CrossRef]
15. Goodfellow, I.; Pouget-Abadie, J.; Mirza, M.; Xu, B.; Warde-Farley, D.; Ozair, S.; Courville, A.; Bengio, Y. Generative adversarial networks. *Commun. ACM* **2020**, *63*, 139–144. [CrossRef]
16. Yan, C.; Zhang, Z.; Nyemba, S.; Malin, B.A. Generating Electronic Health Records with Multiple Data Types and Constraints. In Proceedings of the AMIA 2020, American Medical Informatics Association Annual Symposium, Virtual, 14–18 November 2020.
17. Xu, L.; Skoularidou, M.; Cuesta-Infante, A.; Veeramachaneni, K. Modeling tabular data using conditional gan. *arXiv* **2019**, arXiv:1907.00503. [CrossRef]
18. Pradhan, P.K.; Das, A.; Kumar, A.; Baruah, U.; Sen, B.; Ghosal, P. SwinSight: A hierarchical vision transformer using shifted windows to leverage aerial image classification. *Multim. Tools Appl.* **2024**, *83*, 86457–86478. [CrossRef]
19. Wang, Z.; Sun, J. PromptEHR: Conditional Electronic Healthcare Records Generation with Prompt Learning. In Proceedings of the 2022 Conference on Empirical Methods in Natural Language Processing, EMNLP 2022, Abu Dhabi, United Arab Emirates, 7–11 December 2022; Goldberg, Y., Kozareva, Z., Zhang, Y., Eds.; Association for Computational Linguistics: Stroudsburg, PA, USA, 2022; pp. 2873–2885. [CrossRef]
20. Zhang, Z.; Yan, C.; Lasko, T.A.; Sun, J.; Malin, B.A. SynTEG: A framework for temporal structured electronic health data simulation. *J. Am. Med. Inform. Assoc.* **2021**, *28*, 596–604. [CrossRef]
21. Baowaly, M.K.; Lin, C.; Liu, C.; Chen, K. Synthesizing electronic health records using improved generative adversarial networks. *J. Am. Med. Inform. Assoc.* **2019**, *26*, 228–241. [CrossRef]
22. Lu, C.; Reddy, C.K.; Wang, P.; Nie, D.; Ning, Y. Multi-Label Clinical Time-Series Generation via Conditional GAN. *IEEE Trans. Knowl. Data Eng.* **2024**, *36*, 1728–1740. [CrossRef]
23. Nikolentzos, G.; Vazirgiannis, M.; Xypolopoulos, C.; Lingman, M.; Brandt, E.G. Synthetic electronic health records generated with variational graph autoencoders. *Npj Digit. Med.* **2023**, *6*, 83. [CrossRef]
24. Pang, C.; Jiang, X.; Pavinkurve, N.P.; Kalluri, K.S.; Minto, E.L.; Patterson, J.; Zhang, L.; Hripcsak, G.; Elhadad, N.; Natarajan, K. CEHR-GPT: Generating Electronic Health Records with Chronological Patient Timelines. *arXiv* **2024**, arXiv:2402.04400. [CrossRef]
25. Esteban, C.; Hyland, S.L.; Rätsch, G. Real-valued (Medical) Time Series Generation with Recurrent Conditional GANs. *arXiv* **2017**, arXiv:1706.02633. [CrossRef]
26. Karami, H.; Hartley, M.; Atienza, D.; Ionescu, A. TimEHR: Image-based Time Series Generation for Electronic Health Records. *arXiv* **2024**, arXiv:2402.06318. [CrossRef]
27. Yoon, J.; Mizrahi, M.J.; Ghalaty, N.F.; Jarvinen, T.; Ravi, A.S.; Brune, P.; Kong, F.; Anderson, D.; Lee, G.; Meir, A.; et al. EHR-Safe: Generating high-fidelity and privacy-preserving synthetic electronic health records. *Npj Digit. Med.* **2023**, *6*, 141. [CrossRef]
28. Lee, Y.; Chae, Y.; Jung, K. Leveraging VQ-VAE tokenization for autoregressive modeling of medical time series. *Artif. Intell. Med.* **2024**, *154*, 102925. [CrossRef]
29. Zhou, X.; Jia, Q.; Hu, Y.; Xie, R.; Huang, T.; Yu, F.R. GenG: An LLM-Based Generic Time Series Data Generation Approach for Edge Intelligence via Cross-Domain Collaboration. In Proceedings of the IEEE INFOCOM 2024—IEEE Conference on Computer Communications Workshops, Vancouver, BC, Canada, 20 May 2024. [CrossRef]
30. Zhang, Y.; Sheng, M.; Zhou, R.; Wang, Y.; Han, G.; Zhang, H.; Xing, C.; Dong, J. HKGB: An inclusive, extensible, intelligent, semi-auto-constructed knowledge graph framework for healthcare with clinicians' expertise incorporated. *Inf. Process. Manag.* **2020**, *57*, 102324. [CrossRef]
31. Borisov, V.; Seßler, K.; Leemann, T.; Pawelczyk, M.; Kasneci, G. Language Models are Realistic Tabular Data Generators. In Proceedings of the The Eleventh International Conference on Learning Representations, ICLR 2023, Kigali, Rwanda, 1–5 May 2023.
32. Hernandez, M.; Epelde, G.; Alberdi, A.; Cilla, R.; Rankin, D. Synthetic data generation for tabular health records: A systematic review. *Neurocomputing* **2022**, *493*, 28–45. [CrossRef]
33. Viana, D.; Teixeira, R.; Baptista, J.; Pinto, T. Synthetic Data Generation Models for Time Series: A Literature Review. In Proceedings of the International Conference on Electrical, Computer and Energy Technologies, ICECET 2024, Sydney, Australia, 25–27 July 2024. [CrossRef]

34. Inan, M.S.K.; Hossain, S.; Uddin, M.N. Data augmentation guided breast cancer diagnosis and prognosis using an integrated deep-generative framework based on breast tumor's morphological information. *Inform. Med. Unlocked* **2023**, *37*, 101171. [CrossRef]
35. Chu, Z.; Chen, J.; Chen, Q.; Yu, W.; He, T.; Wang, H.; Peng, W.; Liu, M.; Qin, B.; Liu, T. Navigate through enigmatic labyrinth a survey of chain of thought reasoning: Advances, frontiers and future. *arXiv* **2023**, arXiv:2309.15402.
36. Lee, J.; Yoon, W.; Kim, S.; Kim, D.; Kim, S.; So, C.H.; Kang, J. BioBERT: A pre-trained biomedical language representation model for biomedical text mining. *Bioinformatics* **2020**, *36*, 1234–1240. [CrossRef] [PubMed]
37. Peng, Y.; Yan, S.; Lu, Z. Transfer learning in biomedical natural language processing: An evaluation of BERT and ELMo on ten benchmarking datasets. *arXiv* **2019**, arXiv:1906.05474. [CrossRef]
38. Wei, J.; Wang, X.; Schuurmans, D.; Bosma, M.; Xia, F.; Chi, E.; Le, Q.V.; Zhou, D. Chain-of-thought prompting elicits reasoning in large language models. *Adv. Neural Inf. Process. Syst.* **2022**, *35*, 24824–24837.
39. Lin, B.Y.; Chen, X.; Chen, J.; Ren, X. Kagnet: Knowledge-aware graph networks for commonsense reasoning. *arXiv* **2019**, arXiv:1909.02151.
40. Soman, K.; Rose, P.W.; Morris, J.H.; Akbas, R.E.; Smith, B.; Peetoom, B.; Villouta-Reyes, C.; Cerono, G.; Shi, Y.; Rizk-Jackson, A.; et al. Biomedical knowledge graph-optimized prompt generation for large language models. *Bioinformatics* **2024**, *40*, btae560. [CrossRef]
41. Zhao, R.; Zhao, F.; Wang, L.; Wang, X.; Xu, G. Kg-cot: Chain-of-thought prompting of large language models over knowledge graphs for knowledge-aware question answering. In Proceedings of the Thirty-Third International Joint Conference on Artificial Intelligence (IJCAI-24), Jeju, Republic of Korea, 3–9 August 2024; pp. 6642–6650.
42. Matsumoto, N.; Moran, J.; Choi, H.; Hernandez, M.E.; Venkatesan, M.; Wang, P.; Moore, J.H. KRAGEN: A knowledge graph-enhanced RAG framework for biomedical problem solving using large language models. *Bioinformatics* **2024**, *40*, btae353. [CrossRef]
43. Shao, S.; Lin, S.; Huang, Z. A Medical Consultation System for Geriatric Disease Based on Multi-agent Architecture and Knowledge Graph. In *Proceedings of the Health Information Science-13th International Conference, HIS 2024, Hong Kong, China, 8–10 December 2024*; Proceedings; Siuly, S., Xing, C., Li, X., Zhou, R., Eds.; Lecture Notes in Computer Science; Springer: Singapore, 2024; Volume 15336, pp. 313–325. [CrossRef]
44. Rae, J.W.; Borgeaud, S.; Cai, T.; Millican, K.; Hoffmann, J.; Song, F.; Aslanides, J.; Henderson, S.; Ring, R.; Young, S.; et al. Scaling language models: Methods, analysis & insights from training gopher. *arXiv* **2021**, arXiv:2112.11446.
45. Ma, K.; Cheng, H.; Liu, X.; Nyberg, E.; Gao, J. Open-domain question answering via chain of reasoning over heterogeneous knowledge. *arXiv* **2022**, arXiv:2210.12338. [CrossRef]
46. Xia, Y.; Wang, R.; Liu, X.; Li, M.; Yu, T.; Chen, X.; McAuley, J.; Li, S. Beyond chain-of-thought: A survey of chain-of-x paradigms for llms. *arXiv* **2024**, arXiv:2404.15676.
47. Lewis, P.; Perez, E.; Piktus, A.; Petroni, F.; Karpukhin, V.; Goyal, N.; Küttler, H.; Lewis, M.; Yih, W.t.; Rocktäschel, T.; et al. Retrieval-augmented generation for knowledge-intensive nlp tasks. *Adv. Neural Inf. Process. Syst.* **2020**, *33*, 9459–9474.
48. Zhao, X.; Liu, S.; Yang, S.Y.; Miao, C. Medrag: Enhancing retrieval-augmented generation with knowledge graph-elicited reasoning for healthcare copilot. In Proceedings of the Proceedings of the ACM on Web Conference 2025, Sydney, Australia, 28 April–2 May 2025; pp. 4442–4457.
49. Jiang, J.; Zhou, K.; Zhao, W.X.; Li, Y.; Wen, J.R. ReasoningLM: Enabling structural subgraph reasoning in pre-trained language models for question answering over knowledge graph. *arXiv* **2023**, arXiv:2401.00158.
50. Kang, M.; Kwak, J.M.; Baek, J.; Hwang, S.J. Knowledge graph-augmented language models for knowledge-grounded dialogue generation. *arXiv* **2023**, arXiv:2305.18846.
51. Reimers, N.; Gurevych, I. Sentence-bert: Sentence embeddings using siamese bert-networks. *arXiv* **2019**, arXiv:1908.10084. [CrossRef]
52. Luo, Y.; Sheng, M.; Liu, X.; Wang, K.; Zhang, Y.; Zhao, H. Ltgan: Multi-label time-series gan with constraints for electronic health records generation. In Proceedings of the International Conference on Health Information Science, Hong Kong, China, 8–10 December 2024; Springer: Singapore, 2025; pp. 36–47.
53. Kim, Y.; Xu, X.; McDuff, D.; Breazeal, C.; Park, H.W. Health-llm: Large language models for health prediction via wearable sensor data. *arXiv* **2024**, arXiv:2401.06866. [CrossRef]
54. Jin, M.; Yu, Q.; Shu, D.; Zhang, C.; Fan, L.; Hua, W.; Zhu, S.; Meng, Y.; Wang, Z.; Du, M.; et al. Health-LLM: Personalized retrieval-augmented disease prediction system. *arXiv* **2024**, arXiv:2402.00746.
55. Harutyunyan, H.; Khachatrian, H.; Kale, D.C.; Ver Steeg, G.; Galstyan, A. Multitask learning and benchmarking with clinical time series data. *Sci. Data* **2019**, *6*, 96. [CrossRef] [PubMed]
56. Hao, R.; Sheng, M.; Zhang, Y.; Zhao, H.; Hao, C.; Li, W.; Wang, L.; Li, C. Enhancing clustering performance in sepsis time series data using gravity field. In Proceedings of the International Conference on Health Information Science, Melbourne, Australia, 23–24 October 2023; Springer: Singapore, 2023; pp. 199–212.

57. Wang, Z.; Zhao, H.; Ren, P.; Zhou, Y.; Sheng, M. Learning optimal treatment strategies for sepsis using offline reinforcement learning in continuous space. In Proceedings of the International Conference on Health Information Science, Virtual, 28–30 October 2022; Springer: Cham, Switzerland, 2022; pp. 113–124.
58. Le, S.; Pellegrini, E.; Green-Saxena, A.; Summers, C.; Hoffman, J.; Calvert, J.; Das, R. Supervised machine learning for the early prediction of acute respiratory distress syndrome (ARDS). *J. Crit. Care* **2020**, *60*, 96–102. [CrossRef]
59. Zheng, H.; Zhu, J.; Xie, W.; Zhong, J. Reinforcement learning assisted oxygen therapy for COVID-19 patients under intensive care. *BMC Med. Inform. Decis. Mak.* **2021**, *21*, 350. [CrossRef]
60. Touvron, H.; Martin, L.; Stone, K.; Albert, P.; Almahairi, A.; Babaei, Y.; Bashlykov, N.; Batra, S.; Bhargava, P.; Bhosale, S.; et al. Llama 2: Open foundation and fine-tuned chat models. *arXiv* **2023**, arXiv:2307.09288. [CrossRef]
61. Team, G.; Georgiev, P.; Lei, V.I.; Burnell, R.; Bai, L.; Gulati, A.; Tanzer, G.; Vincent, D.; Pan, Z.; Wang, S.; et al. Gemini 1.5: Unlocking multimodal understanding across millions of tokens of context. *arXiv* **2024**, arXiv:2403.05530. [CrossRef]
62. Brown, T.; Mann, B.; Ryder, N.; Subbiah, M.; Kaplan, J.D.; Dhariwal, P.; Neelakantan, A.; Shyam, P.; Sastry, G.; Askell, A.; et al. Language models are few-shot learners. *Adv. Neural Inf. Process. Syst.* **2020**, *33*, 1877–1901.
63. Hao, C.; Hao, R.; Zhao, H.; Zhang, Y.; Sheng, M.; An, Y. Identification and validation of sepsis subphenotypes using time-series data. *Heliyon* **2024**, *10*, e28520. [CrossRef] [PubMed]

sensors

MDPI

Article

The Impact of Various Cockpit Display Interfaces on Novice Pilots' Mental Workload and Situational Awareness: A Comparative Study

Huimin Tang [1], Boon Giin Lee [1,2], Dave Towey [1] and Matthew Pike [1,*]

[1] School of Computer Science, University of Nottingham Ningbo China, 199 Taikang East Road, Ningbo 315100, China; huimin.tang@nottingham.edu.cn (H.T.); boon-giin.lee@nottingham.edu.cn (B.G.L.); dave.towey@nottingham.edu.cn (D.T.)

[2] Nottingham Ningbo China Beacons of Excellence Research and Innovation Institute, Ningbo 315101, China

* Correspondence: matthew.pike@nottingham.edu.cn

Abstract: Future airspace is expected to become more congested with additional in-service cargo and commercial flights. Pilots will face additional burdens in such an environment, given the increasing number of factors that they must simultaneously consider while completing their work activities. Therefore, care and attention must be paid to the mental workload (MWL) experienced by operating pilots. If left unaddressed, a state of mental overload could affect the pilot's ability to complete his or her work activities in a safe and correct manner. This study examines the impact of two different cockpit display interfaces (CDIs), the Steam Gauge panel and the G1000 Glass panel, on novice pilots' MWL and situational awareness (SA) in a flight simulator-based setting. A combination of objective (EEG and HRV) and subjective (NASA-TLX) assessments is used to assess novice pilots' cognitive states during this study. Our results indicate that the gauge design of the CDI affects novice pilots' SA and MWL, with the G1000 Glass panel being more effective in reducing the MWL and improving SA compared with the Steam Gauge panel. The results of this study have implications for the design of future flight deck interfaces and the training of future pilots.

Keywords: observation study; mental workload; situational awareness; physiological data; instrumentation design

1. Introduction

The number of in-service airplanes has increased in the past 20 years, according to data from GAMA (https://gama.aero/wp-content/uploads/2021ShipmentReport-2022-05-16-1.pdf (accessed on 6 July 2022)). It follows that future air traffic will be more congested, increasing the number of factors that could lead to accidents (https://www.iata.org/en/iata-repository/publications/economic-reports/global-outlook-for-air-transport---december-2023---report/ (accessed on 19 April 2024)). In addition, increased congestion will lead to pilots having to process more information from air traffic controllers, which will increase the number of parallel tasks they need to perform [1]. Mental workload (MWL) is a psychological construct modeled on real activity [2] to describe how much mental effort a given task requires at a given time [3]. The majority of recent aviation accidents have seen high MWL levels [4], highlighting the importance of pilots' MWL in Flight Operation Safety Assessment (FOSA). Situational awareness (SA) [5] is another important factor in ensuring flight safety [6]. It represents operators' understanding of a specific situation and can be viewed as a three-step cognitive concept comprised of (1) *perception*, (2) *comprehension* and (3) *projection* [2]. SA-friendly systems improve pilot efficiency and increase flight safety [7].

The airplane's cockpit display interface (CDI) is one of the most important devices for pilot–airplane interaction [8]. The CDI provides critical airplane state information, such as the heading, altitude and speed. Endsley [5] pointed out that the CDI needs to support pilots in building sufficient SA to maximize flight safety. Existing studies have shown the importance of interface design on operators' MWL [9]. However, current studies mainly focus on how to design the CDI to better support pilots' SA [10] or focus on how individual parts of the CDI, such as the navigation system, influence pilots' MWL [11,12], and rarely is the relationship between a CDI's support for SA and pilots' MWL considered.

Two research questions were proposed for this study to study how different subparts of the CDI influence pilots' SA and how the impact on SA influences their MWL:

RQ1:Do different designs of CDIs that provide the same information have different effects on a pilot's SA?

RQ2:What impact do different designs of CDIs with the same information have on pilots' MWL?

This study focuses on a specific combination of several displays on a CDI can provide pilots with spatial awareness to study the proposed RQs. Spatial awareness has been noted as one of the most important components of SA for aviation [13], which can be provided by three orientation variables—pitch, row and heading—and three position variables: altitude, lateral path deviation and flight path location. In this study, the impacts of the selected displays' layouts on two specific Cessna-172 series CDI types, the Steam Gauge (SG) panel and the G1000 Glass (GG) panel, on novice pilots' SA and MWL were compared with each other using objective and subjective measures. The GG panel is a digital flight display that integrates all key flight data into a single display, while the SG panel displays information using gyroscopes. The experiment asked the participants to complete a flight task as a primary task and finish a parallel subtask to simulate a real-world flight scenario. The 4D multiple resources model [14] and the resources–performance relationship [15] were adopted to interpret the impact of the CDI design on participants' MWL and SA. Our study reaffirms the previous finding that the CDI design influences pilots' SA [8,16,17] and reveals how differently designed CDIs will influence pilots' MWL. Additionally, this work provides a further demonstration of applying MWL measures in the evaluation of real-world flight displays. Future flight displays can be informed by the results of this study, and CDI designers can understand the impact of different display designs on pilots' SA and MWL using the methodology employed in this study.

2. Related Work

Two aspects will be discussed in this section: (1) situational awareness (SA) and mental workload (MWL) and (2) information processing models.

2.1. Situational Awareness and Mental Workload

SA is the assessment of operators' correctness in comprehending a given situation [2]. It has been modeled as a three-step process: (1) perceiving the elements in an environment during a period (*perception*), (2) comprehending their meaning (*comprehension*) and (3) projecting their following status based on the understanding of the current situation (*projection*) [5]. Existing studies have identified a relationship between the CDI design and pilots' SA [8,16,17]. During a computer-based flight simulation, Andre et al. [16] presented navigation displays in three display formats and two color-coding types (six trials). They investigated the influence of an inside-out pilot-centered navigation display, an outside-in world-centered navigation display and a 3D outside-in navigation system on novice pilots' SA. The navigation displays were shown both in monochrome and color. Participants were instructed to accumulate as many points as possible in each of the six trials. The study found that the outside-in navigation display had a more positive effect on pilots' SA compared with the other display formats, and the use of color in displays improved situational awareness, particularly during emergency situations. Wei et al. [8] evaluated participants' SA levels for three differently designed GG panels by having them monitor an aircraft's instrument panel for 12 min, during which a complete flight was simulated (take-off, climb, cruise, descent and landing). The instrument panel was programmed to freeze at preset intervals during the task, and the participants were queried on their awareness of the aircraft's current status through a series of questions. They found that the CDI design had a significant impact on pilots' SA, especially during the *projection* and *comprehension* stages. The SA under Boeing's CDI was significantly higher than that under Airbus's at the *projection* stage and significantly lower at the *comprehension* stage. Their findings suggested that although Boeing's CDI helped participants perceive more information, it was more challenging for them to combine the gathered information for understanding.

MWL can be used in evaluating the performance of the operator and the system [3] to improve the system design. Longo [18] emphasized that excessively high or low MWL would negatively influence human performance. Studies have demonstrated a similar relationship between flight performance and MWL [19–21]. Morris and Leung [19] asked different groups of novice pilots to finish different numbers of parallel tasks on a computer-based simulator that simulated a real-world flight task's different demand levels. The primary task required the participants to keep two crosshairs in the center of the screen using a joystick. The study found a decline in the participants' primary task performance under a higher MWL.

As reported by Wei et al. [20], the accuracy of manipulating the flight simulator decreased, while the reaction time increased as the pilots' MWL increased. A limited number of studies have examined the relationship between CDI design and pilots' MWL, with a focus on the navigation system (primary flight display) on the CDI. For example, Beringer and Ball [12] and Uenking and Hughes [22] investigated the optimal placement of the primary flight display, and Uenking and Hughes [22] and Davis et al. [23] examined ways to design it to minimize MWL. Both Davis et al. [23] and Uenking and Hughes [22] found that the new primary flight displays resulted in a lower MWL compared with conventional round dial instrumentation. Beringer and Ball [12] suggested that the field of view used for the primary flight display should not be less than 40 degrees. However, few studies have examined the effect of the entire CDI design on a pilot's MWL. Casner [11] compared the impact of the CDI with round dial and electronic instruments on pilots' MWL, but the small sample size precluded clear conclusions.

Studies have highlighted the interaction between SA and MWL and its effect on pilots' performance [8,24]. Lin and Lu [24] designed a simulated helicopter rescue task. The task difficulty was induced by different emergencies, such as some suddenly broken instruments or random alarm lights on the panel. They found that low SA led to a high MWL, and decreased pilots' performance, especially for unskilled pilots.

2.2. Information Processing Models

The 4D multiple resources model (MRM) proposed by Wickens [25] describes three stages of information processing: *perception*, *cognition* and *responding*. Different dimensions, such as the *visual* and *auditory* dimensions, might be involved in information processing, depending on the task demands, and consume the corresponding types of cognitive resources. This model is especially suitable for predicting probable inter-task interference during multitasking [25], meaning designers can predict such interference in advance [26] and identify resources demand conflicts [27]. Existing studies have also established that this model can accurately represent changes in MWL during multitasking [28]. When the tasks require resources from the same dimension, such as several tasks requiring *audio* resources to handle different sounds, the operator's MWL will be greater relative to task demands from differing resource dimensions. Resource competition happens frequently during piloting [14]. For example, the competition for *visual* resources is intense due to the need to process information from a large number of instruments [29] or the requirements when simultaneously perceiving information from both the instruments and the outside environment [30]. The MRM has been widely used in exploring the relationship between the pilots' workload and their performance [31]. Thorpe et al. [31] used the MRM to explain their findings that the negative effect of two parallel tasks requiring different resources (visual aviation task and auditory secondary task) was less than that of two parallel tasks requiring the same resources (visual aviation task and visual secondary task) on pilots' flight performance.

The resources–performance relationship adapted by Sharples and Megaw [15] (p. 520) showed that the human capacity for processing information is limited (Figure 1). When the spare capacity approaches zero, primary task performance will decline. Mental overload occurs if the required resources exceed the available resources ([15], p. 543). Combined with the MRM [25], when several tasks (including the primary task) require resources from the same dimension, the remaining capacity in this dimension declines, increasing the possibility of primary task performance degradation. Therefore, resource competition in one dimension could induce both a higher MWL and worse performance. This has been shown in existing pilot-related studies [8,32].

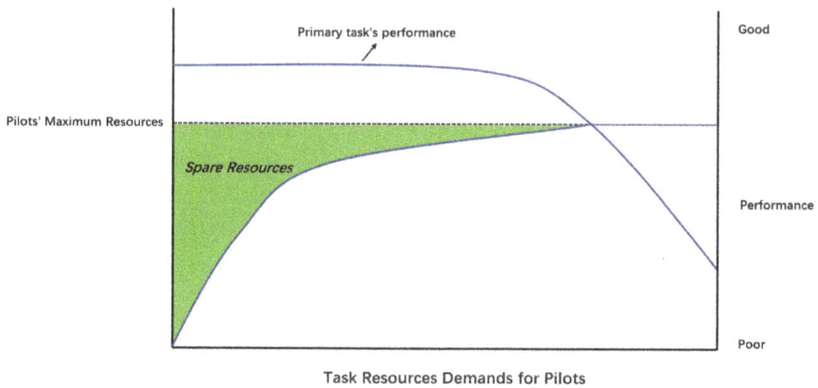

Figure 1. Resources–performance relationship, adapted from Sharples and Megaw [15] (p. 520).

3. Methods

3.1. Hypotheses

Pilots will consume fewer cognitive resources on building up spatial awareness if the displays related to these variables are appropriately constructed. Wickens [13] stated that maintaining SA can be viewed as a cognitive task that requires cognitive resources to process selective information and subsequently may impact pilots' MWL. The Cessna 172 SG panel (Figure 2) and the Cessna 172s GG panel (Figure 3) were chosen for this study because of their widespread use and distinctive differences in spatial awareness-related instrumentation design, including the airspeed indicator, altimeter, attitude indicator and heading indicator ((a–d) in Figures 2 and 3).

The following hypotheses were developed based on the research questions:

- **HPSA** (Operator's SA)
  Under the same task difficulty, there will be a significant difference in SA between two different CDIs.
- **HPMWL** (Operator's MWL)
  For the same task, the operator's MWL will be significantly lower with a CDI that has a higher SA level than with a CDI that has a lower SA level.

HPSA is proposed to address RQ1 and is based on the findings of Wei et al. [8], who identified a relationship between different CDI designs and their impact upon pilots' SA (discussed in Section 2.1). In addition, Endsley [33] pointed out that the system interfaces that were designed based on SA-oriented interface design principles [33] have been shown to better support operators' SA, suggesting the interface design's impact on the operator's SA. Compared with the digital display, the pointer-type ones used on the SG panel have a higher possibility of misreading, especially when the operator is asked to make quantitative readings from the display [34]. Such displays have a higher possibility of depleting the operator's resources for maintaining sufficient SA.

Figure 2. The Cessna 172 Steam Gauge (SG) panel.

Figure 3. The Cessna G1000 Glass (GG) panel.

HPMWL was drawn from the resources–performance relationship ([15], p. 520) and MRM [25] for exploring RQ2. Facing limited resources, operators will theoretically have more spare capacity for other parallel cognitive tasks and have less intense competition in resources when SA can be maintained from the CDI using fewer resources, especially when the parallel tasks require resources from the same dimension as maintaining SA. This decreases the possibility of a high MWL and mental overload. A lower MWL may result from the CDI that better supports the pilot's SA.

3.2. *Task*

3.2.1. Primary Task

A flight simulator running *Microsoft Flight Simulator 2020* was provided by the School of Aerospace at the University of Nottingham Ningbo China (UNNC) and was used to conduct the primary task. Participants were required to pilot the airplane, maintaining a heading of 90° E at an altitude of 2500 feet (0.76 km) for 4 minutes. The task was completed at nighttime in the simulation environment, meaning that there were little-to-no visual landmark cues for participants to utilize. Instead, the participants had to rely solely on the readings from the CDI (Figure 4) to complete the task.

There were several considerations in the primary task design to reduce the influence of factors unrelated to the CDI. The task tried to reduce extra operations from the task that might distract participants, such as communicating with the traffic controller. Because this study focused on the instruments that provide spatial awareness of the CDI, the tasks were designed such that the participants were required to regularly consult the CDI.

Figure 4. Flight simulator and environment for primary task.

3.2.2. Secondary Task

The N-back task [35] was the secondary task, which required the participants to speak out the number that previously appeared n items ago in the task. At the **perception** stage, referring to the MRM [25], both the primary and N-back tasks needed to use **visual** resources, making the competition in the *visual* dimension become intense and resulting in a significant increase in the participants' MWL. Based on the resources–performance relationship ([15], p. 520), the use of *visual* resources by the N-back task resulted in a decrease or even a complete consumption of spare resources for the primary task, leading to a degradation in the primary task's performance. Based on the resources–performance relationship ([15], p. 520), the utilization of *visual* resources by the N-back task led to a partial or total depletion of spare resources for the primary task, resulting in a decline in primary task performance. At the *cognition* stage, the competition for *verbal* resources between the N-back task and the primary task became more intense as the N-back task's difficulty increased. This implies that as the difficulty of the N-back task increases, primary task performance may degrade, and mental workload (MWL) may increase. Hence, we expected the following:

- As the difficulty of the N-back task increased, the performance of the primary task degraded.
- As the difficulty of the N-back task increased, the MWL of the participants rose.

Participants were required to complete three rounds of tasks with each CDI:

- Single flight;
- Flight + one-back task;
- Flight + two-back task.

A "baseline" task of sitting still for one minute was conducted at the beginning of the study to record the participants' physiological data that could reflect their MWL (discussed further below) during a resting state. The order of the six tasks for each participant was arranged using Latin square rotation to minimize the effects of the task order on the results [36] (see Table 1). The number of participants was larger than six. Thus, the task order started from the beginning of the table again after one round.

Table 1. Experiment order.

	Order of the Experiment					
Participant	**1**	**2**	**3**	**4**	**5**	**6**
1	single-flight digital	1 back digital	2 back analogue	2 back digital	1 back analogue	single-flight analogue
2	1 back digital	2 back digital	single-flight digital	single-flight analogue	2 back analogue	1 back analogue
3	2 back digital	single-flight analogue	1 back digital	1 back analogue	single-flight digital	2 back analogue
4	single-flight analogue	1 back analogue	2 back digital	2 back analogue	1 back digital	single-flight digital
5	1 back analogue	2 back analogue	single-flight analogue	single-flight digital	2 back digital	1 back digital
6	2 back analogue	single-flight digital	1 back analogue	1 back digital	single-flight analogue	2 back digital

3.3. Participants

Ten participants (5 male and 5 female) with an average age of 21 years were recruited to take part in the experiment. The participants were all from UNNC. All participants had normal or corrected vision and reported no history of head trauma or brain damage. All the participants had less than 20 h of flight experience on flight simulators but no experience with real-world piloting to ensure there would not be confounding variables caused by different experience levels or different educational backgrounds. The study was approved by the faculty's ethics committee, and all participants provided informed consent prior to beginning the study.

3.4. Data Collection

PsychoPy (https://www.psychopy.org/ (accessed on 21 April 2021)) was used to present the N-back task and record associated performance data as well as to stream sensor data using the lab streaming layer (LSL) protocol (https://github.com/sccn/labstreaminglayer (accessed on 21 April 2021)). *LabRecorder* (https://github.com/labstreaminglayer/App-LabRecorder (accessed on 21 April 2021)) was used to receive the streams transferred via the LSL and record them together into an XDF file with time synchronization between them onto the researcher's laptop as the study was being completed.

Changes in MWL have been shown to be reflected in changes in electroencephalography (EEG) components. Among them, the θ band (4∼7 Hz) and α band (8∼12 Hz) [37] have been successfully used in many pilots' MWL-related experiments (e.g., [38]). Several studies have found that the α band decreased in the frontal and central regions of the cerebral cortex [39] and parietal regions [40], whereas the θ band increased as the MWL increased, especially in the frontal regions [41]. *Muse2* (https://choosemuse.com/muse-2/ (accessed on 21 April 2021)), a noninvasive EEG headband with four sensors on the forehead and ears, and a sampling rate of 256 Hz were used in this study. EEG measures electrical activity originating from different brain regions by placing sensors over the scalp [37]. The headband sits on the subject's forehead (locations AF7 and AF8) and rests on the ears (TP9 and TP10), as shown in Figure 5, using the international 10-20 EEG placement system [42]. Several studies have raised concerns regarding the accuracy and correctness of *Muse2*'s readings (e.g., [43]), but the device remains popular because it is highly portable, minimally invasive, easy to use and low in cost, especially when compared with clinical EEG systems [43]. *Muse2* has been proven to be sensitive in monitoring users' brain activities [44]. *BlueMuse* (https://github.com/sccn/xdf/releases/tag/v1.12 (accessed on 21 April 2021)) was used to collect data from *Muse2* and to interface with the LSL.

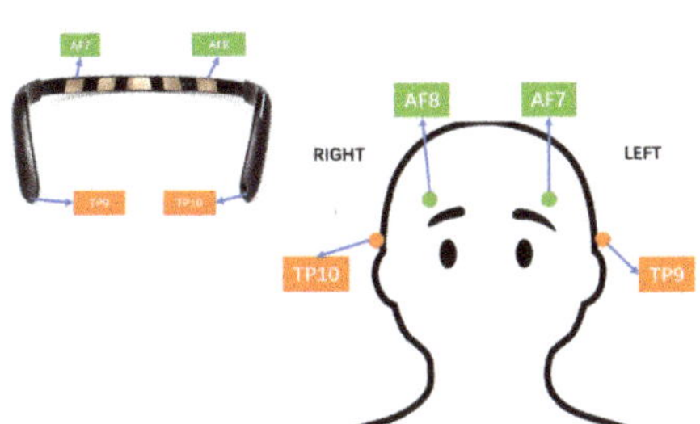

Figure 5. *Muse2* sensor locations using the 10-20 EEG placement system.

Heart rate variability (HRV) is another common physiological measure and has a known relationship with MWL [45]. The results of spectral analysis on HRV show that the low-frequency band (LF, 0.04∼0.15 Hz) represents cardiac sympathetic nervous activities, which are more active under heavy workload and stress, while the high-frequency band (HF, 0.15∼0.40 Hz) relates to the cardiac vagal nervous activities, which are in charge of cardiac activities in normal cases [37]. The LF/HF ratio, representing the balance of the sympathetic tone and vagal tone, can be used as a reliable index to reflect the changes in MWL [46].

According to Murata [47], the LF/HF ratio increases as the MWL increases. A *Polar H10* (https://www.polar.com/uk-en/sensors/h10-heart-rate-sensor (accessed on 21 April 2021)), a sports chest strap, was used to capture the participants' HRV during the study. The H10 is an ECG device that has been validated in previous studies measuring HRV under naturalistic task conditions [48]. *BLEPolarDirect* (https://github.com/roberttwomey/nosc-toolkit (accessed on 21 April 2021)) was used to pass the data returned (via Bluetooth) from the *Polar H10* to *PsychoPy* via the LSL.

3.5. Procedure

Before the start of the study, the participants were given 20 minutes to familiarize themselves with the flight simulator. At the beginning of the study, the participants were asked to sit still and relax for one minute while their physiological data were recorded using *Muse2* and the *Polar H10*. The baseline data were used to ensure the validity of the subsequent data collected and to eliminate noise in the EEG data. At the beginning of each task in the study, the airplane in the flight simulator was set to a starting altitude of 2500 feet (0.76 km) and an airspeed of 163 knots, heading in the direction of 90 degrees east. This airspeed is typical for a Cessna 172 series airplane [49]. The participants were expected to maintain the same altitude and direction during the experiment (Figure 6). Maintaining the airspeed requires pilots to consider various factors such as outside weather conditions, wind speed, air friction and the airplane's engine power. Although the weather and wind speed were constant in this study, changes in altitude have an impact on air drag and engine power and vice versa, resulting in changes in airspeed. As the participants in the study were novice pilots and may not have had the ability to maintain both the airspeed and altitude, they were not required to control the airspeed during the experiment.

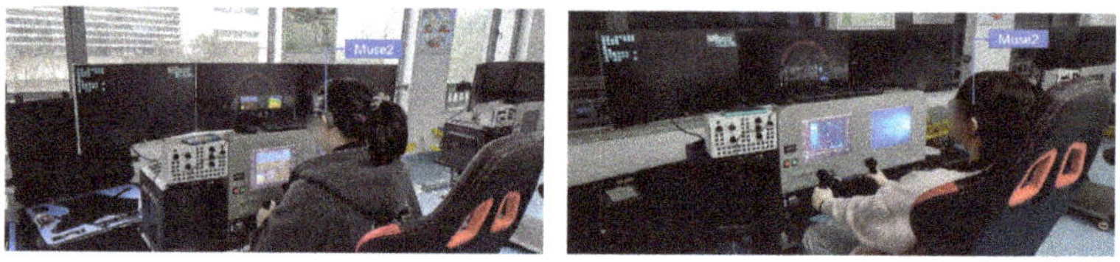

(**a**) During study 1 (**b**) During study 2

Figure 6. During study.

The participants were required to perform an N-back task concurrently with a primary flight task. The N-back task was initiated at the same time as the flight task, and the numbers for the task were displayed on a separate screen as the participant started to operate the flight simulator. There was no set time for when the participants should start the N-back task, and they were able to begin it whenever they deemed their airplane to be stable. The participants were asked to verbally provide the answer to the N-back task. When they completed the N-back task, the flight task ended (Figure 7). The N-back answers, EEG data and HRV data were recorded using *PsychoPy*, while the flight performance data were recorded by the flight simulator.

After each task, the participants completed the NASA Task Load Index (NASA-TLX) assessment to evaluate their experience during the task. The NASA-TLX is a commonly used subjective MWL measure [50]. It has been shown to be quite reliable [51] and has been used extensively in aviation contexts [1,32]. The NASA-TLX predicts MWL by asking the participants to subjectively evaluate six linear scales: mental demand, physical demand, temporal demand, performance, effort and frustration.

After completing the assessment, the participants took one minute of relaxation time before beginning the next task. This process was repeated for all subsequent tasks.

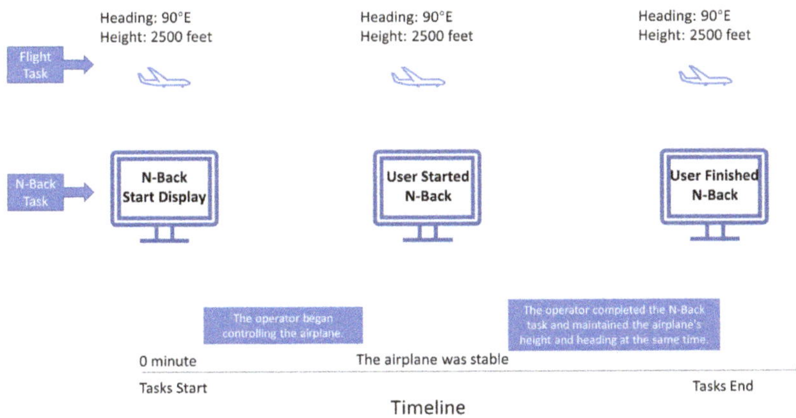

Figure 7. Procedure.

4. Results

In this study, all of the dependent variables (flight performance, NASA-TLX score, HRV and EEG) were determined to be non-parametric data based on the normal distribution of their values. All data were determined to be distribution-free using a Shapiro–Wilk test ($p < 0.05$). All dependent variables were either ordinal (NASA-TLX score) or continuous (flight performance, HRV and EEG) data. Therefore, the Wilcoxon test was used to examine the impact of the type of CDI (two levels) on the dependent variables, while the Friedman test was applied to address the influence of the difficulty of the N-back task (three levels) on the dependent variables [52]. Median rather than mean values were used in the data analysis because of the relatively small sample size and to reduce outlier effects [53].

4.1. Flight Performance

The participants were required to maintain the airplane's altitude and heading. Consequently, the flight performance criteria were based on the time the operators kept the airplane at different altitudes and headings during the task (Table 2).

Table 2. Flight performance criteria.

	Heading (°)	Altitude (Feet)
Perfect (5 points)	90 ± 10	2500 ± 100
Good (3 points)	65–80, 100–115	2250–2400, 2600–2750
Normal (2 points)	50–65, 115–130	2100–2250, 2750–2900
Poor (0 point)	Others	Others

After applying the scoring criteria, a final score for each participant was calculated as follows:

$$scoreList = [5, 3, 2, 0] \tag{1}$$

$$performanceList = [Perfect, Good, Normal, Poor] \tag{2}$$

$$rating = \frac{time\ of\ keeping\ at\ this\ performance\ level}{total\ flight\ time} \tag{3}$$

$$finalScore = \sum_{i=1}^{4} rating_{performanceList[i]} \times scoreList[i] \tag{4}$$

The flight task required the operator to have sufficient SA about the airplane because its altitude and heading would continually change based on real-time spatial awareness. For example, the operators needed to understand that the current data on the CDI indicated a "downtrend", and they needed to make the right decision based on the information on the CDI to keep the altitude from changing. Therefore, flight performance acted as the mapping of the SA level in this study.

The flight performance with the GG panel was significantly better than on that with the SG panel ($Z = -3.816, p < 0.001$), which supports the hypothesis that the CDI type would have an impact on flight performance (**HPSA**) and was consistent with previous findings [8,16,17]. The overall better flight performance with the GG panel (Figure 8a) suggests that the GG panel is the more SA-friendly CDI compared with the SG panel. No significant impact from the difficulty of the N-back task on flight performance was detected. To further examine the impact of the CDI type on flight performance under different N-back conditions, we conducted follow-up Wilcoxon tests on flight performance grouped by CDI type under each of the three N-back conditions. The results revealed significantly better flight performance with the GG panel under the single-flight ($Z = -2.395, p = 0.017$) and one-back ($Z = -2.497, p = 0.013$) conditions, but there was no significant difference between the CDI types under the two-back condition. This suggests that when the multi-task condition does not exceed the operator's limited cognitive resources, the interface design has a significant influence on their SA level. However, when participants experience an extremely high MWL or mental overload (such as in the two-back condition in this study), they may make task management errors [54], causing the flight task to become their secondary task and resulting in a lack of cognitive resources. This may explain the non-significant difference in SA level between the two CDI types under the two-back condition. Despite this, the median scores still showed that the SA level was higher when using the GG panel compared with the SG panel under the two-back condition (Figure 8b), suggesting that the GG panel is the more SA-friendly CDI.

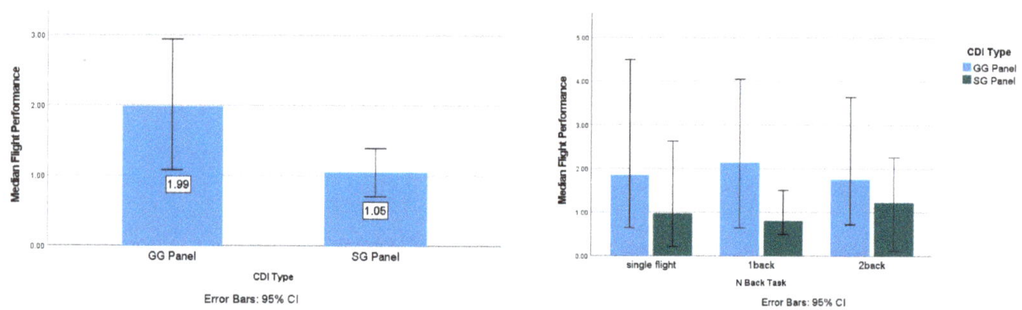

(**a**) (**b**)

Figure 8. Median flight score under different conditions. (**a**) Median flight scores for different CDIs; (**b**) Median flight scores grouped by CDI and N-back conditions.

4.2. MWL Measures

4.2.1. HRV Data

The *hrvanalysis* library in Python [55] was used to calculate the LF/HF ratio. No significant differences in the LF/HF ratio were found between the different CDI types or N-back conditions. There was also no clear trend in the HRV data between the different conditions (Figure 9). This may be because the flight task in this study was designed to be a "cruise" stage, which is typically associated with relatively low cognitive demands. According to Wilson [21], HRV data tend to have more noticeable changes during the takeoff and final landing stages of a flight, while the HRV data during other stages of the flight may only exhibit slight changes. HRV may not be sensitive enough to identify differences in MWL under different task demands if the demand changes are not large enough [56].

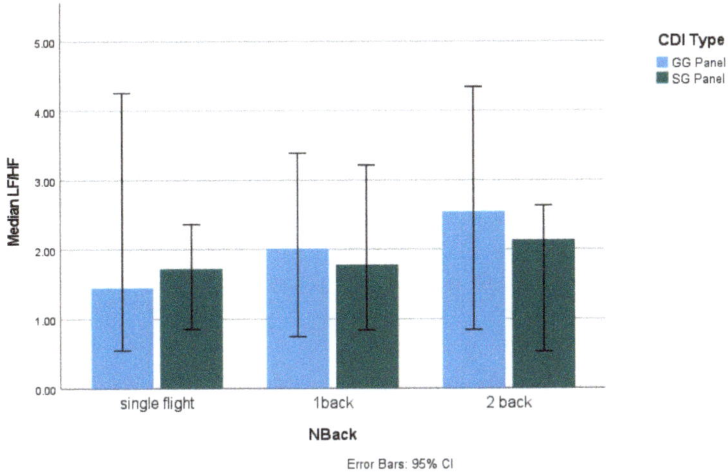

Figure 9. Median LF/HF ratio under different conditions.

4.2.2. EEG Data

The EEG data were first channel re-located based on the international 10–20 system using *EEGLab* (https://github.com/sccn/eeglab (accessed on 21 April 2021)) to ensure the channel names in the data file had the correct correspondence with the locations on the scalp. Then, the signals were re-referenced using EEGLab to ensure the signals in each channel were not impacted by the different distances between the sensors and the reference sensor. The mean mastoids (MMs) of TP9 and TP10 were used as the re-referencing point, which is a widely used re-referencing method [57]. Then, after filtering the EEG data using a 1–30 Hz filter, a wavelet transform [58] was applied to extract the α band and θ band required for this study from the original EEG signal, which is a method for decomposing the EEG signal into its frequency components through dilatations, contractions and shifts [59]. Additionally, the wavelet transform is suitable for the spectral analysis of each decomposed frequency band, making it easier to calculate the relative energy of the bands.

The θ Band

The statistical analysis revealed that the CDI type and N-back task did not have a significant effect on the relative energy of the θ band in either the AF7 and AF8 channels. The overall relative energy of the θ band was lower under the GG panel and highest under the two-back condition in the AF7 channel (Figure 10a,c). Previous research has shown that when individuals engage in mental activities that require higher cognitive resources, the θ power tends to increase more over the frontal cortex area of the brain [60], which is the area that the AF7 and AF8 electrodes of *Muse2* correspond to. Therefore, the lower relative energy of the θ band in the AF7 channel when operating the GG panel observed in this study may indicate that the operators had a lower MWL under this condition. This finding supports the conclusion from the flight performance analysis that the GG panel is the more SA-friendly CDI in this study, even though it does not fully support the hypothesis that the MWL would be lower under the GG panel (**HPMWL**). The higher relative energy of the θ band in the AF7 channel under the two-back condition suggests that the two-back task induced a higher MWL than the single-flight and one-back conditions. This is consistent with the conclusion from the flight performance analysis that the non-significant difference between the CDI types under the two-back condition may have been caused by the participants experiencing extremely high MWLs or mental overload. However, the findings in the AF8 channel showed a trend that was opposite to that in the AF7 channel (Figure 10b,d). The relative energy of the θ band was higher for the GG panel, and the single flight induced the highest MWL among the three N-back conditions.

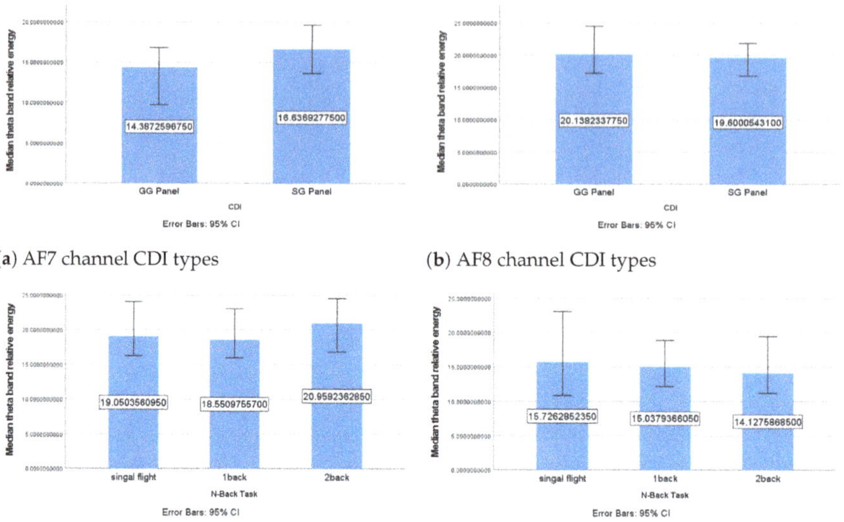

(**a**) AF7 channel CDI types (**b**) AF8 channel CDI types

(**c**) AF7 channel N-back difficulty (**d**) AF8 channel N-back difficulty

Figure 10. Median θ band's relative energy under different conditions.

The α Band

When using the GG panel, the relative energy of the α band was significantly higher in the AF8 channel ($Z = -2.211, p = 0.027$) and marginally significantly higher in the AF7 channel ($Z = -1.8, p = 0.072$) compared with the SG panel (Figure 11a,b). When the Wilcoxon test was applied to the relative energy of the α band in the AF8 channel grouped by CDI type under each of the three N-back conditions, a significantly higher α band relative energy for the GG panel was found under the single-flight condition ($Z = -2.293, p = 0.022$), and a marginally significantly higher α band relative energy for the GG panel was found under both the one-back ($Z = -2.211, p = 0.062$) and two-back conditions ($Z = -1.886, p = 0.059$).

The relative energy of the α band showed a significant difference between the different N-back conditions in both the AF7 ($\chi^2 = 6.4, p = 0.041$) and AF8 ($\chi^2 = 6.3, p = 0.043$) channels. After applying a Bonferroni correction [61], the α band's relative energy under the single-flight condition was significantly higher than under the one-back condition in the AF7 channel ($p = 0.034$) and significantly higher than under the two-back condition in the AF8 channel ($p = 0.018$) (Figure 11c,d). This suggests that the N-back tasks increased the MWL of the participants compared with the single-flight condition.

The results of this study suggest that the GG panel may be designed to be more spatial awareness-friendly, inducing a lower MWL compared with the SG panel under the same conditions. This was inferred from the decrease in the overall relative energy of the α band in both the AF7 and AF8 channels while using the SG panel, as the α waves' increment has been shown to reflect a relaxed mental state [62]. The N-back tasks also increased the MWL of the participants compared with the single-flight condition, as indicated by the significantly increased relative energy of the α band in the AF7 and AF8 channels. It is worth noting that there were some marginally significant results ($p < 0.075$), and increasing the sample size may further strengthen these results [63].

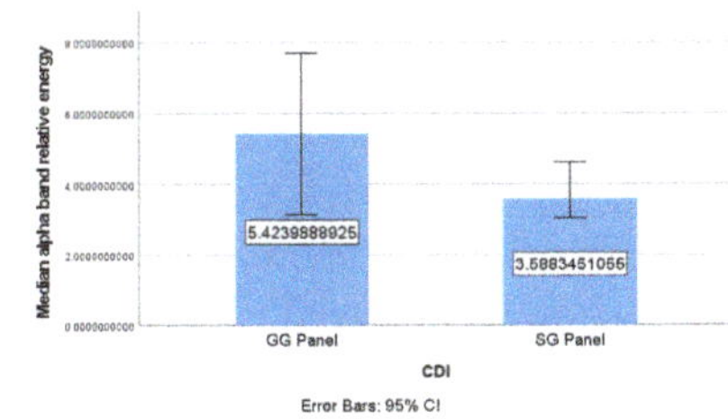

(**a**) AF7 channel CDI types

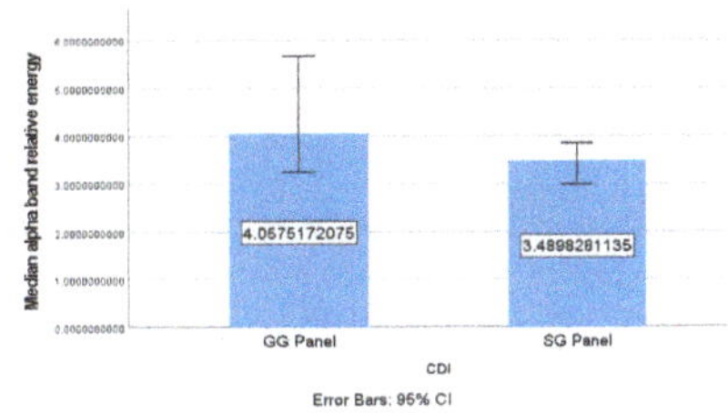

(**b**) AF8 channel CDI types

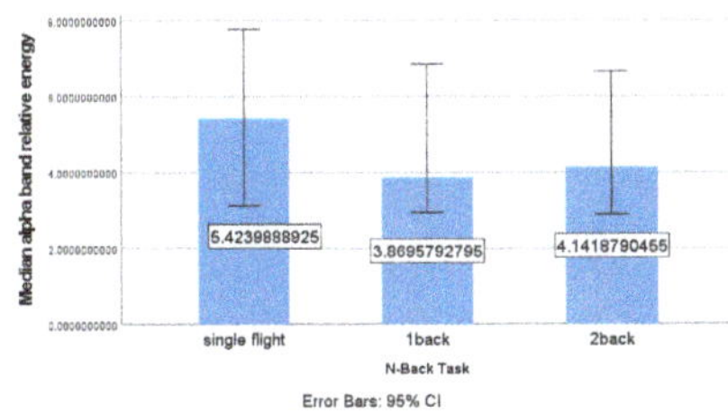

(**c**) AF7 channel N-back difficulty

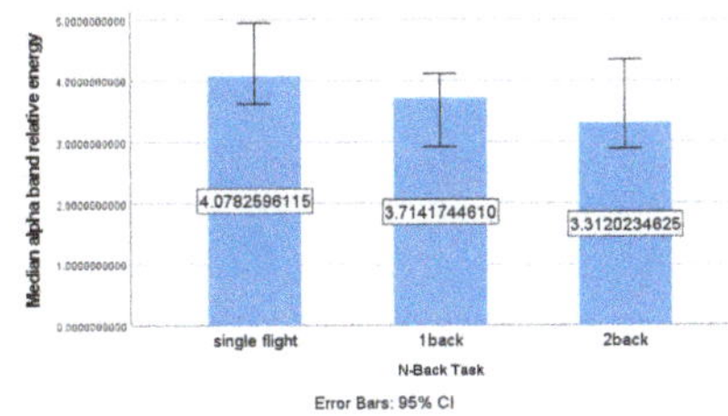

(**d**) AF8 channel N-back difficulty

Figure 11. Median α band's relative energy under different conditions.

The θ/α Ratio

No significant impact from the CDI types on the θ/α ratio was identified in either channel. The overall θ/α ratio under the SG panel was higher in both channels (Figure 12a,b). Since the increase in MWL can be reflected by the increment of the θ/α ratio [64], we interpreted that the participants' MWL under the SG panel was descriptively higher than that under the GG panel, which could support the trend in **HPMWL**.

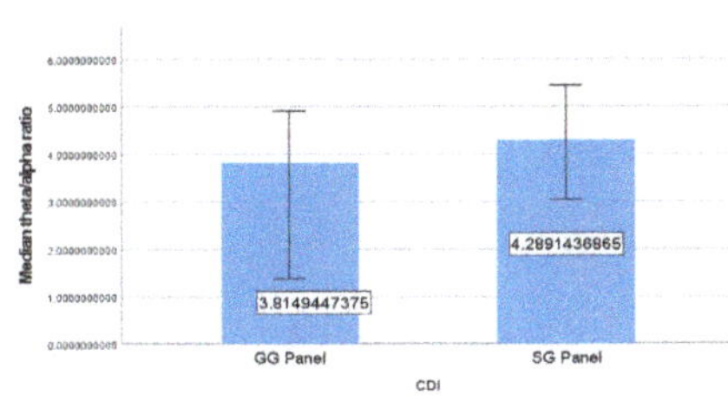

(**a**) AF7 channel CDI types

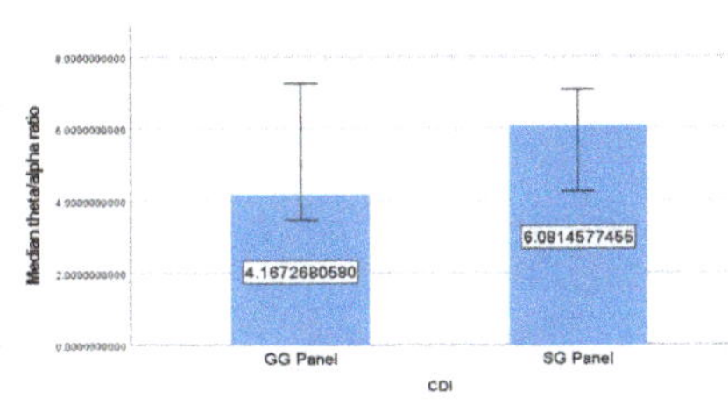

(**b**) AF8 channel CDI types

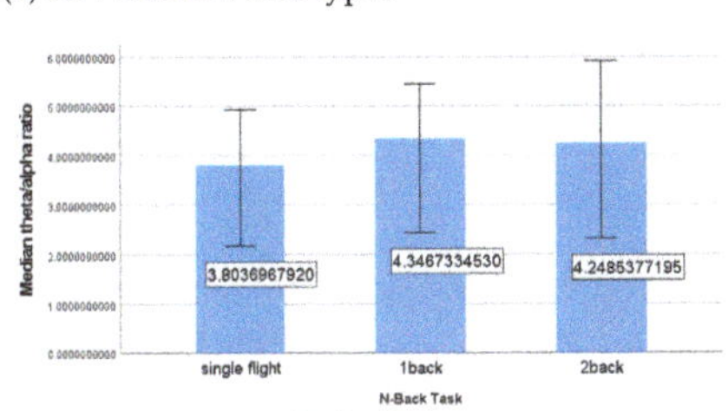

(**c**) AF7 channel N-back difficulty

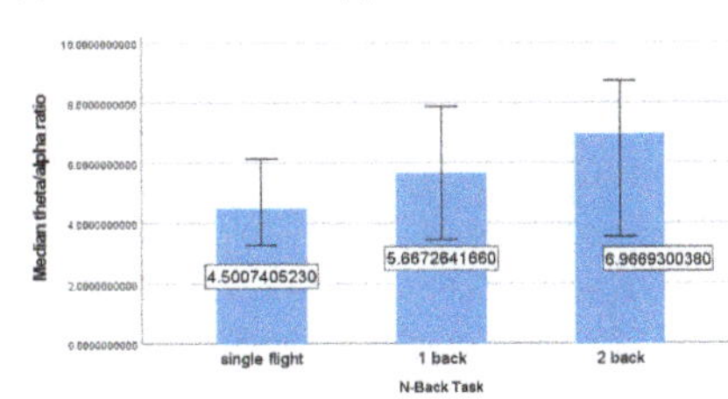

(**d**) AF8 channel N-back difficulty

Figure 12. Median θ/α ratio under different conditions.

There was no significant difference in the θ/α ratios between the different CDI types in either the AF7 or AF8 channels. The overall θ/α ratio was higher under the SG panel in both channels (Figure 12a,b). Since an increase in MWL can be reflected by an increase in the θ/α ratio [64], it can be concluded that the participants' MWL was descriptively higher while using the SG panel compared with the GG panel. This finding supports **HPMWL**.

There was no significant difference in the θ/α ratio between the different N-back conditions in the AF7 channel (Figure 12c), but a significant difference was found in the AF8 channel ($\chi^2 = 9.1, p = 0.011$). After applying a Bonferroni correction [61], a significant difference was found between the single-flight and one-back conditions ($p = 0.034$) and between the single-flight and two-back conditions ($p = 0.022$) in the AF8 channel. The θ/α ratio under the single-flight condition was significantly lower than the one-back and two-back conditions (Figure 12d). This suggests that the N-back tasks increased the MWL of the participants during the flight task, which is consistent with the findings obtained from the analysis of the α band's relative energy.

4.2.3. NASA-TLX Score

The overall NASA-TLX score was used for analysis. It is worth noting that while there was no significant difference in NASA-TLX scores across the CDI types, there was a slightly higher median score when the participants used the SG panel. This suggests that the SG panel may have caused a slightly higher MWL compared with the GG panel, although the difference was not statistically significant. A significant impact of the N-back condition's difficulty on the NASA-TLX score was found ($\chi^2 = 29.1, p < 0.001$). After applying a Bonferroni correction [61], the NASA-TLX score under the two-back condition was significantly higher than that under the single-flight ($p < 0.001$) and one-back conditions ($p < 0.001$). This suggests that the N-back tasks successfully induced different levels of pressure on the operators, with the overall NASA-TLX score increasing with the difficulty of the N-back task (Figure 13b). This is consistent with the findings in the α band's relative energy and the θ/α ratio, which also showed an increase in MWL with increasing N-back condition's difficulty.

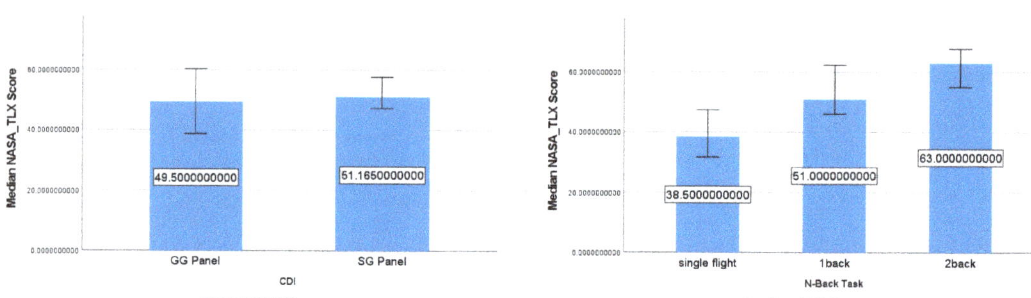

(**a**) Median NASA-TLX scores for different CDIs (**b**) Median NASA-TLX scores under different N-back tasks

Figure 13. Median NASA-TLX scores under different conditions.

4.3. Limitations

The participants in this study were novice pilots, which could lead to easier mental overload than for experienced pilots under similar circumstances. To minimize the possibility of mental overload in situations requiring relatively low cognitive demands due to the participants' unskilled operations, we designed the tasks as cruise tasks and N-back tasks, which were easier to complete compared with other flight stages' tasks. The impact of the same CDI on pilots might also be different depending on their familiarity level with the cockpit. In future work, the participants can be pilots with various experience levels, the primary task can be a more cognitive resource-dependent one, such as takeoff, and the secondary tasks can be real parallel tasks, such as "communicating with an air traffic controller". This would be more similar to today's aviation and improve the study's universality.

The physiological evaluation tools for MWL could be more comprehensive in future studies. As reflected in the current results, HRV is not sensitive enough to detect changes in MWL when there are no significant changes. The eye-tracking technique can be considered a future evaluation tool. The pupil size change has been highlighted by previous studies as being sensitive enough for MWL measurement [65,66]. The participants' focused areas can be revealed by the areas of interest captured by the eye tracker [67,68], making it suitable for investigating the specific parts of the CDI that have a noticeable impact on participants' SA and MWL.

5. Conclusions

In this study, we compared the effect of using the G1000 Glass (GG) and Steam Gauge (SG) panels in the Cessna 172 series as cockpit display interfaces (CDIs). The results showed that the GG panel was more situational awareness (SA)-friendly, as the participants had lower mental workloads (MWLs) when using it. This emphasizes the importance of designing SA-friendly interfaces in the aviation industry. However, it is worth noting that MWL can also be affected by other parallel flight tasks involving a CDI.

This study also found that the N-back tasks used in the experiment were successful in inducing different levels of MWL in the participants. These findings have implications for the design of CDIs and the management of MWL in the aviation industry and contribute to all the areas that involve interface design and MWL-SA evaluation.

We believe that the results of this study provide a clear direction for future work, which can be replicated on a larger scale with more diverse groups to improve the representativeness of the study. Our next step is to investigate how the layout design affects MWL under different flight tasks such as takeoff and landing. Additionally, this study successfully demonstrated that measuring SA and MWL simultaneously is viable, providing a direction for future interface design and improvement by using these two aspects as evaluation tools together.

Author Contributions: Conceptualization, H.T. and M.P.; methodology, H.T. and M.P.; formal analysis, H.T.; data curation, H.T.; writing—original draft preparation, H.T.; writing—review and editing, B.G.L., D.T. and M.P.; supervision, M.P. All authors have read and agreed to the published version of the manuscript.

Funding: This work was supported by the UNNC Education Foundation through Li Dak Sum Innovation Fellowship under Grant LDS202307.

Institutional Review Board Statement: This study was conducted in accordance with the Declaration of Helsinki and approved by the UNNC Faculty of Science and Engineering's ethics committee.

Informed Consent Statement: Informed consent was obtained from all subjects involved in the study.

Data Availability Statement: The raw data supporting the conclusions of this article will be made available by the authors on request.

Acknowledgments: The authors acknowledge the support of the University of Nottingham Ningbo China (UNNC) Artificial Intelligence and Optimisation (AIOP) research group and the Sensors, Sensor Networks and Instrumentation (SSNI) research group. The authors also thank the School of Aerospace at the University of Nottingham Ningbo China, who provided access to the flight simulators used in this research.

Conflicts of Interest: The authors declare no conflicts of interest.

References

1. Li, W.; Li, R.; Xie, X.; Chang, Y. Evaluating mental workload during multitasking in simulated flight. *Brain Behav.* **2022**, *12*, e2489. [CrossRef] [PubMed]
2. Wickens, C.D. Workload and situation awareness. In *Stress, Workload, and Fatigue*; Lawrence Erlbaum Associates, Inc.: Mahwah, NJ, USA, 2001; pp. 443–450.
3. Cain, B. *A Review of the Mental Workload Literature*; Technical Report; Defence Research And Development Toronto (Canada): North York, ON, Canada, 2007.

4. Wickens, C.D.; Hollands, J.G.; Banbury, S.; Parasuraman, R. *Engineering Psychology and Human Performance*; Psychology Press: London, UK, 2015; pp. 480–507.
5. Endsley, M.R. Toward a theory of situation awareness in dynamic systems. *Hum. Factors* **1995**, *37*, 32–64. [CrossRef]
6. Selcon, S.J.; Hardiman, T.D.; Croft, D.G.; Endsley, M.R. A Test-Battery Approach to Cognitive Engineering: To Meta-Measure or not to Meta-Measure, that is the Question! In Proceedings of the 40th Annual Meeting of Human Factors and Ergonomics Society, Santa Monica, CA, USA, 1 October 1996; SAGE Publications Sage CA: Los Angeles, CA, USA, 1996; Volume 40, pp. 228–232.
7. Stanton, N.A.; Plant, K.L.; Roberts, A.P.; Allison, C.K. Use of Highways in the Sky and a virtual pad for landing Head Up Display symbology to enable improved helicopter pilots situation awareness and workload in degraded visual conditions. *Ergonomics* **2019**, *62*, 255–267. [CrossRef] [PubMed]
8. Wei, H.; Zhuang, D.; Wanyan, X.; Wang, Q. An experimental analysis of situation awareness for cockpit display interface evaluation based on flight simulation. *Chin. J. Aeronaut.* **2013**, *26*, 884–889. [CrossRef]
9. Yan, S.; Wei, Y.; Tran, C.C. Evaluation and prediction mental workload in user interface of maritime operations using eye response. *Int. J. Ind. Ergon.* **2019**, *71*, 117–127. [CrossRef]
10. Feng, C.; Wanyan, X.; Liu, S.; Chen, H.; Zhuang, D. An approach to situation awareness (SA) assessment in flight simulation: SA dynamic circulation (SADC) model. *Hum. Factors Ergon. Manuf. Serv. Ind.* **2021**, *31*, 559–569. [CrossRef]
11. Casner, S.M. Perceived vs. measured effects of advanced cockpit systems on pilot workload and error: Are pilots' beliefs misaligned with reality? *Appl. Ergon.* **2009**, *40*, 448–456. [CrossRef] [PubMed]
12. Beringer, D.B.; Ball, J.D. A comparison of pilot navigation performance using conventional instrumentation, head-down, and head-up highway-in-the-sky primary flight displays. In Proceedings of the Human Factors and Ergonomics Society 45th Annual Meeting, Minneapolis, MN, USA, 8–12 October 2001; SAGE Publications Sage CA: Los Angeles, CA, USA, 2001; Volume 45, pp. 16–20.
13. Wickens, C.D. Situation awareness and workload in aviation. *Curr. Dir. Psychol. Sci.* **2002**, *11*, 128–133. [CrossRef]
14. Wickens, C.D. Multiple resources and performance prediction. *Theor. Issues Ergon. Sci.* **2002**, *3*, 159–177. [CrossRef]
15. Sharples, S.; Megaw, T. The definition and measurement of human workload. *Eval. Hum. Work.* **2015**, 516–544.
16. Andre, A.D.; Wickens, C.D.; Moorman, L. Display formatting techniques for improving situation awareness in the aircraft cockpit. *Int. J. Aviat. Psychol.* **1991**, *1*, 205–218. [CrossRef]
17. Wu, X.; Wanyan, X.; Zhuang, D.; Liu, S. Pilot situational awareness modeling for cockpit interface evaluation. In Proceedings of the International Conference on Engineering Psychology and Cognitive Ergonomics, Toronto, ON, Canada, 17–22 July 2016; Springer: Berlin/Heidelberg, Germany, 2016; pp. 476–484.
18. Longo, L. Formalising Human Mental Workload as a Defeasible Computational Concept. Ph.D. Thesis, The University of Dublin, Trinity College, Ireland, 2014.
19. Morris, C.H.; Leung, Y.K. Pilot mental workload: How well do pilots really perform? *Ergonomics* **2006**, *49*, 1581–1596. [CrossRef] [PubMed]
20. Wei, Z.; Zhuang, D.; Wanyan, X.; Liu, C.; Zhuang, H. A model for discrimination and prediction of mental workload of aircraft cockpit display interface. *Chin. J. Aeronaut.* **2014**, *27*, 1070–1077. [CrossRef]
21. Wilson, G.F. An analysis of mental workload in pilots during flight using multiple psychophysiological measures. *Int. J. Aviat. Psychol.* **2002**, *12*, 3–18. [CrossRef]
22. Uenking, M.D.; Hughes, M.F. The Efficacy of using synthetic vision terrain-Textured images to improve pilot situation awareness. *SAE Trans.* **2002**, *111*, 572–583.
23. Davis, R.C.; Wilt, D.W.; Henion, J.; Alter, K.; Snow, P.; Deaton, J.E. Formal tests for LLM approaches using refined cockpit display technology. In *Enhanced and Synthetic Vision 2005*; SPIE: Bellingham, WA, USA, 2005; Volume 5802, pp. 231–238.
24. Lin, L.W.; Lu, M.S. Empirical research on the relationship between Helicopter Pilots' mental workloads and situation awareness levels. *J. Am. Helicopter Soc.* **2016**, *61*, 1–8. [CrossRef]
25. Wickens, C.D. Multiple resources and mental workload. *Hum. Factors* **2008**, *50*, 449–455. [CrossRef] [PubMed]
26. Horrey, W.J.; Wickens, J.D. Multiple resource modeling of task interference in vehicle control, hazard awareness and in-vehicle task performance. In Proceedings of the Driving Assesment Conference, Park City, UT, USA, 21–24 July 2003; Volume 2, pp. 7–12.
27. Wickens, C.D. Multiple resource time sharing models. In *Handbook of Human Factors and Ergonomics Methods*; CRC Press: Boca Raton, FL, USA, 2004; pp. 427–434.
28. Ryu, K.; Myung, R. Evaluation of mental workload with a combined measure based on physiological indices during a dual task of tracking and mental arithmetic. *Int. J. Ind. Ergon.* **2005**, *35*, 991–1009. [CrossRef]
29. Liggett, K.K.; Reising, J.M.; Hartsock, D.C. Development and evaluation of a background attitude indicator. *Int. J. Aviat. Psychol.* **1999**, *9*, 49–71. [CrossRef]
30. Xiao, X.; Wanyan, X.; Zhuang, D. Mental workload prediction based on attentional resource allocation and information processing. *Bio-Med. Mater. Eng.* **2015**, *26*, S871–S879. [CrossRef]
31. Thorpe, A.; Estival, D.; Molesworth, B.; Eidels, A. Pilot errors: Communication comes last. *Saf. Sci.* **2022**, *149*, 105686. [CrossRef]
32. Rainieri, G.; Fraboni, F.; Russo, G.; Tul, M.; Pingitore, A.; Tessari, A.; Pietrantoni, L. Visual scanning techniques and mental workload of helicopter pilots during simulated flight. *Aerosp. Med. Hum. Perform.* **2021**, *92*, 11–19. [CrossRef] [PubMed]
33. Endsley, M.R. Situation awareness. *Handb. Hum. Factors Ergon.* **2021**, 434–455. [CrossRef]

34. Grether, W.F. Instrument reading. I. The design of long-scale indicators for speed and accuracy of quantitative readings. *J. Appl. Psychol.* **1949**, *33*, 363. [CrossRef]
35. Jaeggi, S.M.; Buschkuehl, M.; Perrig, W.J.; Meier, B. The concurrent validity of the N-back task as a working memory measure. *Memory* **2010**, *18*, 394–412. [CrossRef] [PubMed]
36. Anderson, V.L.; McLean, R.A. *Design of Experiments: A Realistic Approach*; CRC Press: Boca Raton, FL, USA, 2018; Chapter 8.
37. Lean, Y.; Shan, F. Brief review on physiological and biochemical evaluations of human mental workload. *Hum. Factors Ergon. Manuf. Serv. Ind.* **2012**, *22*, 177–187. [CrossRef]
38. Hankins, T.C.; Wilson, G.F. A comparison of heart rate, eye activity, EEG and subjective measures of pilot mental workload during flight. *Aviat. Space Environ. Med.* **1998**, *69*, 360–367. [PubMed]
39. Fink, A.; Grabner, R.; Neuper, C.; Neubauer, A. EEG alpha band dissociation with increasing task demands. *Cogn. Brain Res.* **2005**, *24*, 252–259. [CrossRef] [PubMed]
40. Brouwer, A.M.; Hogervorst, M.A.; Van Erp, J.B.; Heffelaar, T.; Zimmerman, P.H.; Oostenveld, R. Estimating workload using EEG spectral power and ERPs in the n-back task. *J. Neural Eng.* **2012**, *9*, 045008. [CrossRef]
41. Esposito, F.; Aragri, A.; Piccoli, T.; Tedeschi, G.; Goebel, R.; Di Salle, F. Distributed analysis of simultaneous EEG-fMRI time-series: Modeling and interpretation issues. *Magn. Reson. Imaging* **2009**, *27*, 1120–1130. [CrossRef]
42. Jasper, H.H. The ten-twenty electrode system of the International Federation. *Electroencephalogr. Clin. Neurophysiol.* **1958**, *10*, 370–375.
43. Jahn, N.T.; Meshi, D.; Bente, G.; Schmälzle, R. Media neuroscience on a shoestring: Examining electrocortical responses to visual stimuli via mobile EEG. *J. Media Psychol. Theor. Methods Appl.* **2022** . [CrossRef]
44. Pu, L.; Lion, K.M.; Todorovic, M.; Moyle, W. Portable EEG monitoring for older adults with dementia and chronic pain-A feasibility study. *Geriatr. Nurs.* **2021**, *42*, 124–128. [CrossRef] [PubMed]
45. Jaiswal, D.; Chowdhury, A.; Banerjee, T.; Chatterjee, D. Effect of Mental Workload on Breathing Pattern and Heart Rate for a Working Memory Task: A Pilot Study. In Proceedings of the 2019 41st Annual International Conference of the IEEE Engineering in Medicine and Biology Society (EMBC), Berlin, Germany, 23–27 July 2019; IEEE: Piscataway, NJ, USA, 2019; pp. 2202–2206.
46. Murai, K.; Hayashi, Y.; Okazaki, T.; Stone, L.C.; Mitomo, N. Evaluation of ship navigator's mental workload using nasal temperature and heart rate variability. In Proceedings of the 2008 IEEE International Conference on Systems, Man and Cybernetics, Singapore, 12–15 October 2008; IEEE: Piscataway, NJ, USA, 2008; pp. 1528–1533.
47. Murata, A. Experimental discussion on measurement of mental workload-evaluation of mental workload by HRV measures. *IEICE Trans. Fundam. Electron. Commun. Comput. Sci.* **1994**, *77*, 409–416.
48. Chattopadhyay, S.; Das, R. Comparing heart rate variability with polar H10 sensor and pulse rate variability with LYFAS: A novel study. *J. Biomed. Eng. Technol.* **2021**, *9*, 1–9. [CrossRef]
49. H-Aviation. Cessna 172S Skyhawk—FAQ's. 2020. Available online: https://www.h-aviation.com/cessna-172s---skyhawk.html#:~:text=How%20fast%20does%20the%20Cessna,a%20maximum%20of%20302km%2Fh (accessed on 10 November 2020).
50. Hart, S.G. NASA-task load index (NASA-TLX); 20 years later. In Proceedings of the Human Factors and Ergonomics Society Annual Meeting, San Fransisco, CA, USA, 16–20 October 2006; Sage Publications Sage CA: Los Angeles, CA, USA, 2006; Volume 50, pp. 904–908.
51. Hart, S.G.; Staveland, L.E. Development of NASA-TLX (Task Load Index): Results of empirical and theoretical research. In *Advances in Psychology*; Elsevier: Amsterdam, The Netherlands, 1988; Volume 52, pp. 139–183.
52. Shankar, S.; Singh, R. Demystifying statistics: How to choose a statistical test? *Indian J. Rheumatol.* **2014**, *9*, 77–81. [CrossRef]
53. Khorana, A.; Pareek, A.; Ollivier, M.; Madjarova, S.J.; Kunze, K.N.; Nwachukwu, B.U.; Karlsson, J.; Marigi, E.M.; Williams III, R.J. Choosing the appropriate measure of central tendency: Mean, median, or mode? *Knee Surgery Sport. Traumatol. Arthrosc.* **2023**, *31*, 12–15. [CrossRef]
54. Wickens, C.D. Attentional tunneling and task management. In Proceedings of the 2005 International Symposium on Aviation Psychology, Dayton, OH, USA, 18–21 April 2005; p. 812.
55. Champseix, R. Hrv-Analysis, 2021. Version 1.0.4. Github. 2021. Available online: https://github.com/Aura-healthcare/hrv-analysis (accessed on 4 November 2020).
56. Veltman, J.; Gaillard, A. Physiological workload reactions to increasing levels of task difficulty. *Ergonomics* **1998**, *41*, 656–669. [CrossRef] [PubMed]
57. Yao, D.; Wang, L.; Oostenveld, R.; Nielsen, K.D.; Arendt-Nielsen, L.; Chen, A.C. A comparative study of different references for EEG spectral mapping: The issue of the neutral reference and the use of the infinity reference. *Physiol. Meas.* **2005**, *26*, 173. [CrossRef]
58. Mallat, S.G. A theory for multiresolution signal decomposition: The wavelet representation. *IEEE Trans. Pattern Anal. Mach. Intell.* **1989**, *11*, 674–693. [CrossRef]
59. Hazarika, N.; Chen, J.Z.; Tsoi, A.C.; Sergejew, A. Classification of EEG signals using the wavelet transform. *Signal Process.* **1997**, *59*, 61–72. [CrossRef]
60. Tsang, P.; Psychologist, M. *Mental Workload and Situation Awareness*; Wiley: Hoboken, NJ, USA, 2006; Chapter 9, pp. 243–268.
61. Napierala, M.A. What is the Bonferroni correction? *Aaos Now* **2012**, 40–41. Available online: http://www.aaos.org/news/aaosnow/apr12/research7.asp (accessed on 6 June 2022).

62. Kandel, E.R.; Schwartz, J.H.; Jessell, T.M.; Siegelbaum, S.; Hudspeth, A.J.; Mack, S. *Principles of Neural Science*, 6th ed.; McGraw-Hill: New York, NY, USA, 2021; p. 1450.
63. Pritschet, L.; Powell, D.; Horne, Z. Marginally significant effects as evidence for hypotheses: Changing attitudes over four decades. *Psychol. Sci.* **2016**, *27*, 1036–1042. [CrossRef] [PubMed]
64. Stipacek, A.; Grabner, R.; Neuper, C.; Fink, A.; Neubauer, A. Sensitivity of human EEG alpha band desynchronization to different working memory components and increasing levels of memory load. *Neurosci. Lett.* **2003**, *353*, 193–196. [CrossRef] [PubMed]
65. Iqbal, S.T.; Zheng, X.S.; Bailey, B.P. Task-evoked pupillary response to mental workload in human-computer interaction. In Proceedings of the CHI'04 Extended Abstracts on Human Factors in Computing Systems, Vienna, Austria, 24–29 April 2004; pp. 1477–1480.
66. Marquart, G.; Cabrall, C.; De Winter, J. Review of eye-related measures of drivers' mental workload. *Procedia Manuf.* **2015**, *3*, 2854–2861. [CrossRef]
67. Jacob, R.J. Eye tracking in advanced interface design. *Virtual Environ. Adv. Interface Des.* **1995**, *258*, 2.
68. Poole, A.; Ball, L.J. Eye tracking in HCI and usability research. In *Encyclopedia of Human Computer Interaction*; IGI Global: Hershey, PA, USA, 2006; pp. 211–219.

sensors

MDPI

Article

Relabeling for Indoor Localization Using Stationary Beacons in Nursing Care Facilities

Christina Garcia * and Sozo Inoue

Graduate School of Life Science and Systems Engineering, Kyushu Institute of Technology, 2-4 Hibikino, Wakamatsu Ward, Kitakyushu 808-0135, Japan; sozo@brain.kyutech.ac.jp

* Correspondence: garcia.christina-alvarez442@mail.kyutech.jp or alvarez7.christina@gmail.com

Abstract: In this study, we propose an augmentation method for machine learning based on relabeling data in caregiving and nursing staff indoor localization with Bluetooth Low Energy (BLE) technology. Indoor localization is used to monitor staff-to-patient assistance in caregiving and to gain insights into workload management. However, improving accuracy is challenging when there is a limited amount of data available for training. In this paper, we propose a data augmentation method to reuse the Received Signal Strength (RSS) from different beacons by relabeling to the locations with less samples, resolving data imbalance. Standard deviation and Kullback–Leibler divergence between minority and majority classes are used to measure signal pattern to find matching beacons to relabel. By matching beacons between classes, two variations of relabeling are implemented, specifically full and partial matching. The performance is evaluated using the real-world dataset we collected for five days in a nursing care facility installed with 25 BLE beacons. A Random Forest model is utilized for location recognition, and performance is compared using the weighted F1-score to account for class imbalance. By increasing the beacon data with our proposed relabeling method for data augmentation, we achieve a higher minority class F1-score compared to augmentation with Random Sampling, Synthetic Minority Oversampling Technique (SMOTE) and Adaptive Synthetic Sampling (ADASYN). Our proposed method utilizes collected beacon data by leveraging majority class samples. Full matching demonstrated a 6 to 8% improvement from the original baseline overall weighted F1-score.

Keywords: oversampling; data augmentation; machine learning; signal measurement; signal pattern; relabeling; indoor localization; beacon; nursing care

1. Introduction

The increasing population of older adults is impacting the nursing workforce, leading to a shortage of skilled staff [1]. As the demand for services grows, the use of nursing homes is also escalating, resulting in a rise in the patient-to-caregiver ratio [2,3]. Research efforts to better comprehend activity patterns during patient assistance are significant to leverage staff and improve elderly care delivery [4].

Indoor Positioning System (IPS) allows localization within enclosed spaces, facilitating navigation and the tracking of individuals or objects through a network of transmitters and receivers [5,6]. Employing indoor positioning to monitor care routine and patient assistance is helpful to support nursing records and to optimize care response [7–9]. Several key positioning techniques exist, but the standard approach includes trilateration, triangulation, multilateration, and fingerprinting [10]. However, these techniques are limited by environmental factors such as hardware requirements, setup complexity, signal obstruction, and time synchronization.

IPS differ in technology, signal measurement, and localization techniques tailored to suit the environment and particular use-case requirements [8]. Radiofrequency-based technologies [11], specifically Wi-Fi [12–16], Radio Frequency Identification (RFID) [17,18], and Bluetooth Low Energy (BLE) [19–21], are typically preferred considering flexibility to setup and integration with IoT. Recently, studies on Wi-FI have been optimizing line of sight (LOS) and non-line of sight (NLOS) [22,23] approaches, including combination with vision [24] for localization. Furthermore, sensors such as IMU [25] and Geomagnetism [26] are integrated in movement and positioning. However, considering restrictive data collection in caregiving activities, beacons are preferred in nursing homes with straightforward installation and lesser complexity imposed on participants [3,27]. The presence of unavoidable equipment in the facility can also hamper the collection of geomagnetic data. Given that Indoor Positioning Systems (IPS) are contingent upon the specific use case and environment, Bluetooth Low Energy (BLE) has been selected based on target availability, straightforward deployment, and privacy considerations pertinent to elderly individuals. Additionally, the energy-efficient power consumption of BLE allows longer data collection with minimal disturbance in the facility monitoring the device remotely.

A common challenge to indoor localization accuracy is low quality of signals. Outlier detection [28] and filtering methods are applied to resolve this issue. Moving average [29], weighted average [30], Bayesian sequential Monte Carlo [20], and Gaussian filtering [31] has been proposed for signal smoothing to achieve better positioning. Kalman filter [21,32] is a widely used method for filtering signals in IPS. In addition, machine learning [7,10,14,29,33] and deep learning [12,30,34] methods are being employed to optimize signal features which are not covered by standard localization techniques. Network issues, environmental factors affecting signal quality, and hardware malfunction can impact collected data. Moreover, data imbalance due to unequal representation of different areas or activities within the facility in a real-world setting affects indoor positioning accuracy [2].

Data augmentation methods are utilized to address the issue of class imbalance, including in IPS applications. Data augmentation is a method to artificially increase the training dataset by creating a modified version of the current data. To increase real data, Random Sampling (RS) [17] is commonly applied to minority class samples by duplication. On the other hand, creating synthetic samples is an alternative method. Adaptive Synthetic Sampling (ADASYN) [35] and Synthetic Minority Over-sampling Technique (SMOTE) [36] both generate synthetic data points for the minority class by interpolation based on the nearest neighbors principle. Currently, existing oversampling techniques suffer from repeating data, resulting in overfitting, while synthetic samples are prone to noise and data misrepresentation. Since augmented data are intended to reflect actual samples, oversampling should be carefully considered. At present, augmentation methods do not optimize real data from the majority classes, and few studies exist that investigate the use of signal patterns for augmenting training data in the context of indoor localization [2].

This paper introduces a data augmentation approach to address the challenge of unequally represented locations in beacon-based indoor localization. By analyzing signal patterns in different rooms, we successfully implement a relabeling strategy, utilizing Received Signal Strength Indicator (RSSI) values from one location as a proxy in another. As more nursing homes adopt IPS with IoT, this research aims to leverage beacon data collected in real-world environments to improve indoor localization. The main objective of this research is to develop an augmentation method by increasing training samples with beacon data from different rooms to improve location detection of minority classes. We focus on identifying semantic locations where rooms are preferred over exact geographical positions to support caregiving records. Specifically, this study aims to address the following research questions:

1. How do we identify and match signal patterns between different locations in the facility? We utilize standard deviation and Kullback–Leibler divergence to analyze signal patterns, facilitating the identification of similar beacons across minority and majority classes.
2. How can we use the samples from other beacons to augment location with less data? We propose a novel relabeling approach that reassigns signals from matching beacons to areas with fewer samples, thereby addressing data imbalance issues.

3. What is the performance of oversampling based on signal pattern relabeling compared to existing augmentation methods? To evaluate, we compare the performance of relabeling with Random Sampling, ADASYN, and SMOTE applied to our collected real-world data from a nursing facility. To further assess, we test relabeling to different rooms to validate its effectiveness.

We employ different augmentation techniques in the training set comprising 3.5 days of data and perform indoor localization using 1.5 days of test data. Overall and target class F1-scores are measured to account for the data imbalance, including precision, and recall. Our proposed method achieves an improved target class F1-score of 27% to 40% compared to the baseline data. With full matching, the relabeling method demonstrates a 6 to 8% improvement from the original baseline overall weighted F1-score. KL divergence results in a better F1-score than the standard deviation for both full and partial matching. This method effectively expands the training set, enhancing model accuracy, as demonstrated in a nursing care facility where beacon devices and mobile applications are employed for data collection.

The subsequent sections of this paper are structured as follows: Section 2 covers the relevant research on indoor localization with BLE technology, including the distinction of our work over other localization and oversampling techniques. In Section 3, we elaborate on the proposed method, which encompasses signal pattern and relabeling approach, while Section 4 covers the data collection process, including the evaluation of proposed relabeling in comparison to baseline and other augmentation strategies. A comprehensive analysis of the results follows in Section 5 emphasizing the significant insights from our research. Section 6 concludes the paper, highlighting the main contributions and outlining potential directions of future work.

2. Related Literature

In this chapter, we cover relevant works on the design of indoor positioning implemented in the nursing facility. With the advent of the Internet of Things (IoT), devices that easily connect to the network such as beacons, tags, and mobile devices [2,37–40] are preferred in system design. Signal measurements are generally based on time, angle, and received signal strength (RSS) [28,29,31,38,41].

2.1. Indoor Localization with Beacons

Beacons are battery-driven radio transmitters used in indoor positioning with proximity sensors that emit BLE signals [42,43]. Unlike the classic Bluetooth, which is connection-oriented, BLE has advertising functionality and does not necessarily have to pair [44,45]. Beacons are preferred in IPS since they are flexible, easy to deploy, and cost-effective, with low power consumption that can last up to a year [46].

BLE beacons are used to track nursing activities to better understand care routines, which is crucial to optimize workload distribution considering the low staff-to-patient ratio [47]. Existing systems combine beacons with Wireless Local Area Networks (WLAN) for tracking both patients and nurses by analyzing RSSI, Time of Arrival (ToA), and the Angle of Arrival (AoA) [7]. This involves the setup of more hardware as receiver and transmitter modules. BLE beacons are often paired with smartphones used by caregivers for detection. To automatically record daily caregiving routines, beacons are used for time-spatial recognition [5,9].

In some nursing facilities, multiple patients share a common room. Considering the effects of setting up indoor positioning on the privacy of the elderly, which can affect their social interaction, the placement of beacons should be carefully considered in planning IPS [27]. Studies on nursing homes commonly use RSS and BLE for indoor localization [3,9,27] with its energy-efficient power consumption. However, noise and multipath can affect the signal quality and positioning performance.

2.2. Data Augmentation

The uneven distribution of classes within a dataset is referred to as data imbalance, which is a prevalent problem in machine learning. This indicates that the classifier performs well for the majority class but poorly for the minority class. Since minority classes are crucial to solving many real-world issues, correctly categorizing samples from this class is similarly crucial [48–50].

Data augmentation is significant in improving indoor positioning systems. The technical and non-technical challenges of IPS in real-world environments [51] have impacted the precision of indoor positioning with uncontrolled variables such as signal interference, physical obstructions, and hardware breakdown. Data collection in on-site scenario [52] affects data quality and labeling, resulting in an imbalanced and unequal representation of classes in dataset [2,53]. In order to create a robust and unbiased model for IPS, data augmentation shows up as a crucial and imperative approach to address data imbalance. By applying several changes to the current dataset, data augmentation creates synthetic data that diversify the training set and lessens the effects of class imbalance [54–56].

Three basic approaches dominate data-level solutions to the class imbalance issue: Random Sampling [57,58], Synthetic Minority Over-Sampling Technique (SMOTE) [56], and Adaptive Synthetic Sampling (ADASYN) [35]. By randomly adding additional copies of selected minority classes to the training data, Random Over-Sampling balances class distribution [57]. On the other hand, instead of oversampling by replacement, ADASYN and SMOTE oversample the minority class by producing synthetic instances [36]. However, these techniques have several limitations.

Overfitting can result from Random Over-Sampling since duplicates of minority-class samples are added to the dataset without adding new information. If the original data are already highly dimensional, this increases the calculation cost and lengthen the classifier's training period. Conversely, Random Under-Sampling randomly removes instances from the majority class, potentially leading to the neglect of essential data.

When training a model, applying SMOTE creates a linear mapping of the data, which can lead to overfitting issues. Also, there is a risk of overlap because the SMOTE method does not account for the position of general data close to the uncommon class data [49,56,59,60] Similarly with ADASYN, possible class overlap occurs in boundary areas as oversampling targets to resample between neighboring minority and majority classes. In general, both synthetic approaches are sensitive to noise and need parameter tuning as both are dependent on data distribution [61].

A combination of augmentation techniques has been applied in indoor localization [62,63] including adding information to the reference dataset and deep learning-based approach [34,64], yet these involve computational complexities. Few studies have focused on using methods based on signal patterns between sensors to augment training data specifically for indoor localization purposes in nursing homes.

Accurate labels ensure that models learn the correct patterns and relationships in the data. In machine learning, relabeling is employed in the data augmentation process to handle proper class labeling, which is crucial for the accuracy of supervised learning models. In existing studies, relabeling is commonly applied to preserve the original class labels [65]. Recently, relabeling has been used to address class imbalance when using logistic regression by assigning new labels to classes with fewer instances [66]. Data augmentation leading to the loss of label information can reduce model performance.

2.3. Signal Pattern

Signal patterns refer to the specific way that radio frequencies, like Bluetooth, behave in an indoor setting. Fluctuations in the signal are essential metrics providing valuable insights into the alterations of signal intensity that occur within indoor settings [67]. Trilateration and triangulation are traditional localization techniques that utilize geometric principles to assess signal behavior, measuring distance and angle from reference points.

In the non-conventional approach, statistical techniques are optimized to handle inherent variability and uncertainty in the signal-based analysis. The received signal's standard deviation reflects the signal strength fluctuations at different positions or distances within an indoor environment [68–70]. On the other hand, variance measures the spread of the RSSI values around the mean, offering perspectives on the consistency of the signal strength measurements at different locations [68]. KL divergence is a measure of the difference between two probability density functions. In IPS, it is calculated to measure similarity in the online phase of unlabeled data in existing database and up-to-date data [10].

Integrating multiple signal sources, sensor fusion techniques, and machine learning provides more robust and reliable localization. Implementing adaptive algorithms that can learn and adjust to environmental changes further improves positioning performance. In this study, we delve into the signal patterns between locations by employing statistical techniques, specifically standard deviation and KL divergence, to measure features to find matching patterns for oversampling.

In this work, we propose to compare the signal pattern features of labeled beacon data from different locations and determine the divergence between minority and majority classes as a foundation for data augmentation. By considering the layout of the care facility, we focus on leveraging the current data from stationary beacons by using the majority class to oversample locations with less BLE data. We utilize relabeling to update labels of the augmented data derived from other locations, aligning them with the labels of the minority class. Specifically, signal patterns observed from selected positioned beacons from various rooms are utilized to guide the relabeling process. Full and partial matching represent two distinct relabeling variations that take into account the comprehensive arrangement of installed beacons.

This research strategically deploys BLE beacons to enable indoor localization of nursing staff, ensuring that the architecture of the nursing home remains unaltered and that caregiving services are delivered without interruption. This work prioritizes identifying semantic room locations over exact coordinates, considering the relevance to caregiving records. To protect the privacy of patients, all devices are mounted outside the doors of patient rooms.

3. Material and Methods

This section introduces our developed relabeling method depicted in Figure 1 for beacon data augmentation.

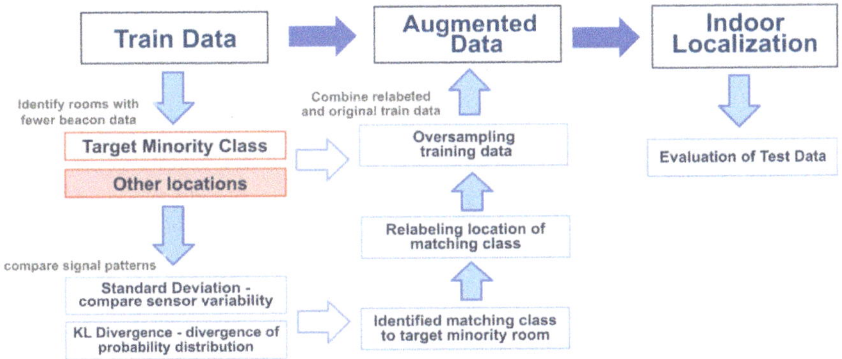

Figure 1. Proposed oversampling approach based on signal pattern relabeling for indoor localization.

In this approach, we aim to increase the training data sample, particularly locations with less beacon data. Initially, training data are analyzed to identify rooms with fewer beacon signals, termed as the "Target Minority Class". To oversample, we identify matching locations to minority classes from other locations by comparing signal patterns using standard deviation and KL divergence. To proceed with data augmentation, we apply relabeling to the identified matching class by updating the location label similar to the target minority. The relabeled set is then added to the original training data, increasing the set. The augmented data are then used to enhance the indoor localization employing machine learning, with the model's efficacy evaluated using test data.

The subsection commences with information on the beacon data from the site. Understanding the signals detected from beacons to comprehend variability is necessary to execute relabeling. Figure 2 describes the overview of the indoor localization employed in the nursing facility from data collection to the use of the relabeling approach.

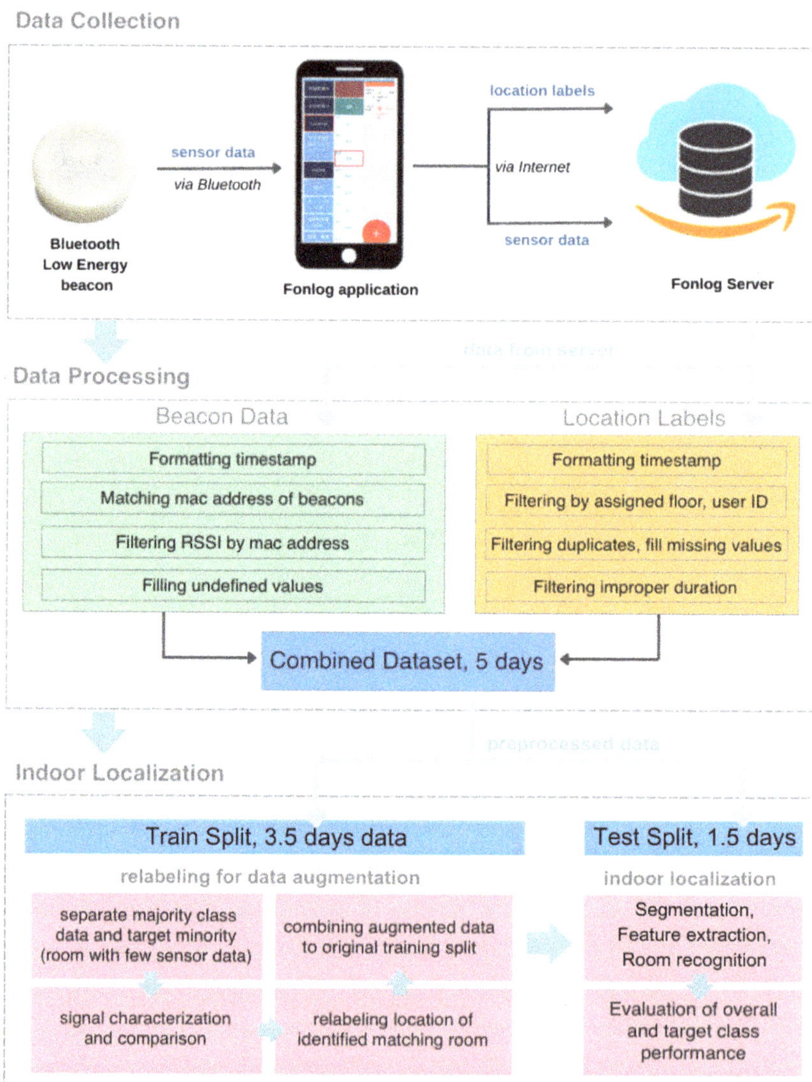

Figure 2. An overview of the proposed indoor positioning system in nursing care facility.

3.1. Signal from Detected Beacons

Prior to deployment, the respective MAC addresses of the beacons are recorded to match the raw data from the server, as shown in Table 1. After pre-processing the raw data, the MAC addresses of detected devices are filtered to match the beacons. The final dataframe of RSSI values is shown in Figure 3.

Table 1. Collected raw beacon data stored in the server.

User_id	Timestamp	Mac Address	RSSI
90	2023-04-10 10:22:55.589+0900	FD:07:0E:D5:28:AE	−75
90	2023-04-10 10:22:55.599+0900	D2:1C:25:72:FB:E3	−62

Based on Figure 3, we can suppose $RSSI_t$ represents the detected RSSI at timestamp t at location m. We then define the RSSI measurement for all beacons n, and the corresponding location label as in Equation (1). Overall, the signal database can be expressed as Equation (2) and the labels as Equation (3).

$$RSSI_t = [RSSI_{t1}, \ldots, RSSI_{1n}], label_t = m_t, \tag{1}$$

$$RSSI = \{[RSSI_{11}, \ldots, RSSI_{1n}], \ldots, [RSSI_{t1}, \ldots, RSSI_{tn}]\}, \tag{2}$$

$$label = \{m_1, m_2, m_3, \ldots, m_t\}, \tag{3}$$

where t represents the timestamp, n denotes the total number of beacons installed on site, and m refers to the associated elderly room. With location label m, training data are observed to identify rooms with fewer detected signals.

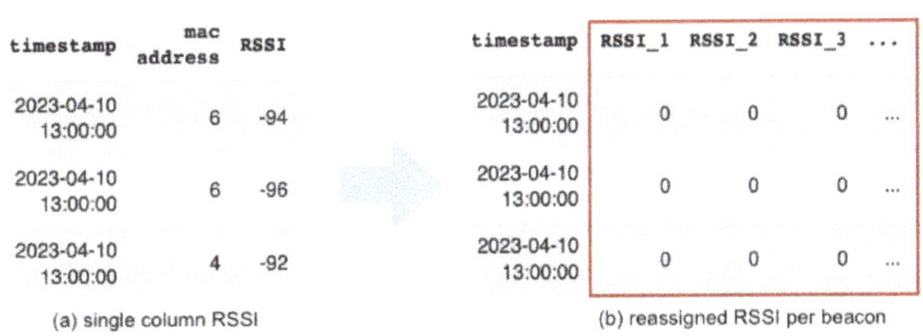

Figure 3. Matrix of the received signal strength from beacons. (**a**) Single column RSSI. (**b**) The corresponding RSSI for each beacon, repositioned into separate columns.

3.2. Matching

Generally, the accuracy of IPS decreases as the distance between the transmitter and receiver increases [71]. Observing this from the histogram of detected beacons and considering the setting of installed devices covering up to a 5 m range, we limit the analysis to six beacons to better comprehend signal patterns between rooms. Specifically, we only focus on elderly rooms installed with stationary beacons following a similar layout. Figure 4 shows the targeted surrounding beacons investigated for each room.

Prior to calculating the signal pattern feature, we classify rooms based on the completeness of the selected surrounding beacons. For each location, six BLE transmitters are filtered such that

$$f_m = [fl_m, sl_m, s_m, f_m, sr_m, fr_m]. \tag{4}$$

In Equation (4), s is the beacon on the location (source) room, f is the beacon on the room in front, fl and sl are the beacons on the left side rooms while sr and fr are beacons on the right side, respectively. In full matching, we only consider elderly rooms forming complete six beacons as candidate match to the minority class. On the other hand, with partial matching, all elderly rooms are considered as possible match to the minority class regardless of incomplete sensors forming the targeted six. As an example, in Figure 4, Room 515 is considered only in partial matching where $f_{515} = [null, 13, 15, 1, 16, 2]$. Room 521, on the other hand, is also a candidate in full matching, where $f_{521} = [8, 20, 21, 10, 22, 11]$. For locations with incomplete beacons resulting in null values representing no signal detected, we replace null with zeroes for the model to better differentiate from distant beacons with higher negative values and to calculate the signal pattern feature.

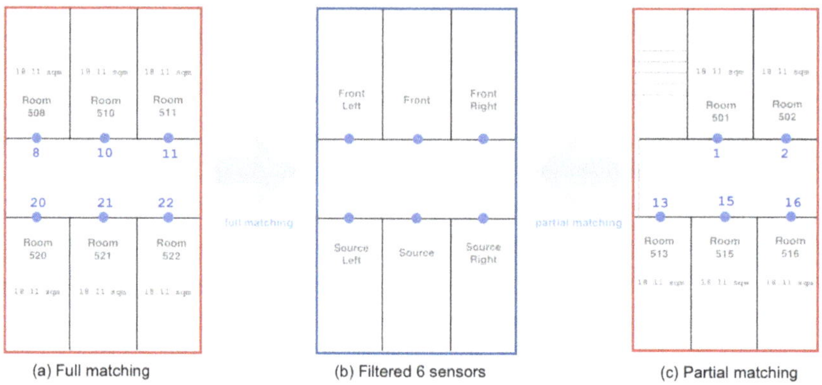

Figure 4. Full and Partial Matching based on selected sensors. (**a**) Full matching with complete 6 sensors. (**b**) A total of 6 sensors surrounding the location. (**c**) Partial matching with incomplete 6 sensors.

3.3. Relabeling Based on Signal Pattern

A signal pattern feature is computed from the six beacons to define the signal variability in different rooms. A sub-dataframe $pattern_m$ containing the RSSI values of the beacons in the list f_m is created for each room in the majority class. The sub-dataframe of the minority class is denoted as $pattern_{min}$. From these sub-dataframes, standard deviation and KL divergence are measured to represent each room's signal pattern feature, which is then used to identify similar locations. The objective is to pinpoint the room that has a matching signal pattern with the minority class in order to move forward with the relabeling process. Initially, the majority class is downsampled to match the number of samples in the minority class as depicted in Figure 5. Outlined in Steps 1 to 4 are the procedures to calculate the signal pattern feature of $pattern_m$ and $pattern_{min}$. See Algorithm 1.

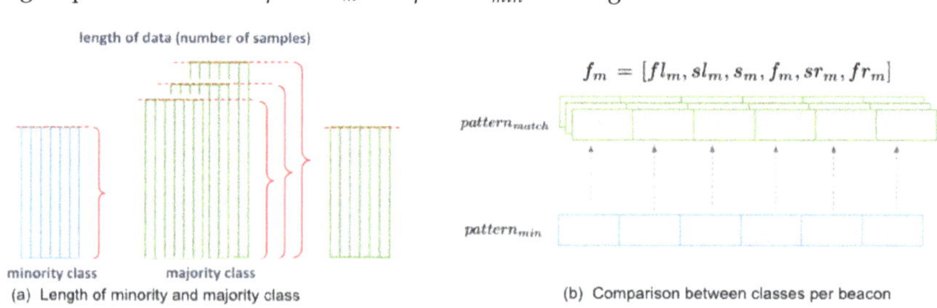

Figure 5. Comparing signal pattern feature between minority and majority class. (**a**) Length of samples considered. (**b**) Comparison between rooms per beacon.

The signal pattern feature based on standard deviation calculates the absolute differences in standard deviation for each beacon between minority and majority classes. This approach considers individually the variability difference for each beacon before summing them. The matching location from the majority class is identified with the least standard deviation difference sum expressed in Equation (5),

$$\Delta\sigma_{\text{total}} = \sum |\sigma_{\text{min}} - \sigma_{\text{match}}|, \quad (5)$$

where $\Delta\sigma_{\text{total}}$ represents the total difference in standard deviation values, σ_{min} is the standard deviation for a specific beacon in the minority subdataframe, σ_{match} is the standard deviation of each beacon in the candidate sub-dataframes $pattern_m$, and $\sum$ represents the summation symbol, indicating the sum of absolute differences across all beacons.

Algorithm 1: Algorithm for comparing the signal pattern feature between rooms

Input:
$m \leftarrow$ room number
$d_{\text{train}} \leftarrow$ train data
Output:
$pattern_{\text{match}} \leftarrow pattern_{\text{m}}$

1. Identify rooms with small sample in train data, df = d_{train}.
2. Define $pattern_{min}$ and $pattern_m$ following Equation (4) such that for i in labels, m
   $d_i = df[df['m']] == "i"]$
   if i == "1": $f_1 = [fl_1, sl_1, s_1, f_1, sr_1, fr_1]$
   ...
   elif i == "25": $f_{25} = [fl_{25}, sl_{25}, s_{25}, f_{25}, sr_{25}, fr_{25}]$.
3. Group $pattern_m$ accordingly as candidates for full and partial match.
4. Calculate the signal pattern feature from $pattern_{min}$ and $pattern_m$.
5. Compare the signal pattern features to identify $pattern_{\text{match}}$.

On the other hand, Kullback–Leibler (KL) Divergence is calculated from minority to majority classes. With this approach, we measure the divergence between the probability distribution represented by minority and majority classes, normalizing the data to represent probability distributions. The KL Divergence applied is expressed in Equation (6),

$$D_{\text{KL}}(P \parallel Q) = \sum P(i) \log\left(\frac{Q(i)}{P(i)}\right), \quad (6)$$

where $P(i)$ represents the probability of observation i in $pattern_{min}$ and $Q(i)$ represents the probability of observation i in candidate match $pattern_m$. After calculating and plotting all the statistical values, we observe the trend of the signals in $pattern_{min}$ and locate similarity in the rest of $pattern_m$ from all other rooms to find $pattern_{match}$. We apply both statistical measures to partial and full matching variations of relabeling forming four candidate matches.

To proceed with relabeling, train data d_{train}, $pattern_{min}$ for the location with low sample and identified $pattern_{match}$ are required. The expected output is the augmented train data $d_{train,new}$ with concatenated original train data d_{train} and relabeled data $d_{relabeled}$. Listed from one to five are the sequential steps followed to execute the relabeling process. See Algorithm 2.

Nursing homes frequently feature an evenly distributed space across rooms. Given the environment's influence on signal behavior, the relabeling method is specifically designed for floors where the rooms share the same geometric layout. The relabeling method is exclusively applied to patient rooms in the original floor plan based on matching as depicted in Figure 4. In this study, relabeling is not applied for signals from beacons positioned in areas with diverse dimensions and layouts, such as open spaces in the cafeteria. It is crucial that the beacon devices are installed in fixed positions. Moreover, as the proposed approach specifically filters data to the six surrounding beacons, Standard Deviation and KL Divergence were chosen to measure signal patterns as they represent the variability of the targeted signals. To apply relabeling to other sensors, the representation of signal patterns can be updated using other statistical measures that better suit the behavior of the data being analyzed.

Algorithm 2: Algorithm for relabeling the location of matching class

Input:
$d_{\text{train}} \leftarrow$ train data
$pattern_{match}$, $pattern_{min}$
Output:
$d_{\text{train, new}} \leftarrow d_{\text{train}}, \quad d_{\text{relabeled}}$

1. Create a dataframe for the relabeled data $d_{\text{relabeled}}$.
2. Populate $d_{\text{relabeled}}$ with values from $pattern_{match}$ following the same columns representing RSSI values of the six beacons.
3. Add remaining columns from d_{train} in $d_{\text{relabeled}}$ for beacons not included in $pattern_{match}$. Fill with zero values.
4. For relabeling, assign the location of the minority class to the labels for $d_{\text{relabeled}}$.
5. To oversample, combine the augmented data $d_{\text{relabeled}}$ to the original train split d_{train} in a new dataframe $d_{\text{train,new}}$.

3.4. Indoor Localization

In this paper, indoor localization is approached as a recognition problem using machine learning [72] to identify semantic room locations. This approach is designed to enhance caregiving records by emphasizing the functional nature of spaces within the caregiving environment rather than their physical coordinates. For room estimation considering attenuation, RSSI values from all rooms are considered in feature extraction. Discrepancies between the timestamps of location labels and RSSI data are resolved by synchronizing data. To proceed with indoor localization, statistical and temporal features are derived from the RSSI matrix. The Random Forest algorithm is employed for identifying locations as this classifier demonstrates effectiveness in handling imbalanced data [73] and preferred in indoor positioning [74–76]. As we deal with a similar scenario of data imbalance between locations, we opt to adopt this model to focus on relabeling.

Five statistical features are extracted, specifically mean, standard deviation, minimum, maximum, and RSSI count. Integrating the quantity of detected signals per device into the feature set improves the model's effectiveness, as observed in prior work. Time-based attributes, specifically hour, minute, and microsecond, are extracted. A non-overlapping time window of 45 s is implemented. The train and test data division is date-based, adhering to train-test ratios recommended by empirical studies. The training set comprises 3.5 days of data, and the remaining 1.5 days are intended for test data. Both training and test sets reflect the caregiving routine in the facility, where nursing staff frequently visited the common area compared to the patient rooms.

To assess the effectiveness of relabeling, the indoor localization performance of baseline train data is compared to the performance of data augmentation. In addition, various data augmentation methods must be applied to the minority class. For cross-validation, the relabeling approach is applied to different rooms and compared with other methods. The next chapter covers the evaluation of real-world data collected in the nursing home.

4. Data Collection and Evaluation

The majority of the caregiving activities are location dependent; hence, monitoring the rooms visited by nursing staff is necessary for an estimation of the delivered workload and assistance given to elderly individuals on the floor. This information is relevant in automatic care recording, necessitating inputted FonLog data. The nursing care tasks executed by the caregiver fall under the category of patient care, medical assistance, documentation, cleaning, and organization. As the movement of the caregiver is dependent on nursing activities, we elaborate the details on caregiving to record the workload of staff performed in respective locations. Patient care covers bathing, excretion assistance, daytime support, meal preparation, and recreation. Medical assistance encompasses checking patient vitals, administering treatment, and overseeing rehabilitation. Cleaning and organization tasks include changing linens and handling laundry. Care record falls under documentation.

4.1. Data Collection

The data collection approach implemented in the nursing facility is depicted in Figure 6. Pre-deployment testing was conducted with the caregiver to ensure the detection of beacons, connection to the server, and correct reading of both location labels and BLE signals. Furthermore, initial data gathering with the placement of beacons was conducted to make adjustments accordingly following the nursing home's protocol. Initially, user feedback was collected via discussion with the participating staff and user study during the pre-deployment setup. A location labeler was employed to address the staff's challenges in recording locations while carrying out their tasks.

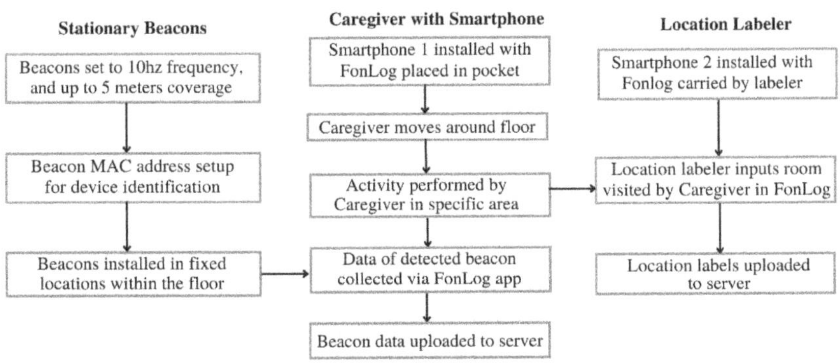

Figure 6. Overview of data collection approach in nursing facility.

Final data collection was performed on the fifth floor of a nursing care facility over a period of five days, from morning to afternoon of April 10 to 14, capturing the locations a caregiver visited during their daily routine. Daily, about seven hours of data collection was performed: three hours in the morning and four hours in the afternoon. The schedule followed the caregiving routine in the facility. In adherence, the placement of beacons outside patient rooms was decided to protect privacy, and data collection was executed following the facility's protocol and proposed schedule.

FonLog, a smartphone application in Figure 7, was installed and used as a data collection tool to log location labels and as a receiver of RSSI signals [39]. In this study, FonLog was explicitly customized to record activities performed on the respective floor of the nursing care facility with strategically positioned beacons.

Two separate smartphones installed with FonLog were set up for data gathering, one for the nursing staff with enabled BLE ID and another for the location labeler. Prior to deployment, the location and Bluetooth settings of the smartphone were set, and beacon detection was verified on the login interface, where the MAC (Media Access Control) addresses are displayed. The nursing staff was requested to bring one smartphone as they performed caregiving across various rooms on the floor. To avoid disturbance during tasks and maintain the quality of nurse care, the device was placed in the front pocket of the uniform.

A common challenge of on-site data collection is a mismatch between beacon data and the corresponding location label, which occurs due to the delay in activity recording. Caregivers usually log the activity after completing various tasks, resulting in discrepancies in timestamps during pre-processing. To mitigate this problem, the data collection process included an observer closely monitoring the caregiver's daily routine. The labeler assigned was positioned in the hallway and recorded locations on the second smartphone, with Fonlog choosing from the list of rooms in the app. Both the carer and labeler were each assigned a unique user ID. Similarly, each patient room and area within the floor was allocated a unique customer ID. All beacon data and labels were temporarily stored in the local storage of each smartphone and uploaded to the same server once internet connection was established. In this study, the data collection involved participating nurses of average height. Any significant height differences should be noted, and the impact on RSSI values should be observed in the case of multiple participants.

(a) Sensor selection screen (b) Activity recording screen (c) Log-in screen with detected MAC address

Figure 7. User interface of the FonLog application. (**a**) Sensor selection screen where BLE ID is enabled. (**b**) Activity recording screen showing respective rooms within the floor. (**c**) Log-in interface reflecting detected mac address of BLE device.

Beacon data recorded from Fonlog contain user_id, timestamp, MAC address, and RSSI, while location data by labeler consists of user_id, activity performed, start and stop time, customer_id, and specified location. In our initial approach, we aligned the timezones of both the location and beacon data, and the timestamps were subsequently reformatted. Duplicates and rows with null values were removed, and entries lacking in start-time and stop-time with undefined duration were filtered from location labels. Figure 8 depicts the map of the target floor in the nursing facility with installed beacons.

Before deployment, beacon frequency and height placement were decided based on experiments performed in the lab and hallway of the university, reflecting the facility's layout. In total, 25 stationary beacons indicated by blue points on the map were installed 2 m from the ground, strategically positioned outside 19 elderly rooms, and in common areas to cover the usual route of the nursing staff. All beacons were configured to a frequency of 10 Hz, where the actual data showed RSSI detection ranging from 3 to 5 Hz.

In medical and elderly care facilities, there are substantial limitations on where BLE beacons can be placed. In scenarios where users carry the beacon tag, scanners are strategically positioned in corners and specific areas, taking into account the geometry of the location [9]. This setup allows for position calculation using methods that do not rely on machine learning. For this research, the placement of beacons was determined in agreement with the nursing home to effectively track the movement of caregivers on the respective floor. Specifically, beacons were mounted at an appropriate height on the doors outside each patient room. This approach was adopted to respect privacy and to minimize disturbance to the elderly during the installation and maintenance of the devices. In contrast to other studies [9], beacons were placed in stationary positions as emitters while Fonlog installed in the caregiver's smartphone served as the scanner.

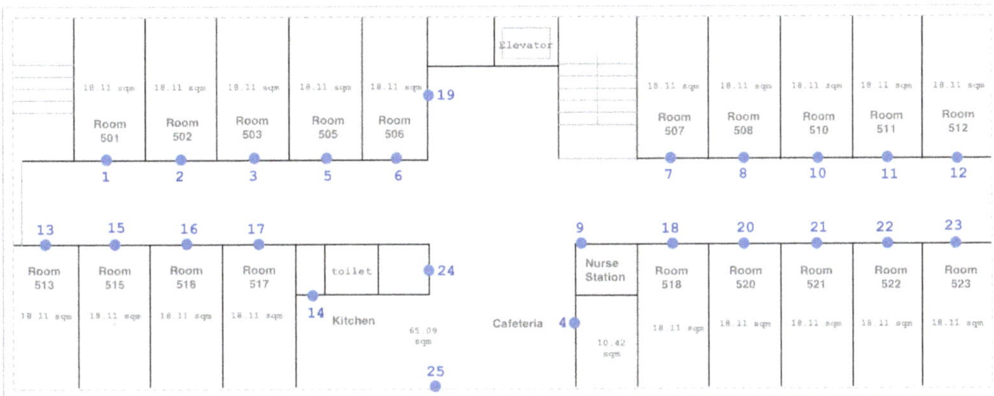

Figure 8. Layout of the facility showing the placement of installed beacons at each location.

The transmission power of beacons varies depending on the calibration of the RSSI at a distance of one meter. Beacons configured for broader coverage have higher transmission power. According to the specification sheet of the BLE device used, an RSSI detection range of 100 m corresponds to a transmission power of +4 dBm. All beacons were set to a detection range of zero to five meters coverage with a transmission power of up to −30 dBm (decibel-milliwatts).

Figure 9 displays the preprocessed beacon data with location labels used in the baseline of the indoor positioning model.

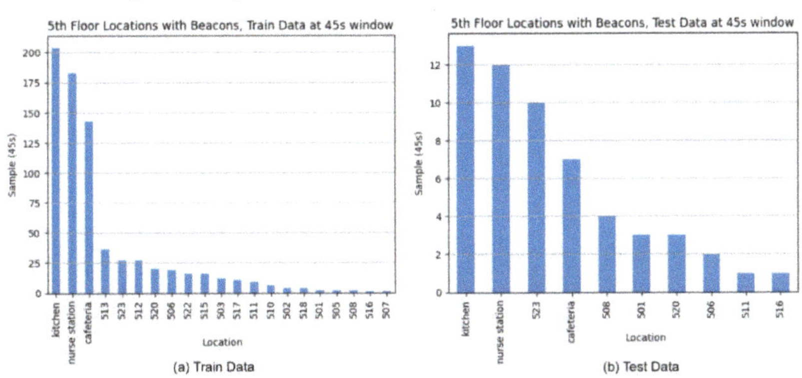

Figure 9. Baseline beacon data with location labels segmented for feature extraction. (**a**) Training split, 45 s window (**b**) Test split, 45 s window.

From the data, we can see an overview of the caregiving routine, which involves more extended activities in the kitchen, nurse station, and cafeteria, where meal assistance and nurse care recording are performed. We can also identify rooms with more extended data, suggesting longer patient assistance is delivered in the respective elderly room. The same test data from the figure are retained to evaluate and compare the performance of different data augmentation approaches in Section 4.3.

4.2. Performance Evaluation

As this study aims to address data imbalance by relabeling the approach, a comparative analysis of signal patterns and an assessment of the efficacy of partial and full room matching are conducted. Furthermore, we evaluate the performance of relabeling against other data augmentation methods. The test data cover 1.5 days, while the training data encompass 3.5 days, both segmented into 45 s windows without overlapping.

The target use case for evaluation in this study is centered on enhancing the performance of indoor localization through the meticulous comparison of signal patterns by specific features. By implementing partial and full matching of rooms, we aim to refine the relabeling process and ensure the most accurate representation of location data. This comparison also extends to assessing the efficacy of our proposed methods against other data augmentation approaches, thus providing a comprehensive analysis of performance improvements in indoor localization systems.

To evaluate the impact of data augmentation to indoor positioning performance, *F1-score*, *Precision* and *Recall* are measured and obtained, respectively, using the following Equations in (7)–(9), where TP is the True Positive, TN is the True Negative, FP is the False Positive, and FN the False Negative values.

$$F1\text{-}score = \frac{TP}{TP + 0.5(FP + FN)}, \quad (7)$$

$$Precision = \frac{TP}{(TP + FP)}, \quad (8)$$

$$Recall = \frac{TP}{(TP + FN)}. \quad (9)$$

We assess the efficacy of applying relabeling to underrepresented locations within the training dataset from the collected data. For this evaluation, reviewing the beacon data for each location in the training set, Rooms 516 and 507 are identified as minority classes. Filtering the device of all the rooms to six beacons following Equation (4) and Figure 4, we proceed with calculating the signal pattern feature. Figure 10 illustrates the resulting standard deviation of each of six beacons in $pattern_{min}$ corresponding to the minority class rooms.

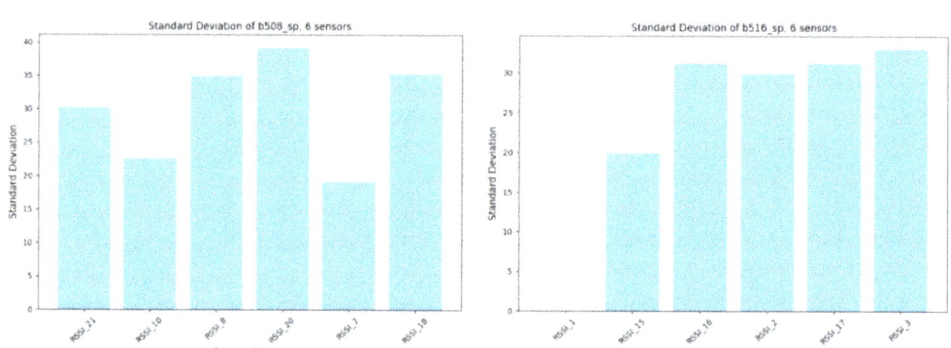

(a) Room 508, filtered to 6 sensors (b) Room 516, filtered to 6 sensors

Figure 10. Standard deviation of selected sensors in target minority class rooms. (**a**) Room 508, filtered to 6 sensors (**b**) Room 516, filtered to 6 sensors.

We then find the match to the minority class by comparing the signal pattern feature of other rooms using standard deviation and KL divergence. Based on the floor layout, corner rooms with only less than six beacons are reserved as candidate matches in partial matching. In comparison, middle rooms with complete six beacons mounted in surrounding locations are considered in full matching. This process allows for two types of relabeling: full and partial matching, by correlating beacons across different classes.

The integration of two variations of matching and two statistical measures for signal pattern features results in four potential scenarios for candidate matching in the relabeling process. To identify the best combination for relabeling, we implement data augmentation to all four use-case scenarios and compare it with other oversampling methods. From this approach, we aim to identify the statistical measure that better represents signal patterns from six beacons, and by comparing full and partial matching, we determine the locations to consider for relabeling.

In the case of full matching using signal patterns determined by standard deviation, as depicted in Figure 11, Room 520 emerges as the closest match to Room 508, and Room 511 is identified as the nearest match to Room 516.

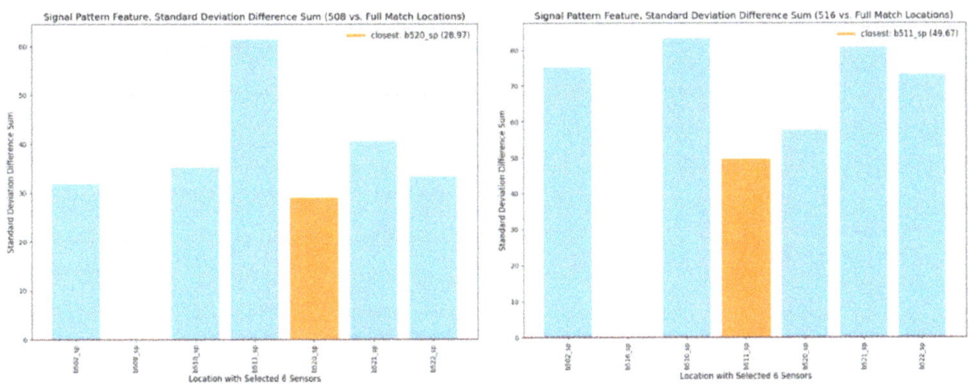

(a) Room 508 with full matching rooms (b) Room 516 with full matching rooms

Figure 11. Full matching with signal pattern feature based on standard deviation. (**a**) Room 508 vs. other locations. (**b**) Room 516 vs. other locations.

Figure 12 reveals that, under partial matching with signal pattern analysis based on standard deviation, Room 520 remains the closest match to Room 508, while Room 518 is identified as the nearest match to Room 516.

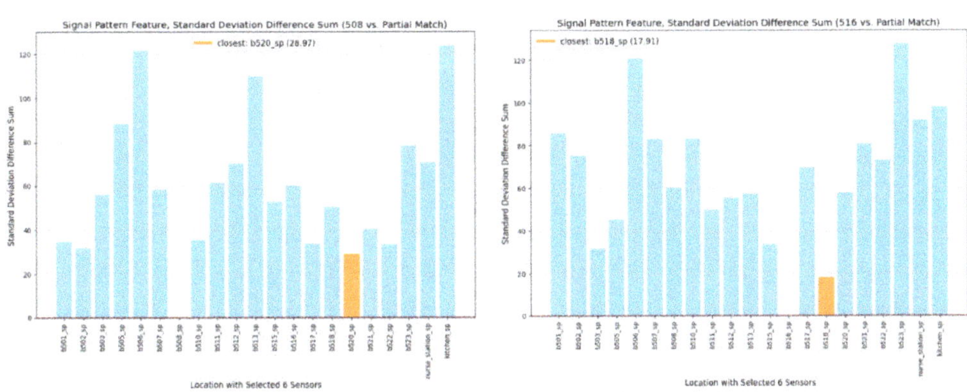

(a) Room 508 with partial matching rooms (b) Room 516 with partial matching rooms

Figure 12. Partial matching with signal pattern feature based on standard deviation. (**a**) Room 508 vs. partial matching locations. (**b**) Room 516 vs. partial matching locations.

Continuing with the assessment of the signal pattern using KL divergence, Figure 13 shows the result of full matching with KL divergence where we identify 522 as the majority class nearest match to 508. At the same time, 520 is the match for 516.

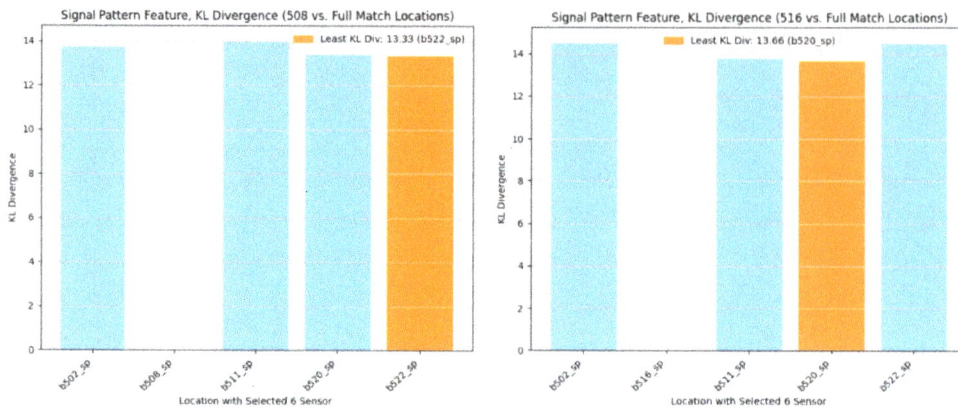

(a) Room 508 with full matching rooms

(b) Room 516 with full matching rooms

Figure 13. Full matching with signal pattern feature based on KL divergence. (**a**) Room 508 vs. other locations. (**b**) Room 516 vs. other locations.

Lastly, partial matching based on KL divergence results in 512 as the nearest match of 508, while 503 is identified as the top match for 516 as reflected in Figure 14.

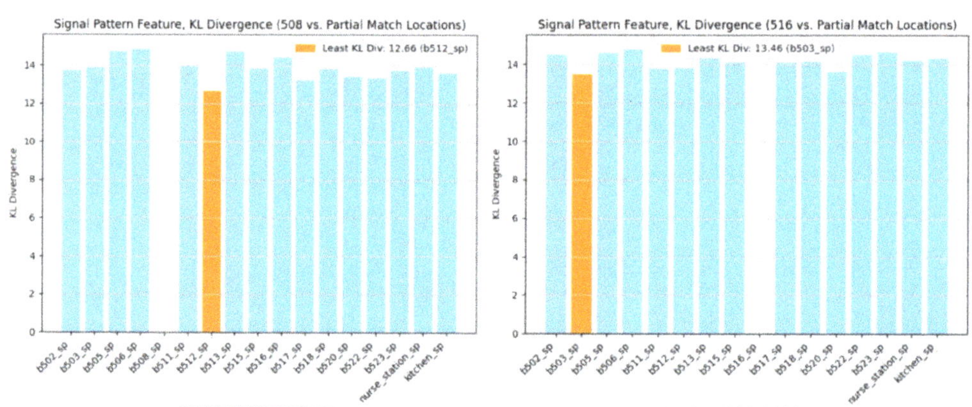

(a) Room 508 with partial matching rooms

(b) Room 516 with partial matching rooms

Figure 14. Signal pattern feature based on KL divergence with partial matching. (**a**) Room 508 vs. partial matching locations. (**b**) Room 516 vs. partial matching locations.

With the matching location identified for each minority class in all four cases, we proceed with relabeling the identified match for each minority location. We concatenate the augmented data to the original training oversampling Rooms 508 and 516. Indoor localization with the Random Forest model is executed using the new training data. Similarly, we apply data augmentation using Random Sampling, ADASYN, and SMOTE to both Room 508 and 516, generating the same length of augmented data as that of the relabeled matches for comparison.

4.3. Results

In our evaluation of indoor localization, we assessed the effectiveness of various combinations of relabeling approaches and statistical measures applied to signal patterns. This assessment aimed to determine the specific combination that yields the most optimal performance in terms of minority class localization.

Comparing the performance of relabeling using full matching based on standard deviation in Table 2, only the proposed relabeling method classified Room 516 after oversampling. Relabeling achieved the highest target class F1-score for both 508 and 516. In the overall model F1-score, relabeling improved model performance by 6%. ADASYN achieved the highest value, followed by the proposed method.

Table 2. Performance of Relabeling with Standard Deviation, Full Matching versus other Sampling.

Oversampling Approach	Target Class Precision	Target Class Recall	Target Class F1-Score	Overall Model Weighted F1-Score
Baseline *	Room 508 = 0.50 Room 516 = 0.00	Room 508 = 0.25 Room 516 = 0.00	Room 508 = 0.33 Room 516 = 0.00	0.60
Random Sampling	Room 508 = 0.67 Room 516 = 0.00	Room 508 = 0.50 Room 516 = 0.00	Room 508 = 0.57 Room 516 = 0.00	0.64
SMOTE	Room 508 = 1.00 Room 516 = 0.00	Room 508 = 0.25 Room 516 = 0.00	Room 508 = 0.40 Room 516 = 0.00	0.63
ADASYN	Room 508 = 0.50 Room 516 = 0.00	Room 508 = 0.50 Room 516 = 0.00	Room 508 = 0.50 Room 516 = 0.00	0.69
Proposed Relabeling	Room 508 = 0.57 Room 516 = 1.00	Room 508 = 1.00 Room 516 = 1.00	Room 508 = 0.73 Room 516 = 1.00	0.66

* Original train data with no augmentation applied.

The performance of relabeling with partial matching based on standard deviation is summarized in Table 3. In target class F1-score, no oversampling method was able to classify Room 516. At the same time, all data augmentation approaches improved the performance for 508, with SMOTE as the highest, followed by relabeling and ADASYN. Applying Random Sampling to the minority class resulted in a lower F1-score. Regarding the overall weighted F1-score, only ADASYN and SMOTE improved the model by 2 to 3%. As for the relabeling approach, the observed decrease in performance can be attributed to incomplete beacon data resulting from partial matching and the possibility of overlapping classes.

Table 3. Performance of Relabeling with Standard Deviation, Partial Matching versus other Sampling.

Oversampling Approach	Target Class Precision	Target Class Recall	Target Class F1-Score	Overall Model Weighted F1-Score
Baseline *	Room 508 = 0.50 Room 516 = 0.00	Room 508 = 0.25 Room 516 = 0.00	Room 508 = 0.33 Room 516 = 0.00	0.60
Random Sampling	Room 508 = 0.67 Room 516 = 0.00	Room 508 = 0.50 Room 516 = 0.00	Room 508 = 0.57 Room 516 = 0.00	0.60
SMOTE	Room 508 = 1.00 Room 516 = 0.00	Room 508 = 0.50 Room 516 = 0.00	Room 508 = 0.67 Room 516 = 0.00	0.63
ADASYN	Room 508 = 0.50 Room 516 = 0.00	Room 508 = 0.75 Room 516 = 0.00	Room 508 = 0.60 Room 516 = 0.00	0.62
Proposed Relabeling	Room 508 = 0.50 Room 516 = 0.00	Room 508 = 0.75 Room 516 = 0.00	Room 508 = 0.60 Room 516 = 0.00	0.59

* Original train data with no augmentation applied.

On the other hand, performance comparison with full matching based on the KL divergence is depicted in Table 4. With this relabeling approach, only the proposed method was able to classify Room 516 after performing data augmentation. Relabeling achieved the highest target class F1-score for both Room 508 and 516. In the overall model F1-score, relabeling improved the model performance by 8% and achieved the highest result.

Table 4. Performance of Relabeling with KL Divergence, Full Matching versus other Oversampling.

Oversampling Approach	Target Class Precision	Target Class Recall	Target Class F1-Score	Overall Model Weighted F1-Score
Baseline *	Room 508 = 0.50 Room 516 = 0.00	Room 508 = 0.25 Room 516 = 0.00	Room 508 = 0.33 Room 516 = 0.00	0.60
Random Sampling	Room 508 = 0.67 Room 516 = 0.00	Room 508 = 0.50 Room 516 = 0.00	Room 508 = 0.57 Room 516 = 0.00	0.66
SMOTE	Room 508 = 0.67 Room 516 = 0.00	Room 508 = 0.50 Room 516 = 0.00	Room 508 = 0.57 Room 516 = 0.00	0.66
ADASYN	Room 508 = 0.60 Room 516 = 0.00	Room 508 = 0.75 Room 516 = 0.00	Room 508 = 0.67 Room 516 = 0.00	0.66
Proposed Relabeling	Room 508 = 0.57 Room 516 = 1.00	Room 508 = 1.00 Room 516 = 1.00	Room 508 = 0.73 Room 516 = 1.00	0.68

* Original train data with no augmentation applied.

Lastly, Table 5 summarizes indoor positioning comparison with partial matching based on KL divergence. Similarly, in target class F1-score, no oversampling method was able to classify Room 516, while relabeling achieved the highest performance for Room 508 at 73%, which is 16% to 29% higher than that of other oversampling methods. In the overall model F1-score, only relabeling resulted in an improved model performance at 62%.

Table 5. Performance of Relabeling with KL Divergence, Partial Matching versus other Oversampling.

Oversampling Approach	Target Class Precision	Target Class Recall	Target Class F1-Score	Overall Model Weighted F1-Score
Baseline *	Room 508 = 0.50 Room 516 = 0.00	Room 508 = 0.25 Room 516 = 0.00	Room 508 = 0.33 Room 516 = 0.00	0.60
Random Sampling	Room 508 = 0.67 Room 516 = 0.00	Room 508 = 0.50 Room 516 = 0.00	Room 508 = 0.57 Room 516 = 0.07	0.60
SMOTE	Room 508 = 0.67 Room 516 = 0.00	Room 508 = 0.50 Room 516 = 0.00	Room 508 = 0.57 Room 516 = 0.00	0.58
ADASYN	Room 508 = 0.40 Room 516 = 0.00	Room 508 = 0.50 Room 516 = 0.00	Room 508 = 0.44 Room 516 = 0.00	0.57
Proposed Relabeling	Room 508 = 0.57 Room 516 = 0.00	Room 508 = 1.00 Room 516 = 0.00	Room 508 = 0.73 Room 516 = 0.00	0.62

* Original train data with no augmentation applied.

5. Discussion

In this section, we outline the findings of our study, building upon the methodologies and approaches discussed previously. We highlight the contribution to the field of indoor localization, particularly in the context of nursing care facilities.

To investigate the impact of relabeling on indoor localization in nursing facilities, we illustrate the performance of the proposed method in Figure 15. In response to the first research question on identifying matching signal patterns between different locations in the facility, we calculate signal pattern features using standard deviation to compare per beacon variability and KL divergence to compare signal distributions between rooms.

To better understand signal patterns, we filter the analysis of beacons to six surrounding devices in the neighboring rooms.

In response to the second research question on how we use samples from other beacons to augment location with fewer data, we apply relabeling to the resulting match from the comparison of signal patterns between rooms. We implement two variations to relabeling based on the completeness of six beacons filtered around the surrounding location. We summarize the resulting performance of applying relabeling to the stationary beacons installed in the nursing facility in Figures 15 and 16.

Between the variations of relabeling, full matching consistently achieves better performance compared to baseline and partial matching in terms of the target class F1-score. As reflected in Figure 15, Room 516 is only classified after employing relabeling with full matching. Matching based on signal pattern features referenced from KL divergence results in a better F1-score than standard deviation. Overall, the proposed method improves the target class F1-score compared to the baseline by 27% to 40%.

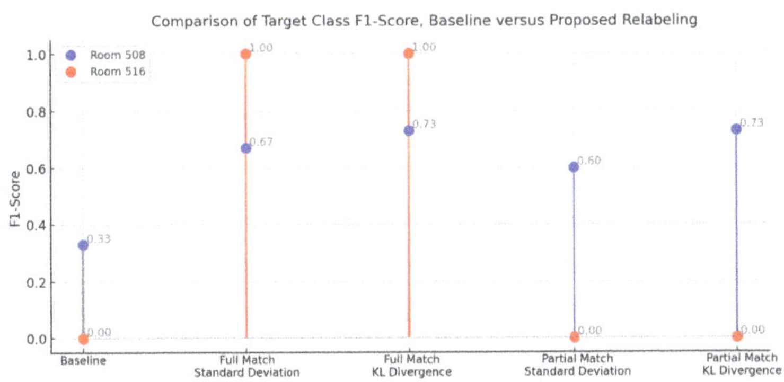

Figure 15. Comparison of Target Class F1-Score, Baseline versus Proposed Relabeling.

By comparing the overall F1-score performance in Figure 16, we determined that full matching achieves better performance than baseline with an increase of 6% to 8%. Moreover, full matching outperformes partial matching by 6% to 7%. For both full and partial matching, KL divergence results in a better F1-score than standard deviation. Overall, the proposed method improves model performance compared to the baseline, except in the case of partial matching with standard deviation. In general, increasing the training data with relabeling based on signal pattern results in improved localization.

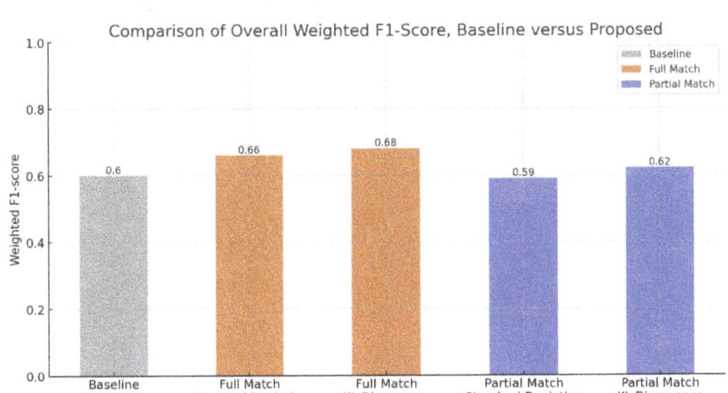

Figure 16. Comparison of Overall Weighted F1-Score, Baseline versus Proposed Relabeling.

Choosing the best combination of relabeling from the matching variation and signal pattern feature, we compare full matching based on KL divergence with the baseline confusion matrix as depicted in Figure 17. For target class F1-scores, relabeling improves the performance of both Room 508 and 516. All samples of 508 and 516 are correctly localized with relabeling. In the overall model F1-score, relabeling improves performance by 8%. On the other hand, a trade-off of the proposed method is potential class overlap, especially from the majority class used in the matching, which can be observed in Room 520 in the confusion matrix affecting the recognition of the room.

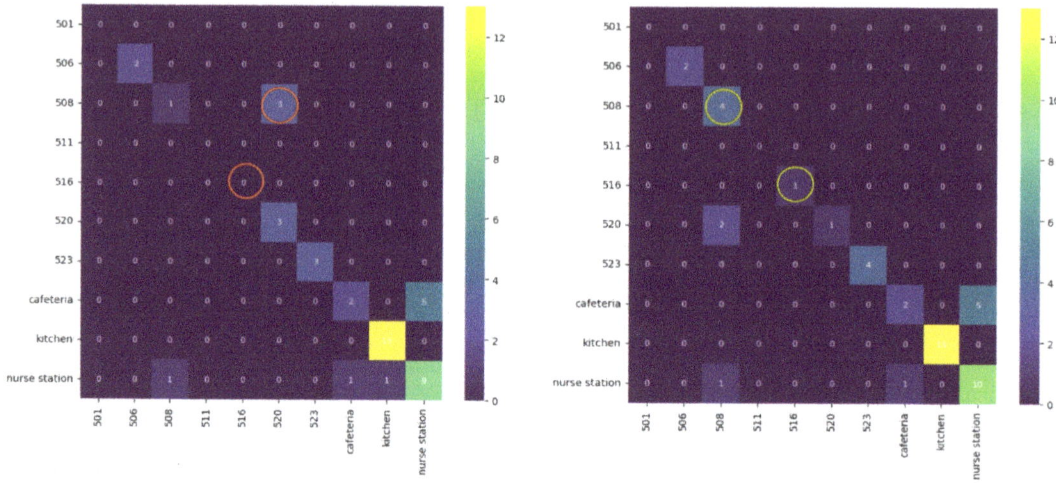

(a) Baseline Train Data (b) Full Matching with KL Divergence

Figure 17. Comparison of indoor positioning with baseline data. (**a**) Confusion matrix, original data. (**b**) Confusion matrix, with relabeled data.

For cross-validation, we test the proposed method by applying data augmentation on other rooms using the same test data. In the original train data, we reduce Rooms 520 and 523 samples and then re-evaluate the performance for baseline before applying augmentation. Table 6 summarizes the result after oversampling Room 520. Relabeling improves both overall and target F1-score of baseline after applying data augmentation and increasing the samples by 96.7%.

Table 6. Comparison of Indoor Localization Performance, Oversampling Room 520.

Oversampling Approach	Train Data Room 520	Room 520 F1-Score	Overall Weighted F1-Score
Baseline	1000	0.00	0.56
Random Sampling	1969	0.50	0.67
SMOTE	1969	0.40	0.67
ADASYN	1969	0.86	0.67
Proposed method	1969	0.40	0.63

For Room 523, the proposed relabeling achieves both the highest overall and target F1-score after applying data augmentation to the room, increasing the samples five-fold the original size as depicted in Table 7.

Given the varying quantities of samples added through oversampling, evaluating and contrasting the augmented data produced by each technique is crucial, as they each function distinctly.

Presently, we remain positive that the application of the proposed method to other rooms relevant insights into the current effectiveness of relabeling. Between the variations in relabeling, full matching effectively utilizes all sensor data from other beacons. Conversely, while partial matching demonstrates some improvement in localization accuracy, its reduced retention of beacon data from other rooms can lead to diminished performance. Overall, as relabeling is based on leveraging samples from other beacons, the proposed data augmentation approach relies on the number of samples in the majority class. In line with other oversampling techniques, the volume of data gathered within the facility for training significantly influences the outcome of data augmentation and the performance of indoor localization.

Table 7. Comparison of Indoor Localization Performance, Oversampling Room 523.

Oversampling Approach	Train Data Room 523	Room 523 F1-Score	Overall Weighted F1-Score
Baseline	2178	0.00	0.53
Random Sampling	11,174	0.00	0.53
SMOTE	11,174	0.00	0.53
ADASYN	11,174	0.33	0.59
Proposed method	11,174	0.33	0.61

6. Conclusions

In this paper, we proposed a relabeling method for data augmentation towards indoor localization in nursing care facilities. Our proposed method addresses the challenge of data imbalance due to unequal representation of locations with low samples in minority classes. By filtering BLE devices to surrounding beacons, we calculated the signal pattern feature of minority and majority classes and identified matching rooms for relabeling. By applying a relabeling approach to identified locations matching the signal patterns of the minority class, indoor localization improved both in the target class and overall model performance. Moreover, we presented a comparative performance of indoor localization using different augmentation techniques and confirmed that our proposed method achieves better indoor localization in nursing homes.

The main contribution of this work is utilizing real data from other beacons referenced from the majority class to augment locations with fewer samples. With our proposed full and partial matching based on calculated signal pattern features, we showed the advantage of having a flexible method where the statistical measure can be varied in accordance with the targeted analyzed data. Moreover, our method displayed model generalization leveraging majority class data. On the other hand, dependence on the number of samples of majority classes should be taken into account when applying relabeling.

We could discern similarities between different rooms by calculating signal pattern features. This enabled us to effectively augment locations with limited data samples by repurposing data from matching rooms. By measuring both KL divergence and standard deviation across classes, we mapped the signal patterns among various locations. The approach based on KL divergence consistently outperforms baseline data and other methods, specifically Random Sampling, SMOTE, and ADASYN. However, it is important to acknowledge the potential for class overlap with the matching class. The augmented data should be carefully examined.

The reliance of the proposed method on the volume of majority class samples and the potential for class overlap necessitates careful application and consideration. To this end, our future work will focus on expanding data collection to enable more robust evaluations. Broadening the scope of data collection to various nursing homes with different layouts is a key component to validate the effectiveness of relabeling in diverse settings. Furthermore, locations within common areas not following specific layouts will be further examined. We aim to refine our relabeling approach to encompass areas within these common spaces with varying geometries and dimensions, such as cafeterias, where nursing activities tend to occur over extended periods. Applying the relabeling approach to other use-case scenarios using different sensors, such as an accelerometer and gyroscope, will be analyzed to understand signal patterns for matching. The aim is to apply the proposed relabeling to other sensor data to assess the performance in improving models after data augmentation.

Currently, the analysis is performed in the offline phase using real-world data. Expanding the system to a more dynamic real-time relabeling and indoor positioning for nursing homes is suggested to adapt to immediate changes in the caregiving environment. Furthermore, elaborating analysis of signal patterns by incorporating additional statistical analysis techniques is suggested to represent signal pattern features of different devices aside from BLE beacons. A hybrid augmentation approach by combining the proposed oversampling method with SMOTE or ADASYN is suggested to resolve the limitations of respective oversampling techniques and further improve indoor localization. Algorithm optimization should be explored to enhance the accuracy and robustness of indoor localization. We plan to investigate the use of additional machine learning algorithms and the implementation of filtering techniques to improve signal quality, potentially increasing the precision of our localization system. On the other hand, the proposed relabeling maintains its effectiveness even with filtered or high-quality signal inputs.

Lastly, as we continue to improve the FonLog application and indoor positioning system, we will integrate user feedback collection via an in-app survey to incorporate suggestions in the system design cycle. This is to ensure adherence to the protocol of the partner facility and adapt to user experience with practical challenges faced in real-world scenarios.

Author Contributions: Conceptualization, C.G. and S.I.; methodology, C.G.; software, C.G. and S.I.; validation, C.G. and S.I.; formal analysis, C.G.; investigation, C.G.; resources, C.G.; data curation, C.G.; writing—original draft preparation, C.G.; writing—review and editing, C.G. and S.I.; visualization, C.G.; supervision, S.I.; project administration, C.G.; funding acquisition, S.I. All authors have read and agreed to the published version of the manuscript.

Funding: This research was funded by the JST-Mirai Program, Creation of Care Weather Forecasting Services in the Nursing and Medical Field, Grant Number JP21473170, and Health Labour Sciences Research, Grant Number 19AC1003 in Japan.

Institutional Review Board Statement: The study was conducted according to the guidelines of the Declaration of Helsinki.

Informed Consent Statement: Informed consent was obtained from all subjects involved in the study.

Data Availability Statement: The data presented in this study are available on request from the corresponding author.

Acknowledgments: We extend our heartfelt gratitude to the managers of Kaikyokan Facility for their invaluable collaboration in securing approval for data collection on site and for assisting with the installation of beacons. This research was made possible with the participating caregiver department, allowing for the staff to carry mobile devices installed with Fonlog during daily nursing routines. The authors extend their gratitude to the participating graduate students from Maekawa Laboratory of Osaka University for their assistance during data collection. Lastly, we thank the Japanese Ministry of Education, Culture, Sports, Science, and Technology (MEXT).

Conflicts of Interest: The authors declare no conflict of interest.

Abbreviations

The following abbreviations are used in this manuscript:

ADASYN	Adaptive Synthetic Sampling
AoA	Angle of Arrival
BLE	Bluetooth Low Energy
HAR	Human Activity Recognition
LOS	Line of Sight
IMU	Inertial Measurement Unit

IoT	Internet of Things
IPS	Indoor Positioning System
KL	Kullback–Leibler
MAC	Media Access Control
NLOS	Non-line of Sight
RF	Random Forest
RFID	Radio Frequency Identification
RS	Random Sampling
RSS	Received Signal Strength
RSSI	Received Signal Strength Indicator
SMOTE	Synthetic Minority Oversampling Technique
ToA	Time of Arrival
Wi-Fi	Wireless Fidelity
WLAN	Wireless Local Area Network

References

1. López, M.P.; Tsitouras, D.J. Flatlining: How the Reluctance to Embrace Immigrant Nurses is Mortally Wounding the U.S. Healthcare System. *J. Health Care Law Policy* **2009**, *12*, 235. [CrossRef]
2. Garcia, C.; Quynh, V.N.P.; Kaneko, H.; Inoue, S. A Relabeling Approach to Signal Patterns for Beacon-based Indoor Localization in Nursing Care Facility. In Proceedings of the 5th International Conference on Activity and Behavior Computing, Kaiserslautern, Germany, 7–9 September 2023.
3. Morita, T.; Taki, K.; Fujimoto, M.; Suwa, H.; Arakawa, Y.; Yasumoto, K. Beacon-Based Time-Spatial Recognition toward Automatic Daily Care Reporting for Nursing Homes. *J. Sens.* **2018**, *2018*, 2625195. [CrossRef]
4. Garcia, C.; Inoue, S. Challenges and Opportunities of Activity Recognition in Clinical Pathways. In Proceedings of the 4th International Conference on Activity and Behavior Computing, London, UK, 27–29 October 2022.
5. Calderoni, L.; Ferrara, M.; Franco, A.; Maio, D. Indoor localization in a hospital environment using random forest classifiers. *Expert. Syst. Appl.* **2015**, *42*, 125–134. [CrossRef]
6. Nguyen, Q.H.; Johnson, P.; Nguyen, T.T.; Randles, M. A novel architecture using iBeacons for localization and tracking of people within healthcare environment. In Proceedings of the 2019 Global IoT Summit (GIoTS), Aarhus, Denmark, 17–21 June 2019; pp. 1–6. [CrossRef]
7. Kanan, R.; Elhassan, O. A combined batteryless radio and wifi indoor positioning for hospital nursing. *J. Commun. Softw. Syst.* **2016**, *12*, 33–44. [CrossRef]
8. Bibbò, L.; Carotenuto, R.; Della Corte, F. An Overview of Indoor Localization System for Human Activity Recognition (HAR) in Healthcare. *Sensors* **2022**, *22*, 8119. [CrossRef] [PubMed]
9. Morita, T.; Taki, K.; Fujimoto, M.; Suwa, H.; Arakawa, Y.; Yasumoto, K. BLE Beacon-based Activity Monitoring System toward Automatic Generation of Daily Report. In Proceedings of the IEEE International Conference on Pervasive Computing and Communications Workshops (PerCom Workshops), Athens, Greece, 19–23 March 2018; pp. 788–793. [CrossRef]
10. Yoo, J.; Park, S. Fingerprint variation detection by unlabeled data for indoor localization. *Pervasive Mob. Comput.* **2020**, *67*, 101219. [CrossRef]
11. Palipana, S.; Pietropaoli, B.; Pesch, D. Recent advances in RF-based passive device-free localisation for indoor applications. *Ad. Hoc. Netw.* **2017**, *64*, 80–98. [CrossRef]
12. Zhu, Q.; Xiong, Q.; Wang, K.; Lu, W.; Liu, T. Accurate WiFi-based indoor localization by using fuzzy classifier and mlps ensemble in complex environment. *J. Franklin Inst.* **2020**, *57*, 1420–1436. [CrossRef]
13. Wang, J.; Tian, Z.; Yang, X.; Zhou, M. TWPalo: Through-the-wall passive localization of moving human with Wi-Fi. *Comput. Commun.* **2020**, *157*, 284–297. [CrossRef]
14. Oussalah, M.; Alakhras, M.; Hussein, M.I. Multivariable fuzzy inference system for fingerprinting indoor localization. *Fuzzy Sets Syst.* **2015**, *8*, 65–89. [CrossRef]
15. Berrueta, B.; Baalaa, O.; Caminadab, A.; Guillet, V. An evaluation method of channel state information fingerprinting for single gateway indoor localization. *J. Netw. Comput. Appl.* **2020**, *159*, 102591. [CrossRef]
16. Fei, H.; Xiao, F.; Huang, H.; Sun, L. Indoor static localization based on Fresnel zones model using COTS Wi-Fi. *J. Netw. Comput. Appl.* **2020**, *167*, 102709. [CrossRef]
17. Xu, B.; Gang, W. Random sampling algorithm in RFID indoor location system. In Proceedings of the Third IEEE International Workshop on Electronic Design, Test and Applications (DELTA'06), Kuala Lumpur, Malaysia, 17–19 January 2006. [CrossRef]
18. Ni, L.; Liu, Y.; Lau, Y.C.; Patil, A. LANDMARC: Indoor Location Sensing Using Active RFID. *Wirel. Netw.* **2004**, *10*, 701–710. [CrossRef]

19. Cortesi, S.; Vogt, C.; Magno, M. Comparison between an RSSI- and an MCPD-Based BLE Indoor Localization System. *J. Comput.* **2023**, *12*, 59. [CrossRef]
20. Daníş, F.S.; Cemgfl, A.T.; Ersoy, C. Adaptive Sequential Monte Carlo Filter for Indoor Positioning and Tracking with Bluetooth Low Energy Beacons. *IEEE Access* **2021**, *9*, 37022–37038. [CrossRef]
21. Zhou, C.; Yuan, J.; Liu, H.; Qiu, J. Bluetooth Indoor Positioning Based on RSSI and Kalman Filter. *Wirel. Pers. Commun.* **2017**, *96*, 4115–4130. [CrossRef]
22. Cao, H.; Wang, Y.; Bi, J.; Zhang, Y.; Yao, G.; Feng, Y.; Si, M. LOS compensation and trusted NLOS recognition assisted WiFi RTT indoor positioning algorithm. *Expert Syst. Appl.* **2024**, *243*, 122867. [CrossRef]
23. Huang, X.; Guo, S.; Wu, Y.; Yang, Y. A fine-grained indoor fingerprinting localization based on magnetic field strength and channel state information. *Pervasive Mob. Comput.* **2017**, *41*, 150–165. [CrossRef]
24. Tang, C.; Sun, W.; Zhang, X.; Zheng, J.; Sun, J.; Liu, C. A Sequential-Multi-Decision Scheme for WiFi Localization Using Vision-Based Refinement. *IEEE Trans Comput.* **2023**. 1–16. [CrossRef]
25. Blandeau, M.; Guichard, R.; Hubaut, R.; Leteneur, S. IMU positioning affects range of motion measurement during squat motion analysis. *J. Biomech.* **2023**, *153*, 111598. [CrossRef]
26. Kuo, Y.-H.; Wu, E.H.-K. Intelligent Geomagnetic Indoor Positioning System. *Electronics* **2023**, *12*, 2227. [CrossRef]
27. Yang, A.C.H.; Lau, N.; Ho, J.C.F. The Role of Bedroom Privacy in Social Interaction among Elderly Residents in Nursing Homes: An Exploratory Case Study of Hong Kong. *J. Sens.* **2020**, *15*, 4101. [CrossRef] [PubMed]
28. Yaro, A.S.; Maly, F.; Prazak, P. Outlier Detection in Time-Series Receive Signal Strength Observation Using Z-Score Method with Sn Scale Estimator for Indoor Localization. *Appl. Sci.* **2023**, *13*, 3900. [CrossRef]
29. Yoo, J.; Kim, H.J. Target Tracking and Classification from Labeled and Unlabeled Data in Wireless Sensor Networks. *J. Sens.* **2014**, *14*, 23871–23884. [CrossRef] [PubMed]
30. Hoang, M.T.; Yuen, B.; Dong, X.; Lu, T.; Westendorp, R.; Reddy, K. Recurrent Neural Networks for Accurate RSSI Indoor Localization. *IEEE Internet Things J.* **2019**, *6*, 10639–10651. [CrossRef]
31. Wang, S. Wireless Network Indoor Positioning Method Using Nonmetric Multidimensional Scaling and RSSI in the Internet of Things Environment. *Math. Probl. Eng.* **2020**, *2020*, 8830891. [CrossRef]
32. Kotanen, A; Hannikainen, M.; Leppakoski, H.; Hamalainen, T.D. Experiments on local positioning with Bluetooth. In Proceedings of the International Conference on Information Technology: Coding and Computing, Las Vegas, NV, USA, 28–30 April 2003; pp. 297–303. [CrossRef]
33. Yaro, A.S.; Maly, F.; Prazak, P. A Survey of the Performance-Limiting Factors of a 2-Dimensional RSS Fingerprinting-Based Indoor Wireless Localization System. *Sensors* **2023**, *23*, 2545. s23052545 [CrossRef]
34. Liu, L.; Zhao, Q.; Miki, S.; Tokunaga, J.; Ebara, H. Indoor Fingerprinting Positioning System Using Deep Learning with Data Augmentation. *Sens. Mater.* **2022**, *34*, 3047–3061. [CrossRef]
35. He, H.; Bai, Y.; Garcia, E.A.; Li, S. ADASYN: Adaptive synthetic sampling approach for imbalanced learning. In Proceedings of the 2008 IEEE International Joint Conference on Neural Networks (IEEE World Congress on Computational Intelligence), Hong Kong, China, 1–8 June 2008; pp. 1322–1328. [CrossRef]
36. Chawla, N.; Bowyer, K.; Hall, L.O.; Kegelmeyer, W.P. SMOTE: Synthetic Minority Over-sampling Technique. *arXiv* **2002**, arXiv:1106.1813.
37. Fikry, M.; Garcia, C.; Quynh, V.N.P.; Inoue, S. Improving Complex Nurse Care Activity Recognition Using Barometric Pressure Sensor. In Proceedings of the 4th International Conference on Activity and Behavior Computing, London, UK, 27–29 October 2022.
38. Sulaiman, B.; Tarapiah, S.; Natsheh, E.; Atalla, S.; Mansoor, W.; Himeur, Y. Radio map generation approaches for an RSSI-based indoor positioning system. *Syst. Soft Comput.* **2023**, *5*, 200054. [CrossRef]
39. Mairittha, N.; Mairittha, T.; Inoue, S. A mobile app for nursing activity recognition. In Proceedings of the 2018 ACM International Joint Conference and 2018 International Symposium on Pervasive and Ubiquitous Computing and Wearable Computers, Singapore, 8–12 October 2018; pp. 400–403.
40. Luschi, A.; Villa, E.A.B.; Gherardelli, M.; Iadanza, E. Designing and Developing a Mobile Application for Indoor Real-time Positioning and Navigation in Healthcare Facilities'. *J. Health Care Technol.* **2022**, *30*, 1371–1395. [CrossRef] [PubMed]
41. Wang, J.; Ghosh, R.; Das, S. A survey on sensor localization. *J. Control Theory Appl.* **2010**, *8*, 2–11. [CrossRef]
42. Using Beacons for Indoor Positioning, Tracking and Indoor Navigation. Available online: www.infsoft.com/basics/positioning-technologies/bluetooth-low-energy-beacons (accessed on 20 November 2023).
43. Locatify. Indoor Positioning Systems based on BLE Beacons. Available online: https://locatify.com/blog/indoor-positioning-systems-ble-beacons (accessed on 26 November 2023).
44. What Is BLE?—Bluetooth Low Energy Basics. Available online: Https://www.rfwireless-world.com/Terminology/what-is-BLE.html (accessed on 24 November 2023).
45. Lee, W. Understanding and Using iBeacons. 2021. Available online: https://www.codemag.com/article/1405051/Understanding-and-Using-iBeacons (accessed on 24 November 2023).
46. Pichler, S. Indoor Positioning with Beacons. 2022. Available online: https://www.esri.com/arcgis-blog/products/arcgis-ips/indoor-gis/indoor-positioning-with-beacons/ (accessed on 26 November 2023).
47. Wichmann, J. Indoor positioning systems in hospitals: A scoping review. *Digit. Health* **2022**, *8*, 20552076221081696. [CrossRef]

48. Singh, A.; Purohit, A. A Survey on Methods for Solving Data Imbalance Problem for Classification. *Int. J. Comput. Appl.* **2015**, *127*, 37–41. [CrossRef]
49. Jiang, X.; Ge, Z. Data Augmentation Classifier for Imbalanced Fault Classification. *IEEE Trans. Autom. Sci. Eng.* **2021**, *18*, 1206–1217. [CrossRef]
50. Fan, J.; Sun, C.; Chen, C.; Jiang, X.; Liu, X.; Zhao, X.; Meng, L.; Dai, C.; Chen, W. EEG data augmentation: Towards class imbalance problem in sleep staging tasks. *J. Neural Eng.* **2020**, *17*, 056017. [CrossRef]
51. Basiri, A.; Lohan, E.S.; Moore, T.; Winstanley, A.; Peltola, P.; Hill, C.; Amirian, P.; Silva, P.F. Indoor location based services challenges, requirements and usability of current solutions. *Comput. Sci. Rev.* **2017**, *24*, 1–12.2 [CrossRef]
52. Ciabattoni, L.; Foresi, G.; Monteriù, A.; Pepa, L.; Pagnotta, D.P.; Spalazzi, L.; Verdini, F. Real time indoor localization integrating a model based pedestrian dead reckoning on smartphone and BLE beacons. *J. Ambient Intell. Human. Comput.* **2019**, *10*, 1–12. [CrossRef]
53. Yoon, J.; Oh, J.; Kim, S. Transfer Learning Approach for Indoor Localization with Small Datasets. *Remote Sens.* **2023**, *15*, 2122. [CrossRef]
54. Maharana, K.; Mondal, S.; Nemade, B. A review: Data pre-processing and data augmentation techniques. *Glob. Transit.* **2022**, *3*, 91–99. [CrossRef]
55. Um, T.T.; Pfister, F.M.J.; Pichler, D.; Endo, S.; Lang, M.; Hirche, S.; Fietzek, U.; Kulić, D. Data augmentation of wearable sensor data for parkinson's disease monitoring using convolutional neural networks. In Proceedings of the 19th ACM International Conference on Multimodal Interaction, Glasgow, UK, 13–17 November 2017; pp. 216–220. [CrossRef]
56. Temraz, M.; Keane, M.T. Solving the class imbalance problem using a counterfactual method for data augmentation. *Mach. Learn. Appl.* **2022**, *9*, 100375. [CrossRef]
57. Ling, C.X.; Li, C. Data mining for direct marketing: Problems and solutions. In Proceedings of the Fourth International Conference on Knowledge Discovery and Data Mining, New York, NY, USA, 27–31 August 1998; pp. 73–79.
58. He, H.; Garcia, E.A. Learning from Imbalanced Data. *IEEE Trans. Knowl. Data Eng.* **2009**, *21*, 1263–1284. [CrossRef]
59. Thabtah, F.; Hammoud, S.; Kamalov, F.; Gonsalves, A. Data imbalance in classification: Experimental evaluation. *Inf. Sci.* **2020**, *513*, 429–441. [CrossRef]
60. Lee, J.; Park, K. GAN-based imbalanced data intrusion detection system. *Pers. Ubiquit Comput.* **2021**, *25*, 121–128. [CrossRef]
61. Brandt, J.; Emil Lanzén, E. *A Comparative Review of SMOTE and ADASYN in Imbalanced Data Classification*; Department of Statistics, Uppsala University: Uppsala, Sweden, 2020.
62. Hilal, A.; Arai, I.; El-Tawab, S. DataLoc+: A Data Augmentation Technique for Machine Learning in Room-Level Indoor Localization. In Proceedings of the IEEE Wireless Communications and Networking Conference (WCNC), Nanjing, China, 29 March–1 April 2021; pp. 1–7. [CrossRef]
63. Yu, S.M.; Park, J.; Ko, S.-W. Combinatorial Data Augmentation for Real-Time Indoor Positioning: Concepts and Experiments. In Proceedings of the IEEE 95th Vehicular Technology Conference: (VTC2022-Spring), Helsinki, Finland, 19–22 June 2022; pp. 1–5. [CrossRef]
64. Liu, A.; Cheng, L.; Yu, C. SASMOTE: A Self-Attention Oversampling Method for Imbalanced CSI Fingerprints in Indoor Positioning Systems. *Sensors* **2022**, *22*, 5677. [CrossRef]
65. Kim, M.; Jeong, C.Y. Label-preserving data augmentation for mobile sensor data. *Multidimens. Syst. Signal Process.* **2021**, *32*, 115–129. [CrossRef]
66. Li, Y.; Adams, N.; Bellotti, T. A relabeling approach to handling the class imbalance problem for logistic regression. *J. Comput. Graph. Stat.* **2022**, *31*, 241–253. [CrossRef]
67. Yoo, J.; Park, J. Indoor Localization Based on Wi-Fi Received Signal Strength Indicators: Feature Extraction, Mobile Fingerprinting, and Trajectory Learning. *Appl. Sci.* **2019**, *9*, 3930. [CrossRef]
68. Bose, A.; Foh, C.H. A practical path loss model for indoor WiFi positioning enhancement. In Proceedings of the 6th International Conference on Information, Communications and Signal Processing, Singapore, 10–13 December 2007; pp. 1–5. [CrossRef]
69. Sahoo, S. Comparing RSSI between Thinly Enclosed and Exposed Devices. Available online: https://ysjournal.com/computer-science/comparing-rssi-between-thinly-enclosed-and-exposed-devices/ (accessed on 26 November 2023).
70. Lee, P.Q.; Seah, W.; Tan, H.; Yao, Z. Wireless sensing without sensors—An experimental study of motion/intrusion detection using RF irregularity. *Meas. Sci. Technol.* **2010**, *21*, 124007. [CrossRef]
71. Yassin, A.; Nasser, Y.; Awad, M.; Al-Dubai, A.; Liu, R.; Yuen, C.; Raulefs, R.; Aboutanios, E. Recent Advances in Indoor Localization: A Survey on Theoretical Approaches and Applications. *IEEE Commun. Surv. Tutorials* **2017**, *19*, 1327–1346. [CrossRef]
72. Thakur, N.; Han, C.Y. Indoor Localization for Personalized Ambient Assisted Living of Multiple Users in Multi-Floor Smart Environments. *Big Data Cogn. Comput.* **2021**, *5*, 42. [CrossRef]
73. Brown, I.; Mues, C. An experimental comparison of classification algorithms for imbalanced credit scoring data sets. *Expert Syst. Appl.* **2012**, *39*, 3446–3453. [CrossRef]
74. Lee, S.; Kim, J.; Moon, N. Random forest and WiFi fingerprint-based indoor location recognition system using smart watch. *Hum. Cent. Comput. Inf. Sci.* **2019**, *9*, 6. [CrossRef]

75. Varma, P.S.; Anand, V. Random Forest Learning Based Indoor Localization as an IoT Service for Smart Buildings. *Wirel. Pers. Commun.* **2021**, *117*, 3209–3227. [CrossRef]
76. Ramadan, M.; Sark, V.; Gutierrez, J.; Grass, E. NLOS Identification for Indoor Localization using Random Forest Algorithm. In Proceedings of the 22nd International ITG Workshop on Smart Antennas, Bochum, Germany, 14–16 March 2018; pp. 1–5.

sensors

MDPI

Article

Deep Learning Models for Stress Analysis in University Students: A Sudoku-Based Study

Qicheng Chen [1] and Boon Giin Lee [2,*]

[1] School of Computer Science, University of Nottingham Ningbo China, Ningbo 315100, China; harrison15120@outlook.com
[2] Nottingham Ningbo China Beacons of Excellence Research and Innovation Institute, School of Computer Science, University of Nottingham Ningbo China, Ningbo 315100, China
* Correspondence: boon-giin.lee@nottingham.edu.cn

Abstract: Due to the phenomenon of "involution" in China, the current generation of college and university students are experiencing escalating levels of stress, both academically and within their families. Extensive research has shown a strong correlation between heightened stress levels and overall well-being decline. Therefore, monitoring students' stress levels is crucial for improving their well-being in educational institutions and at home. Previous studies have primarily focused on recognizing emotions and detecting stress using physiological signals like ECG and EEG. However, these studies often relied on video clips to induce various emotional states, which may not be suitable for university students who already face additional stress to excel academically. In this study, a series of experiments were conducted to evaluate students' stress levels by engaging them in playing Sudoku games under different distracting conditions. The collected physiological signals, including PPG, ECG, and EEG, were analyzed using enhanced models such as LRCN and self-supervised CNN to assess stress levels. The outcomes were compared with participants' self-reported stress levels after the experiments. The findings demonstrate that the enhanced models presented in this study exhibit a high level of proficiency in assessing stress levels. Notably, when subjects were presented with Sudoku-solving tasks accompanied by noisy or discordant audio, the models achieved an impressive accuracy rate of 95.13% and an F1-score of 93.72%. Additionally, when subjects engaged in Sudoku-solving activities with another individual monitoring the process, the models achieved a commendable accuracy rate of 97.76% and an F1-score of 96.67%. Finally, under comforting conditions, the models achieved an exceptional accuracy rate of 98.78% with an F1-score of 95.39%.

Keywords: deep learning; signal processing; stress detection; wearable sensing

1. Introduction

In this current generation, college and university students encounter a multitude of stressors, particularly in China, where the concept of "involution" gained prevalence around 2020. Academic-related stress has emerged as a significant source of anxiety for Chinese students, encompassing pressures to attain top grades, worries about receiving lower scores, concerns regarding the availability of internship opportunities compared to their peers, and anxieties about securing graduate study prospects.

Involution, also known as "Nei Juan" in Chinese pinyin, refers to situations in work and study where individuals put in extra effort, but that effort does not yield a proportional outcome [1]. For instance, the process of securing a graduate study opportunity in a university has become more challenging. In the past, students with relatively high grade point averages (GPAs) would be admitted based on limited quotas. However, admission offices now consider additional factors such as extracurricular activities and published academic papers. Consequently, students are compelled to invest extra effort in internships, extracurricular activities, and academic research to enhance their chances of admission.

This intensifies the competition among students and contributes to a growing sense of stress. A survey conducted by [1] examined 1100 Chinese students and revealed that only 3% of middle school students get a full 8 h of sleep daily, while 19.5% sleep for less than 4 to 5 h. An influential article [2] written by a Chinese mother narrated her struggles to provide her child with top educational opportunities. Despite earning a monthly salary of RMB 30,000, which is considered relatively high in China, it proved insufficient due to the pressure to enroll her daughter in numerous extracurricular classes in order to excel in all exams, even if those classes covered material beyond the scope of the regular school curriculum. The combined pressures from schools, society, and even families have given rise to a highly stressed generation.

Research has indicated that high levels of stress are linked to lower levels of well-being and reduced quality of life. Prolonged exposure to stress can result in severe mental health issues such as anxiety and depression [3]. A survey involving 5551 students [4] found that anxiety is negatively correlated with academic performance, meaning that students with lower anxiety levels tend to have higher GPAs than those with moderate or high levels of anxiety. Furthermore, depression and anxiety can lead to suicide, which is the second most common cause of death among college and university students. According to a report by [5], approximately 1100 students out of 100,000 commit suicide each year. Thus, monitoring stress levels can be extremely beneficial for universities and families in supporting students' academic performance and enhancing their overall well-being.

2. Related Work

Wagh et al. [6] employed Support Vector Machine (SVM), k-Nearest Neighbor (kNN), and Decision Tree (DT) algorithms to classify positive, neutral, and negative emotions using time and time-frequency domain features extracted from various channels of electroencephalogram (EEG) data. The classifiers were trained on the SJTU emotion EEG dataset (SEED), resulting in an accuracy of 72.46% for DT and 60.19% for kNN. Vijayakumar et al. [7] developed a 1D convolutional neural network (CNN) to classify arousal, valence, and liking based on peripheral physiological signals, including blood volume pressure (BVP), horizontal electrooculogram (hEOG), vertical electrooculogram (vEOG), trapezius electromyogram (tEMG), zygomaticus electromyogram (zEMG), respiration rate (RSP), and skin temperature (SKT) data. The CNN model achieved accuracies of 77.03% for arousal, 68.75% for valence, and 74.68% for liking using time and frequency domain features. Miao et al. [8] proposed a parallel spatial-temporal 3D deep residual learning framework called MFBPST-3D-DRLF for emotion recognition using EEG signals. This framework utilized multiple frequency bands (delta, theta, alpha, beta, gamma) of EEG signals to generate a 3D representation of features, which were then trained using a 3D deep residual CNN model. It achieved a classification accuracy of 96.67% on the SEED dataset (positive, neutral, and negative emotions) and 88.21% on the SEED-IV dataset (happy, fear, sad, and neutral).

Montero Quispe et al. [9] employed a novel self-supervised learning approach for emotion recognition, consisting of two stages: self-supervised pre-training and emotion recognition model training. In the pre-training stage, the model learned to recognize six signal variants generated by applying noise, scaling, negation, flipping, permuting, and time warping to the original data. This approach aimed to capture the main characteristics of the data. Results showed that self-supervised learning outperformed fully supervised learning methods in classifying arousal and valence using EEG and electrocardiogram (ECG) data. Tang et al. [10] conducted experiments on emotion recognition using EEG data with a proposed model called Spatial-Temporal Information Learning Network (STILN), which achieved an accuracy of 68.31% for arousal and 67.52% for valence. Choi et al. [11] proposed an attention-LRCN model that reduced motion artifacts in collected photoplethysmography (PPG) data. By combining the attention module with a baseline model and utilizing frequency domain features of PPG as inputs, the proposed model achieved an accuracy of 97.11%. Li et al. [12] developed a 1D convolutional neural network and a multi-layer perceptron to enhance the accuracy of stress detection using the WESAD dataset. They achieved exceptional results, with an accuracy rate of 99.80% for binary classification and an accuracy rate of 99.55% for 3-class classification. Arsalan et al. [13] focused on detecting stress during public speaking activities by collecting EEG, Galvanic Skin Response (GSR), and PPG data. They extracted frequency domain and time domain features and employed support vector machines (SVMs) with radial basis function. Their approach achieved a stress classification accuracy of 96.25% and an impressive F1-score of 95.99%. Han et al. [14] gathered and extracted features from ECG, PPG, and GSR, achieving an accuracy of 94.55% using 10-fold cross-validation and 81.82% in a real-world setting with kNN. Rastgoo et al. [15] developed a stress detection system that dynamically monitors driver stress during driving. They incorporated multiple data sources such as ECG, steering wheel, and weather conditions to predict the driver's stress level. Their approach, combining CNN and long short-term memory (LSTM) networks, achieved an accuracy of 92.80% and an F1-score of 94.56%.

In general, biosignals such as PPG, ECG, and EEG signals have demonstrated their utility in analyzing both physical health status and mental states, including emotions, concentration, and stress. Several public datasets, such as AMIGOS [12], DREAMER [13], SWELL (Smart Reasoning for Well-being at Home and at Work) [14], SEED [15], DEAP [16], and WESAD [17], have been curated specifically for research purposes and contain carefully designed experiments to induce stress in participants while collecting these biosignals. EEG, ECG, and GSR signals were obtained from sets of videos in the AMIGOS dataset. Similarly, the DREAMER dataset and SEED dataset also employed film clips to evoke positive, neutral, and negative emotions, gathering participants' ECG and EEG signals. The DEAP dataset utilized music videos to elicit emotions across different arousal–valence quadrants, capturing EEG and peripheral signals. The WESAD dataset utilized video clips and the Trier Social Stress Test as stimuli, collecting peripheral signals from participants. In the SWELL dataset, participants were exposed to a range of stressors, including time pressure, within a genuine office environment. They were requested to engage in activities such as report writing and giving presentations while their ECG data were systematically gathered. These publicly available datasets have been adapted for a wide range of research focusing on emotion recognition and stress detection. Meanwhile, deep learning methods, including CNN and LSTM, yielded higher accuracies compared to traditional machine learning methods such as SVM and DT.

The majority of public datasets currently available for developing stress detection algorithms utilize video and audio stimuli. The SWELL dataset has published a stress dataset whose data are collected under real-life office scenarios. However, there is no existing stress dataset found that collects data in a school context. Students are often studying and engaging in challenging problem-solving tasks, resulting in physiological signal changes that differ greatly from those observed during passive video viewing. To address this issue, this study devised a stress induction protocol that combined different levels of problem-solving (Sudoku game at medium and hard difficulty levels) with environmental stimuli such as videos and audio, creating a context in which students were required to solve problems amidst various distractions.

3. Materials and Methods

3.1. Experiment Design

In order to replicate a stress-inducing environment, participants were tasked with completing multiple Sudoku games under different scenarios, including both noisy and noise-free conditions. Throughout the experiments, various physiological data, including ECG, PPG, and EEG, were collected from the participants using wearable sensors. Following each Sudoku game, participants were asked to assess their own stress levels by completing a questionnaire. All participants were healthy students from the campus who volunteered for the experiment by filling out a registration form posted on the university's forum. The registration form included a series of questions, such as their medical history related to heart diseases, favorite music, and music that caused discomfort. Students with heart-related conditions were excluded from participating in the experiment. The registration form was created using Wenjuanxing [18], a platform that facilitates the design of questionnaires, exams, voting systems, and rating forms.

A total of 30 participants, consisting of 6 males and 24 females, with a mean age of 20.4, were recruited for the study. Among the participants, 7 were from the Faculty of Science and Engineering (FOSE), 12 were from the Nottingham University Business School (NUBS), and 11 were from the Faculty of Humanity and Social Science (FHSS). In terms of academic status, there were 15 sophomore students, 4 junior students, 8 senior students, and 3 Ph.D. students, as shown in Table 1. Upon registration, participants were informed of the experiment's location and schedule based on their availability. Each experimental session involved only one participant and took place in a small meeting room equipped with one desk, a couple of chairs, and a television with a functioning speaker.

Table 1. Participant information. Long dash (—) means no participant.

Faculty	Sophomore	Junior	Senior	Ph.D.	Total
FOSE	2	1	2	2	7
FHSS	11	—	—	—	11
NUBS	2	3	6	1	12
Total	15	4	8	3	30

The participants were given a brief introduction to the purpose of the experiment and were asked to sign a consent form and an information sheet if they agreed to participate before the experiment began. As the experiment involved solving Sudoku puzzles, the subjects were briefed on the basic rules of Sudoku. An iPad device running a Sudoku app that generated puzzles with the highest difficulty level was used. Following this, the subjects were fitted with sensors to collect physiological data. The experiment commenced only after confirming that the data receiver terminal was functioning correctly.

For each experiment, the participants were tasked with solving three Sudoku puzzles within a time limit of 15 min. The scenarios for each experiment were as follows:

- Scenario 1: The participant was left alone in the room, solving Sudoku puzzles while being exposed to horror or discordant audio, such as white noise, and watching horror videos, such as zombie movies.
- Scenario 2: No music or videos were played during this scenario. Instead, a person was present in the room observing the participant while they solved the Sudoku puzzles.
- Scenario 3: The participant was left alone in the room, solving Sudoku puzzles while being exposed to comforting audio and videos, such as sounds of birds, waterfalls, and rainfall.

The participants were divided into two study groups. The first group consisted of 15 participants who were assigned to solve Sudoku puzzles with a "medium" difficulty level. The second group also had 15 participants, but they were assigned to solve Sudoku puzzles with a "hard" difficulty level. This division was made to investigate the impact of Sudoku difficulty on changes in stress levels. At the end of each trial, participants were asked to indicate their stress levels using the re-designed questionnaire. This questionnaire was designed using Wenjuanxing and was based on the self-report approach proposed by Li et al. [15]. The questionnaire was adapted and revised for this study. The stress levels were assessed on a 3-point scale, where scores of 0 to 4 indicated relaxation or little stress, scores of 5 to 7 indicated medium stress, and scores of 8 indicated high stress. In order to enhance the level of stress and motivation during the Sudoku puzzle completion, incentives in the form of prizes were provided to the participants. The participants had the opportunity to win a prize for each Sudoku puzzle if they were able to successfully complete it within the given time limit of 15 min. Additionally, if a participant successfully completed all the Sudoku puzzles within the designated time frame, they were eligible to receive an additional prize. In total, there were four prizes available to be won.

3.2. Data Collection

The data collection process involved using three separate healthcare devices to gather PPG, ECG, and EEG data from the participants, as shown in Figure 1. PPG data were collected using the Polar Verity Sense (PVS) [19], ECG data were collected using the BMD101 device [20], and EEG data were collected using NeuroSky's MindWave Mobile 2 (MV2) [21]. The sampling rates for data collection were set at 55 Hz for PVS, 512 Hz for BMD101, and 512 Hz for MV2. It is important to note that data collection did not occur during the participant's self-reporting periods.

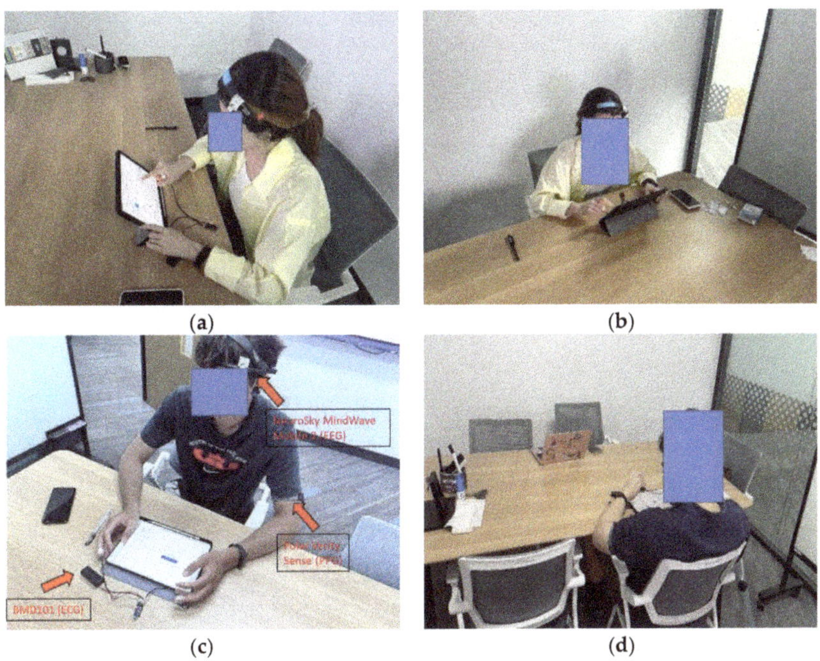

Figure 1. The experiment was designed as a solitary activity, with participants (**a–d**) solving Sudoku puzzles in a designated room. Sensors were attached to the participants during the experiment to collect data.

To capture the signals emitted by BMD101, PVS, and MV2 devices, three distinct programs were employed. The BMD101 package [20] offered a starting program designed specifically for acquiring the ECG signal transmitted by BMD101. Additionally, PVS provided an SDK example for capturing the PPG signal [20]. To collect the EEG signals transmitted by MV2, a ThinkGear socket [21] developed with Node.js on GitHub was utilized. The program for receiving ECG and EEG signals was executed on Windows 11, while the program for capturing the PPG signal ran on an Android platform. All of these programs recorded the raw data in text files.

3.3. Data Pre-Processing

The raw data obtained from the experiment had varying sampling rates, which posed challenges when inputting the data into the classification models without preprocessing. Additionally, there were certain inevitable recording errors during the data collection process. For example, the start time of PPG data collection may have been slightly delayed or advanced compared to the other two types of data, resulting in a few missing data points. Furthermore, there were uncontrollable variables in the experiment, such as instances where participants completed the Sudoku puzzle in less than 15 min, requiring the experiment to be ended earlier. To address these issues, all the data were resampled to a uniform rate of 256 Hz using a resampling method. This resulted in a total of 230,400 data points for each experiment. Subsequently, a Butterworth band-pass filter was applied to remove noise from the data. For PPG data, a low-cut frequency of 0.5 Hz and a high-cut frequency of 5 Hz were utilized. The raw PPG1, PPG2, and PPG3 signals corresponded to the sensor's green, red, and infrared light sources, respectively. For ECG data, the low-cut and high-cut frequencies were set at 5 Hz and 15 Hz, respectively. As for EEG data, a low-cut frequency of 0.1 Hz and a high-cut frequency of 15 Hz were employed. For the segmentation step, a sliding window approach with a duration of 10 s and no overlap was employed on the 15 min raw data. This segmentation divided the original 230,400 data points into 90 segments, with each segment containing 2560 data points. Figure 2 depicts the unprocessed and filtered PPG, ECG, and EEG data. The training pipeline for a model is outlined in Figure 3. On the other hand, Figure 4 illustrates the process of incorporating a collected sample of ECG data, indicating the various steps involved.

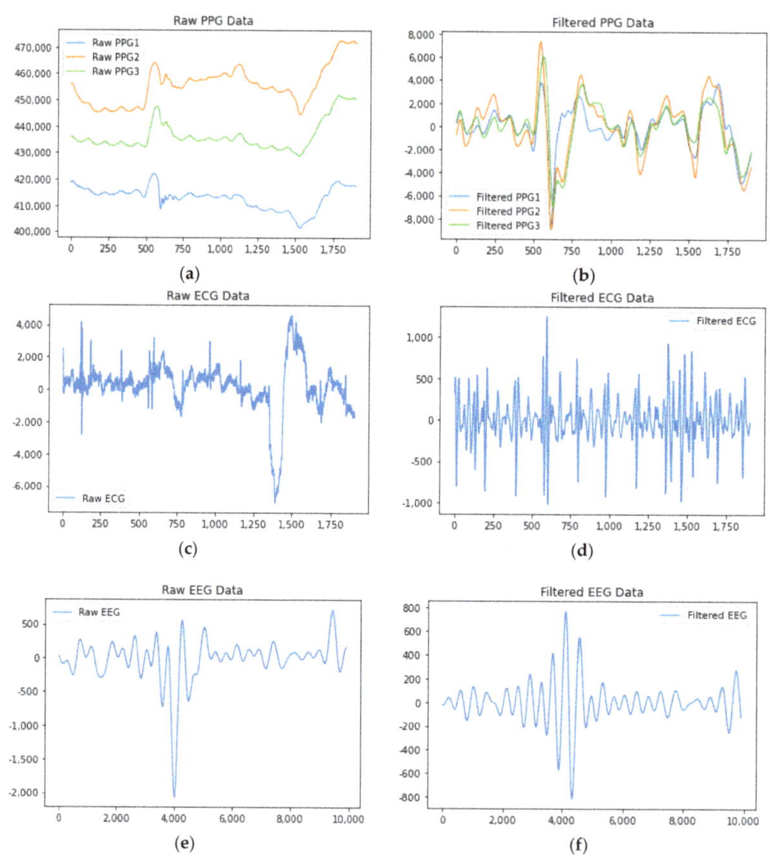

Figure 2. Collected (**a**) raw PPG, (**b**) filtered PPG, (**c**) raw ECG, (**d**) filtered ECG, (**e**) raw EEG, and (**f**) filtered EEG data in this study. The PPG1, PPG2, and PPG3 signals represent the green, red, and infrared light captured by the sensor, respectively. Following the application of filtering techniques, both motion-related noise in PPG and noise in ECG were effectively reduced. However, there were no discernible changes observed in the EEG signal.

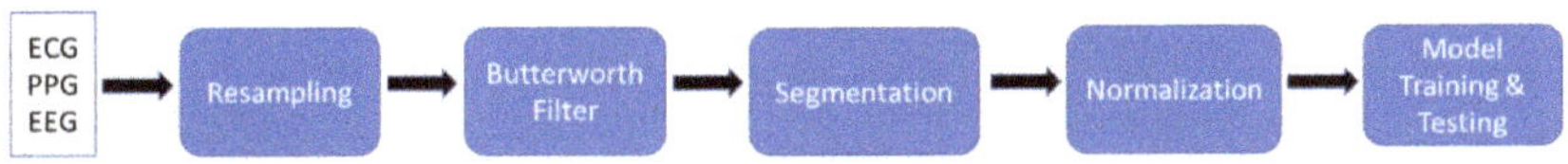

Figure 3. Training pipeline of a model.

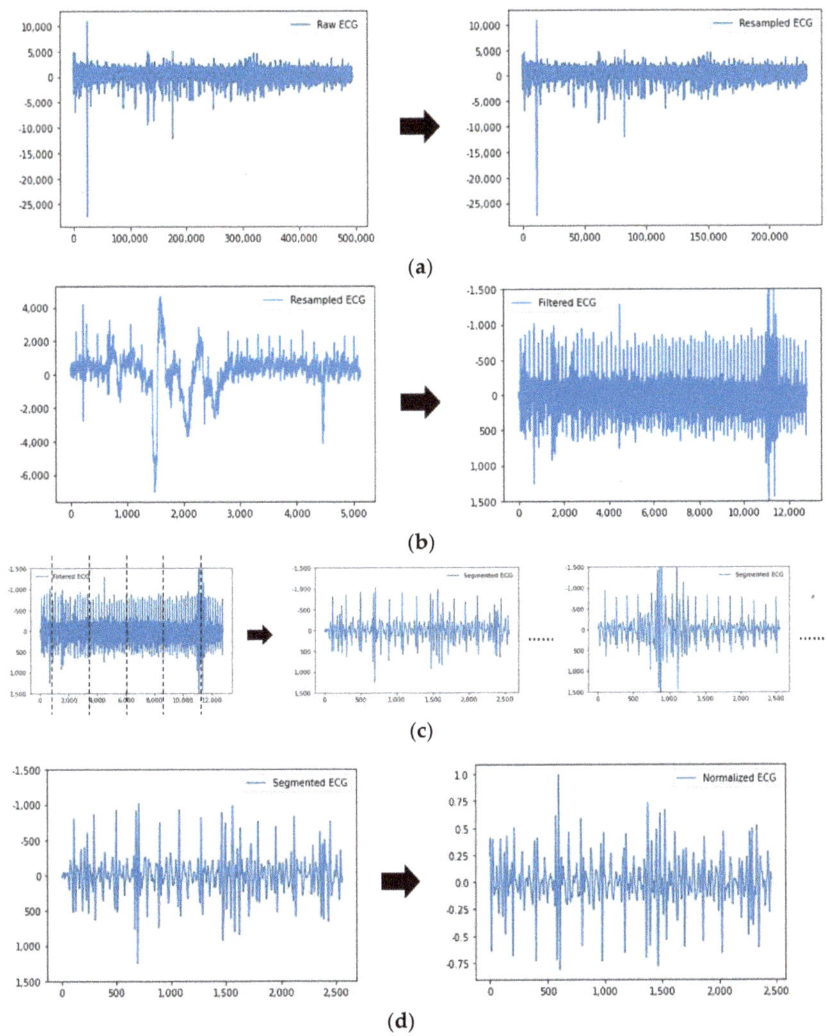

Figure 4. Showcase of data preprocessing procedures for a specific sample of collected ECG data. These steps include (**a**) down-sampling the ECG from 494,033 data points to 230,400 data points, (**b**) employing a Butterworth filter, (**c**) segmenting the data, and (**d**) normalizing the data.

3.4. Models Training

3.4.1. StressNeXt

In the StressNeXt model [22], the parameter quantity is initially reduced using a 1×1 convolutional block. The reduced parameters are then passed through four consecutive multi-kernel blocks. Each multi-kernel block comprises multiple convolutional layers with different sizes, namely 1×1, 1×3, 1×5, and 1×7. Within the multi-kernel block, the input signal is simultaneously processed by these convolutional layers, and instead of concatenating the results at the end, similar to the Inception network [23], the feature maps obtained from these layers are combined by addition. The specific architecture of the StressNeXt model is depicted in Figure 5.

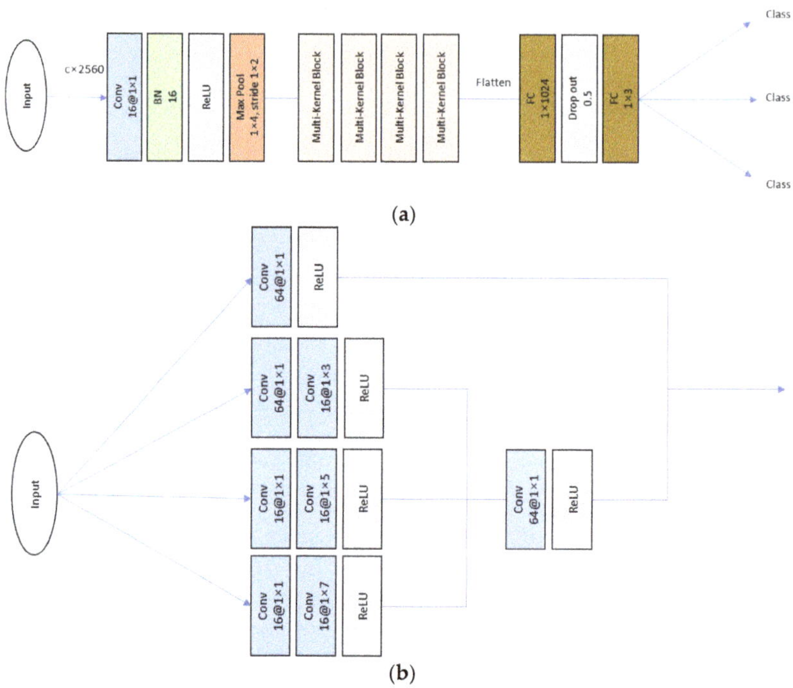

Figure 5. StressNeXt (**a**) model architecture and its (**b**) multi-kernel block design.

3.4.2. LRCN

The attention-LRCN (Long-term Recurrent Convolutional Network) model [14] was originally designed to take the Short-Term Fourier Transform (STFT) of signals as input, which consists of 2 dimensions corresponding to frequency and time. However, in this study, only 1 dimension is used as input. Therefore, the baseline of the attention-LRCN was taken and adapted to perform classification tasks on 1D input signals from the dataset. The model begins with two 1D convolutional layers and a max pooling layer to reduce computation complexity. Next, two residual blocks are applied, each separated by a max pooling layer that further reduces the feature maps. After the residual blocks, two LSTM layers are used to extract time-domain features from the signals and perform the classification task. Figure 6 shows the architecture of the LRCN model.

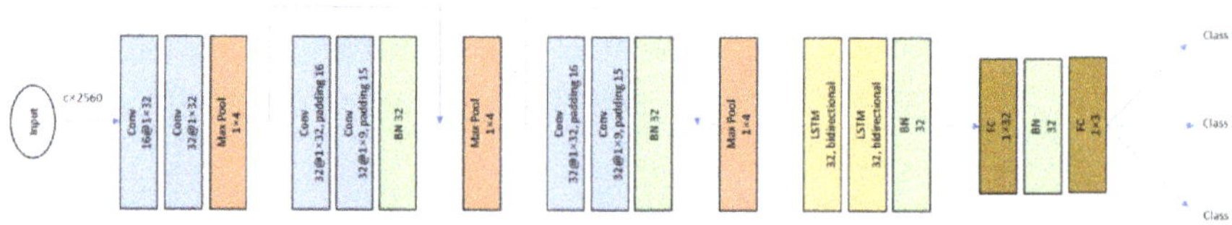

Figure 6. The LRCN architecture consists of 1D convolutional layers followed by ReLU activation.

3.4.3. Self-Supervised CNN

The process of self-supervised learning consists of two stages as illustrated in Figure 7. In the first stage, known as pretraining, a self-supervised CNN [9] is trained to recognize whether an input is a transformed version of the original signals. Various transformations are applied to the signals, including adding noise, scaling, negating, horizontally flipping, and permuting the signal. However, in this study, the time-warped transformation was excluded. Before applying these transformations, the data are segmented and normalized, as shown in Figure 4. For example, Gaussian noise with parameters $\mu = 0$ and $\sigma = 0.01$ is added to the signal, the signal is scaled by a factor of 1.1, and signal permutation involves randomly selecting 20 pieces of length 1/4 of the sampling rate (64 data points) and swapping them to create a permuted version of the original signal. Figure 8 displays both the transformed PPG signal and the original signal as examples for reference. After the pretraining stage, the shared layers (convolutional layers) of the self-supervised CNN are reused for classification tasks.

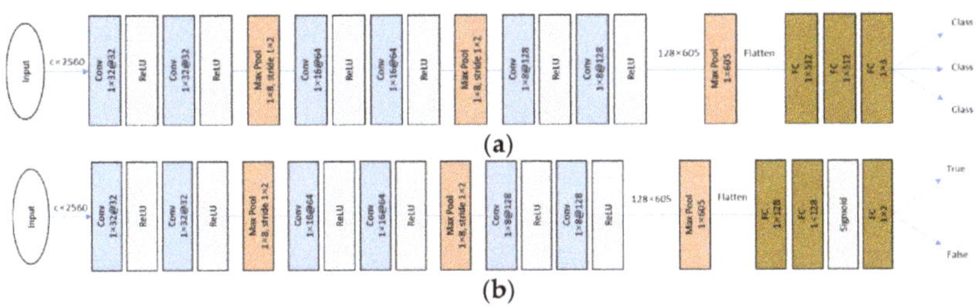

Figure 7. Self-supervised CNN architecture for (**a**) pretrain and (**b**) classifier.

The shared layer of the self-supervised CNN is composed of three convolutional blocks. Each block consists of two 1D convolutional layers and is followed by a max pooling operation. In the first convolutional block, there are 32 filters with a filter size of 32. The second block has 64 filters with a size of 16, and the third block uses 128 filters with a size of 8. The max pooling layers in each block have a size of 8 and a stride of 2. After passing through the shared layer, the feature maps have dimensions of 128×605. These feature maps are then inputted into a global max pooling layer, resulting in an output size of 128×1.

3.4.4. Training Parameters

The models' performance was evaluated using k-fold stratified cross-validation. This approach ensures that each split contains all class labels rather than having one-fold with labels from only one class. The value of k was set to 3 for this training. The optimizer used was Adam, with a learning rate of 0.001. During the training of the models, the total number of epochs was set to 300, while for the self-supervised CNN pretraining stage, it was set to 150 epochs. The validation accuracy of each epoch and F1-score were averaged and returned as the performance for each fold. After completing the cross-validation, the performance of each fold was averaged again to obtain the final model performance.

The models were developed using the PyTorch framework and executed on an HP Shadow Elf 7 laptop running Windows 11. The laptop was equipped with an NVIDIA GeForce RTX 3060 Laptop GPU, an Intel(R) Core (TM) i7-11800H CPU operating at 2.30 GHz, and 16 GB of RAM.

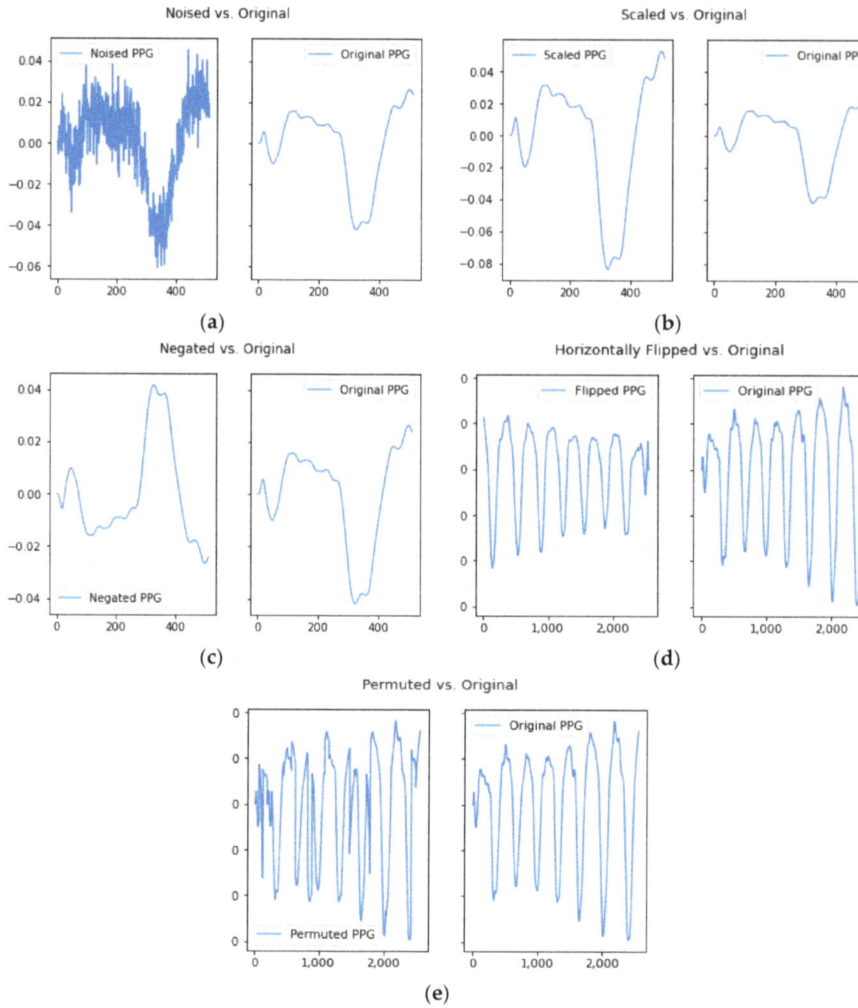

Figure 8. Transformed PPG vs. original PPG: (**a**) noised PPG vs. original PPG (**b**) scaled PPG vs. original PPG (**c**) negated PPG vs. original PPG (**d**) horizontally flipped PPG vs. original PPG (**e**) permuted PPG vs. original PPG.

4. Results and Discussion

Various signal combinations, such as PPG, ECG, EEG, PPG + ECG, PPG + EEG, ECG + EEG, and PPG + ECG + EEG, were explored to determine the most effective signals for stress detection. The performance of the models was evaluated using accuracy and F1-score as metrics. Accuracy represents the percentage of correctly classified data in relation to the entire dataset and serves as a straightforward measure of performance which is represented as follows:

$$accuracy = \frac{1}{n}\sum(p_i == y_i)$$

where p_i is the predicted label of the *ith* sample, and y_i is the true label of the *ith* sample. If p_i is the same as y_i, $p_i == y_i$, it is 1, otherwise, it is 0. F1-score can be interpreted as a harmonic means of precision and recall,

$$F1 = 2 \times \frac{precision \times recall}{precision + recall}$$

where precision assesses the classifier's capacity to avoid mislabeling negative samples as positive, while recall evaluates the classifier's ability to identify all positive samples, with the equations as follows:

$$precision = \frac{TP}{TP + FP}$$

$$recall = \frac{TP}{TP + FN}$$

where *TP* is true positive, *FP* is false positive, and *FN* is false negative.

4.1. Scenario-Based Self-Reporting Stress Analysis

Figures 9 and 10 presented the changes in stress levels across different scenarios while participants solved "medium" and "hard" levels of Sudoku puzzles, respectively. The results indicated that a majority of participants experienced an increase in stress levels in scenario 1, which involved noisy and horror audio/videos. About half of the participants (13 out of 30) reported higher stress levels in scenario 2, where they were monitored by another person. Conversely, in scenario 3, which included comforting audio and video, most participants (22 out of 30) reported a decrease in stress levels. When comparing scenario 3 to scenario 1, 20 out of 30 participants demonstrated a decrease in stress levels in the final scenario. Additionally, in scenario 1, participants solving the "hard" Sudoku puzzles tended to experience higher levels of stress compared to those solving the "medium" Sudoku puzzles. Similarly, in scenario 3, participants solving the "medium" Sudoku puzzles showed a reduction in stress levels compared to those solving the "hard" Sudoku puzzles.

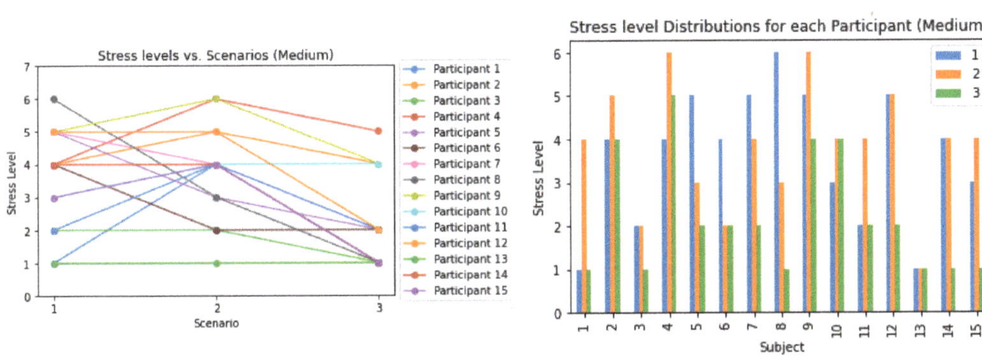

(a) (b)

Figure 9. (**a**) Relationship between stress levels and scenarios, as well as the (**b**) distribution of stress levels for each participant in the medium-level Sudoku where the legend represented the scenarios.

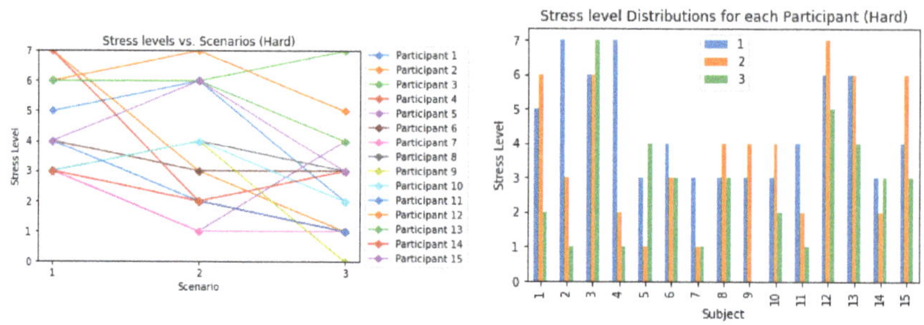

(**a**) (**b**)

Figure 10. (**a**) Relationship between stress levels and scenarios, as well as the (**b**) distribution of stress levels for each participant in the hard level Sudoku where the legend represented the scenarios.

4.2. Classifiers Evaluation

Table 2 displays the accuracy and F1-score results for different models and combinations of biosignals used in the classification. The results revealed that the LRCN model demonstrated the highest effectiveness in predicting stress levels when utilizing ECG data, achieving an average accuracy of 93.42% and an F1-score of 88.11%. Other models also exhibited high levels of accuracy and F1-score, with StressNeXt and self-supervised CNN achieving the best performance when combining ECG and EEG signals. Notably, the utilization of EEG alone resulted in the lowest accuracy and F1-score across all models. This observation could potentially be attributed to participants' suboptimal reduction of movements during the experiments, as proposed by [24,25]. The presence of muscle noise might have compromised the effectiveness of the collected EEG data in this study.

Table 2. Performances of various models and different data combinations.

Model Name	Data	Accuracy	F1-Score
StressNeXt	PPG	83.90%	69.04%
	ECG	85.71%	69.61%
	EEG	66.31%	39.40%
	PPG + ECG	88.22%	77.48%
	PPG + EEG	83.26%	68.35%
	ECG + EEG	90.02%	80.45%
	PPG + ECG + EEG	86.90%	74.26%
LRCN	PPG	84.27%	70.49%
	ECG	93.42%	88.11%
	EEG	80.15%	62.36%
	PPG + ECG	86.77%	74.66%
	PPG + EEG	83.89%	69.67%
	ECG + EEG	91.39%	84.31%
	PPG + ECG + EEG	84.44%	71.35%
Self-Supervised CNN	PPG	81.66%	63.98%
	ECG	90.07%	81.11%
	EEG	74.44%	28.90%
	PPG + ECG	86.05%	71.47%
	PPG + EEG	84.72%	69.90%
	ECG + EEG	90.32%	81.04%
	PPG + ECG + EEG	80.69%	52.70%

Table 3 displays the accuracy and F1-score outcomes for the various model and biosignal combinations utilized in the stress level classification under different experimental scenarios. The findings suggest that the StressNeXt model was most successful in predicting stress levels in scenario 3, while the LRCN model was the most effective in scenarios 1 and 2. ECG was found to be the most effective signal in analyzing stress levels, with all models achieving around 95% accuracy and 90% F1-score when utilizing ECG data. On the other hand, EEG was found to be the least effective signal for detecting stress levels, with all models yielding the lowest accuracies and F1-scores, less than 70% in scenarios 1 and 2 and 90% in scenario 3. In addition, it can be observed that all models were able to predict stress levels rather effectively in scenario 3, with accuracy and F1-score reaching approximately 90% for all signal combinations.

Table 4 displays a comparison of the difficulty levels of Sudoku for each model. The accuracy rates and F1-scores of the StressNeXt model were slightly higher in the "medium" level Sudoku sample group compared to the "hard" level Sudoku sample group. Conversely, the other two models exhibited slightly lower accuracy rates and F1 scores in the "medium" level Sudoku sample group compared to the "hard" level Sudoku sample group. However, these differences were not found to be statistically significant.

The confusion matrices for StressNeXt, LRCN, and self-supervised CNN with different input signals for all scenarios and difficulty levels of Sudokus are displayed in Figure 11, Figure 12, and Figure 13, respectively. The total number of samples after data segmentation was 8100, with 6030 samples for stress level class 0, 1710 samples for class 1, and 360 samples for class 2. The confusion matrices were obtained by averaging the confusion matrix of the validation set in each fold of the 3-fold validation. It is apparent that there is more training data on class 0, and the model seems to lean towards predicting class 0 stress level. StressNeXt was unable to predict class 2 stress levels with ECG, whereas other models can effectively classify most of the class 2 stress using only ECG. The self-supervised CNN failed to distinguish other classes from class 0 with EEG or all signals combined as input and was incapable of classifying any class 2 stress with combined PPG and ECG as input.

Table 3. Performance of various models and different combinations of data in different scenarios.

Model Name	Scenario	Data	Accuracy	F1-Score
StressNeXt	Scenario 1	PPG	84.78%	79.90%
		ECG	93.48%	90.83%
		EEG	63.39%	48.68%
		PPG + ECG	89.52%	85.44%
		PPG + EEG	80.98%	73.95%
		ECG + EEG	92.71%	90.02%
		PPG + ECG + EEG	87.88%	84.37%
	Scenario 2	PPG	86.64%	79.53%
		ECG	96.80%	94.27%
		EEG	64.81%	47.75%
		PPG + ECG	83.73%	65.20%
		PPG + EEG	84.18%	75.01%
		ECG + EEG	95.93%	91.33%
		PPG + ECG + EEG	89.37%	76.86%
	Scenario 3	PPG	95.22%	78.55%
		ECG	97.79%	91.37%
		EEG	86.86%	46.81%
		PPG + ECG	97.79%	91.62%
		PPG + EEG	94.80%	77.17%
		ECG + EEG	98.78%	95.39%
		PPG + ECG + EEG	98.29%	92.88%

Table 3. *Cont.*

Model Name	Scenario	Data	Accuracy	F1-Score
LRCN	Scenario 1	PPG	81.78%	78.10%
		ECG	95.13%	93.72%
		EEG	69.93%	55.68%
		PPG + ECG	88.52%	85.41%
		PPG + EEG	81.15%	77.31%
		ECG + EEG	93.51%	91.17%
		PPG + ECG + EEG	81.51%	77.14%
	Scenario 2	PPG	82.94%	72.06%
		ECG	96.46%	93.90%
		EEG	73.35%	54.19%
		PPG + ECG	85.21%	74.37%
		PPG + EEG	83.31%	73.50%
		ECG + EEG	97.96%	96.67%
		PPG + ECG + EEG	87.84%	75.98%
	Scenario 3	PPG	95.43%	78.90%
		ECG	97.16%	93.63%
		EEG	90.42%	57.76%
		PPG + ECG	95.44%	84.94%
		PPG + EEG	92.14%	68.00%
		ECG + EEG	96.71%	90.31%
		PPG + ECG + EEG	92.00%	67.49%
Self-Supervised CNN	Scenario 1	PPG	92.06%	90.09%
		ECG	95.06%	93.01%
		EEG	63.35%	28.10%
		PPG + ECG	88.22%	82.41%
		PPG + EEG	91.08%	89.11%
		ECG + EEG	87.62%	80.42%
		PPG + ECG + EEG	90.05%	87.98%
	Scenario 2	PPG	92.95%	89.92%
		ECG	96.61%	94.54%
		EEG	69.90%	30.35%
		PPG + ECG	90.53%	79.41%
		PPG + EEG	90.08%	85.39%
		ECG + EEG	96.15%	93.09%
		PPG + ECG + EEG	88.56%	76.57%
	Scenario 3	PPG	95.48%	78.54%
		ECG	98.50%	94.66%
		EEG	89.95%	37.93%
		PPG + ECG	95.14%	77.24%
		PPG + EEG	95.80%	82.97%
		ECG + EEG	97.06%	85.73%
		PPG + ECG + EEG	94.89%	77.83%

Table 4. Performance of various models and different combinations of data was evaluated across different difficulty levels of Sudoku puzzles.

Model Name	Sudoku Difficulty	Data	Accuracy	F1-Score
StressNeXt	Medium	PPG	87.75%	78.58%
		ECG	90.04%	83.72%
		EEG	69.98%	44.72%
		PPG + ECG	83.79%	73.58%
		PPG + EEG	86.06%	75.56%
		ECG + EEG	87.85%	78.48%
		PPG + ECG + EEG	86.25%	76.48%
	Hard	PPG	85.36%	73.37%
		ECG	89.41%	77.91%
		EEG	67.59%	43.92%
		PPG + ECG	84.96%	70.63%
		PPG + EEG	81.80%	67.53%
		ECG + EEG	90.05%	79.82%
		PPG + ECG + EEG	84.52%	69.55%
LRCN	Medium	PPG	85.52%	74.36%
		ECG	91.56%	87.91%
		EEG	79.75%	57.72%
		PPG + ECG	84.90%	74.16%
		PPG + EEG	84.37%	70.64%
		ECG + EEG	82.83%	67.56%
		PPG + ECG + EEG	84.37%	72.07%
	Hard	PPG	82.58%	68.27%
		ECG	92.00%	83.55%
		EEG	77.14%	58.98%
		PPG + ECG	86.98%	75.84%
		PPG + EEG	81.55%	66.90%
		ECG + EEG	87.74%	75.85%
		PPG + ECG + EEG	87.10%	75.75%
Self-Supervised CNN	Medium	PPG	89.70%	81.48%
		ECG	88.71%	80.70%
		EEG	75.55%	29.64%
		PPG + ECG	87.36%	73.33%
		PPG + EEG	92.49%	84.64%
		ECG + EEG	84.92%	70.74%
		PPG + ECG + EEG	89.38%	80.53%
	Hard	PPG	91.29%	83.10%
		ECG	89.74%	81.06%
		EEG	73.34%	28.20%
		PPG + ECG	89.95%	80.51%
		PPG + EEG	84.93%	70.77%
		ECG + EEG	90.63%	82.80%
		PPG + ECG + EEG	89.81%	81.14%

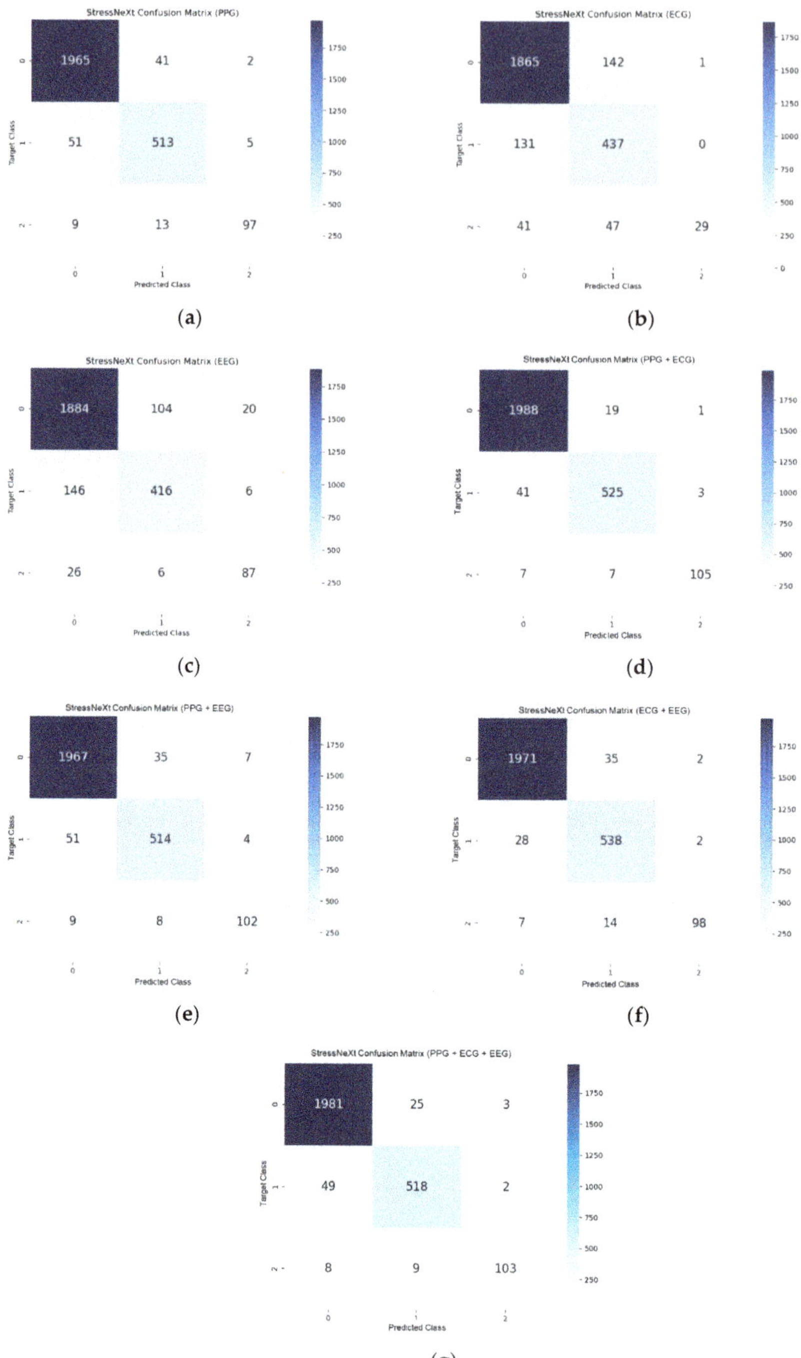

Figure 11. StressNeXt confusion matrix for (**a**) PPG as input, (**b**) ECG as input, (**c**) EEG as input, (**d**) combined PPG and ECG as input, (**e**) combined PPG and EEG as input, (**f**) combined ECG and EEG as input, (**g**) combined PPG, ECG, and EEG as input.

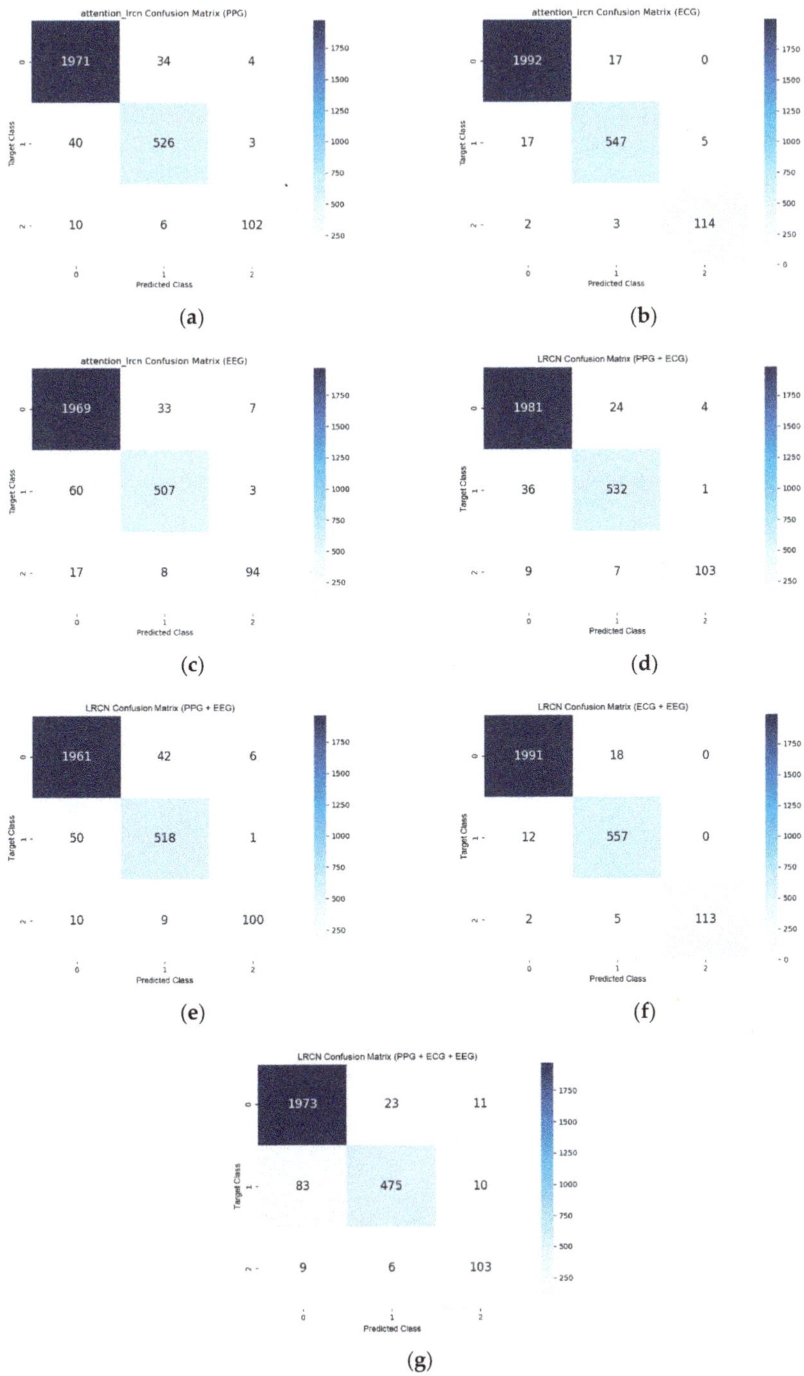

Figure 12. LRCN confusion matrix for (**a**) PPG as input, (**b**) ECG as input, (**c**) EEG as input, (**d**) combined PPG and ECG as input, (**e**) combined PPG and EEG as input, (**f**) combined ECG and EEG as input, (**g**) combined PPG, ECG, and EEG as input.

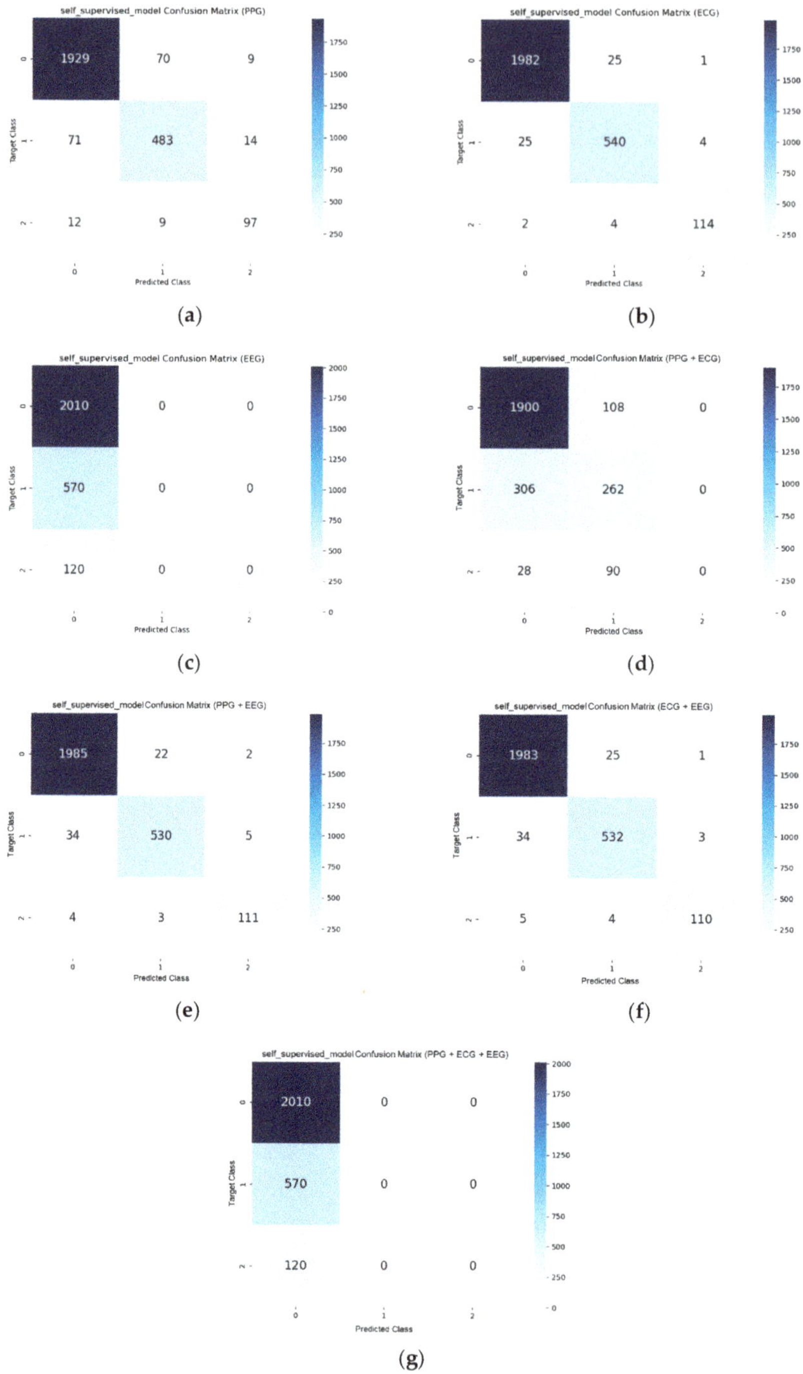

Figure 13. Self-supervised CNN confusion matrix for (**a**) PPG as input, (**b**) ECG as input, (**c**) EEG as input, (**d**) combined PPG and ECG as input, (**e**) combined PPG and EEG as input, (**f**) combined ECG and EEG as input, (**g**) combined PPG, ECG, and EEG as input.

Table 5 illustrates a performance comparison between our proposed model and previous studies on stress detection. The results indicate that our model outperforms others in terms of accuracy in scenario 3, where subjects solve Sudoku under comforting conditions. Similarly, in scenario 2, where subjects solve Sudoku with another individual monitoring, our model achieves the highest F1-score compared to previous models. Overall, our model demonstrates superior accuracy compared to Transformer [26], Random Forest [27], AdaBoost DT [17], DeepER Net [28], Artificial Neural Network [29], Deep ECGNet [30], and CNN-LSTM [31]. The comparison highlights that the effectiveness of these models may be specific to particular datasets, depending on the scenario design and experiment settings. Despite the fact that the study conducted by [12], which utilized Deep 1D-CNN, exhibited superior performance compared to our proposed work, it required more than five sensors as inputs, which may not be feasible for real-world applications. Furthermore, the evaluation of individual sensor data on the performance of our proposed model was not conducted, and the sample size only included 15 participants, which did not adequately represent the majority of human stress levels. Similarly, the study using SVM-RBF [25] demonstrated slightly higher accuracy than our proposed work. However, it is worth noting that our proposed work achieved higher accuracy (93.42%) when utilizing only EEG signals, compared to [13], which utilized PPG (80%) and GSR + PPG (86.25%) signals. Perfect comparison becomes relatively challenging as our current study did not incorporate the GSR signal, as its changes vary based on environmental conditions. In general, most studies included fewer than 20 participants, and variations in task creation and labeling methods likely contribute to performance differences. Nonetheless, our study attains the highest accuracy (98.78%) in scenario 3 and the highest F1-score (96.67%) in scenario 2, surpassing the existing work.

Table 5. Comparative analysis of the performance.

Model	Accuracy	F1-Score	Input Data	Scenarios	Number of Participants
Transformer [26]	71.60%	74.20%	Raw ECG	Participants write reports for each of the two provided topics and make presentation for one of the provided topics (SWELL dataset)	25
Random Forest [27]	78.80%	88.80%	Extracted features of GSR, heart rate	Students perform multiple tasks, including sing-a-song, emails, color-word test, game, arithmetic question, social conversation, eating, homework, put hands in ice bucket	9
AdaBoost DT (3-class classification) [17]	80.34%	72.51%	Extracted features of PPG, EDA, SKT	Participants read magazines, take TSST, and watch amusing videos (WESAD dataset)	17
DeepER Net [28]	83.90%	81.00%	Extracted features of ECG and RSP	University students solve math tasks or color-word test	18
Artificial Neural Network (ANN) [29]	84.32%	78.71%	Extracted features of ACC, PPG, EDA, TEMP, RESP, EMG, and ECG	Participants read magazines, take TSST, and watch amusing videos (WESAD dataset)	17
Deep ECGNet [30]	87.39%	73.96%	Extracted features of ECG	Students take multiple tasks, including arithmetic problems, color-word test, interview	30

Table 5. *Cont.*

Model	Accuracy	F1-Score	Input Data	Scenarios	Number of Participants
CNN-LSTM Network [31]	92.80%	94.56%	Raw ECG, vehicle dynamic data, environmental parameters	Participants drive on a simulator with different scenarios, including urban, highway, city	17
Multi-layer Perceptron [24]	93.64%	92.44%	Raw PPG, EDA, SKT	Participants read magazines, take TSST, and watch amusing videos (WESAD dataset)	17
SVM-RBF [25]	96.25%	96.00%	Extracted features of PPG, GSR, EEG	Participants prepare a talk and speak in front of real audience	40
Deep 1D-CNN [12]	97.48%	96.82%	ECG, EDA, EMG, RESP, TEMP, TEMP, ACC	Participants watched a series of videos	15
Proposed model (general)	93.42%	88.11%	ECG + EEG	Students solve Sudoku puzzles under different distractions, including noisy environment, another individual monitoring, comforting conditions	30
Proposed model (scenario 1)	95.13%	93.72%			
Proposed model (scenario 2)	97.76%	96.67%			
Proposed model (scenario 3)	98.78%	95.39%			

5. Conclusions

In this study, a novel experimental approach was used to evaluate stress levels in participants through the use of Sudoku puzzles as problem-solving tasks. The study utilized PPG, ECG, and EEG data to extract features and assess stress levels. Various scenarios were created using audio and video components to examine the impact of environmental stimuli on stress levels. The results indicated that noisy environments tended to cause higher stress levels, and higher difficulty level Sudoku puzzles may lead to a higher average stress level. The stress detection models used in the experiments showed high effectiveness, with the LRCN model achieving a stress level detection accuracy of 93.42% and an F1-score of 88.11% when ECG was used as an input signal. When analyzing different scenarios, the StressNeXt model demonstrated exceptional effectiveness in predicting stress levels under comforting conditions, achieving an accuracy rate of 98.78% and an F1-score of 95.39%. Conversely, the LRCN model was most effective in predicting stress levels in two scenarios, achieving accuracy rates of 95.13% and 97.96%, with F1 scores of 93.72% and 96.67%, respectively. However, the study has some limitations, such as the subjectivity of self-reporting, which can result in variations in reported stress levels. To improve the reliability of this study, more robust self-reporting methods, such as using the Self-Assessment Manikin (SAM), can be employed to validate the accuracy and reliability of stress level assessment. In addition, this study aims to explore the potential inclusion of other physiological signals, such as GSR or respiratory rate, to examine their effectiveness in stress detection. Furthermore, the study plans to incorporate another dataset containing diverse scenarios and environments to validate the performance of the proposed deep learning models.

Author Contributions: Conceptualization, Q.C. and B.G.L.; methodology, Q.C. and B.G.L.; software, Q.C.; validation, Q.C. and B.G.L.; formal analysis, Q.C. and B.G.L.; investigation, Q.C.; resources, Q.C. and B.G.L.; data curation, Q.C. and B.G.L.; writing, original draft preparation, Q.C.; writing—review and editing, B.G.L.; visualization, Q.C.; supervision, B.G.L.; project administration, B.G.L.; funding acquisition, B.G.L. All authors have read and agreed to the published version of the manuscript.

Funding: This research was funded by Zhejiang Provincial Natural Science Foundation of China, grant number LQ21F020024 and Ningbo Science and Technology Bureau under Major Programme (CM2025), grant number Z2021Z037.

Institutional Review Board Statement: The study was conducted in accordance with the Code of Research Conduct and Research Ethics and approved by the Ethics Committee of the University of Nottingham Ningbo China (7 September 2022).

Informed Consent Statement: Informed consent was obtained from all subjects involved in the study.

Data Availability Statement: Data available upon request due to ethical restrictions.

Conflicts of Interest: The authors declare no conflict of interest. The funders had no role in the design of the study; in the collection, analyses, or interpretation of data; in the writing of the manuscript; or in the decision to publish the results.

References

1. Li, C. From Involution to Education: A Glance to Chinese Young Generation. In Proceedings of the 2021 4th International Conference on Humanities Education and Social Sciences (ICHESS 2021), Xishuangbanna, China, 29–31 October 2021; Atlantis Press: Amsterdam, The Netherlands, 2021; pp. 1884–1887.
2. Ponzini, A. Educating the new Chinese middle-class youth: The role of quality education on ideas of class and status. *J. Chin. Sociol.* **2020**, *7*, 1. [CrossRef]
3. Pascoe, M.C.; Hetrick, S.E.; Parker, A.G. The impact of stress on students in secondary school and higher education. *Int. J. Adolesc. Youth* **2020**, *25*, 104–112. [CrossRef]
4. Chapell, M.S.; Blanding, Z.B.; Silverstein, M.E.; Takahashi, M.; Newman, B.; Gubi, A.; McCann, N. Test anxiety and academic performance in undergraduate and graduate students. *J. Educ. Psychol.* **2005**, *97*, 268–274. [CrossRef]
5. College Student Suicide: Failures and Potential Solutions. Available online: https://www.brainsway.com/knowledge-center/college-student-suicide-failures-and-potential-solutions/#:~:text=How%20Many%20College%20Students%20Commit,for%20death%20among%20college%20students (accessed on 19 February 2023).
6. Wagh, K.P.; Vasanth, K. Performance evaluation of multi-channel electroencephalogram signal (EEG) based time frequency analysis for human emotion recognition. *Biomed. Signal Process. Control* **2022**, *78*, 103966. [CrossRef]
7. Vijayakumar, S.; Flynn, R.; Corcoran, P.; Murray, N. CNN-based Emotion Recognition from Multimodal Peripheral Physiological Signals. In Proceedings of the IMX'22: ACM International Conference on Interactive Media Experiences, Aveiro, Portugal, 22–24 June 2022.
8. Miao, M.; Zheng, L.; Xu, B.; Yang, Z.; Hu, W. A multiple frequency bands parallel spatial–temporal 3D deep residual learning framework for EEG-based emotion recognition. *Biomed. Signal Process. Control* **2023**, *79*, 104141. [CrossRef]
9. Montero Quispe, K.G.; Utyiama, D.M.; Dos Santos, E.M.; Oliveira, H.A.; Souto, E.J. Applying Self-Supervised Representation Learning for Emotion Recognition Using Physiological Signals. *Sensors* **2022**, *22*, 9102. [CrossRef] [PubMed]
10. Tang, Y.; Wang, Y.; Zhang, X.; Wang, Z. STILN: A Novel Spatial-Temporal Information Learning Network for EEG-based Emotion Recognition. *arXiv* **2022**, arXiv:2211.12103. [CrossRef]
11. Choi, J.; Lee, J.S.; Ryu, M.; Hwang, G.; Hwang, G.; Lee, S.J. Attention-LRCN: Long-term Recurrent Convolutional Network for Stress Detection from Photoplethysmography. In Proceedings of the 2022 IEEE International Symposium on Medical Measurements and Applications (MeMeA), Messina, Italy, 22–24 June 2022; IEEE: New York, NY, USA, 2022; pp. 1–6.
12. Miranda-Correa, J.A.; Abadi, M.K.; Sebe, N.; Patras, I. Amigos: A dataset for affect, personality and mood research on individuals and groups. *IEEE Trans. Affect. Comput.* **2018**, *12*, 479–493. [CrossRef]
13. Katsigiannis, S.; Ramzan, N. DREAMER: A database for emotion recognition through EEG and ECG signals from wireless low-cost off-the-shelf devices. *IEEE J. Biomed. Health Inform.* **2017**, *22*, 98–107. [CrossRef] [PubMed]
14. Koldijk, S.; Sappelli, M.; Verberne, S.; Neerincx, M.A.; Kraaij, W. The swell knowledge work dataset for stress and user modeling research. In Proceedings of the 16th International Conference on Multimodal Interaction, Istanbul, Turkey, 12–16 November 2014; pp. 291–298.
15. Zheng, W.-L.; Lu, B.-L. Investigating critical frequency bands and channels for EEG-based emotion recognition with deep neural networks. *IEEE Trans. Auton. Ment. Dev.* **2015**, *7*, 162–175. [CrossRef]
16. Koelstra, S.; Muhl, C.; Soleymani, M.; Lee, J.-S.; Yazdani, A.; Ebrahimi, T.; Pun, T.; Nijholt, A.; Patras, I. Deap: A database for emotion analysis; using physiological signals. *IEEE Trans. Affect. Comput.* **2011**, *3*, 18–31. [CrossRef]

17. Schmidt, P.; Reiss, A.; Duerichen, R.; Marberger, C.; Van Laerhoven, K. Introducing wesad, a multimodal dataset for wearable stress and affect detection. In Proceedings of the 20th ACM International Conference on Multimodal Interaction, Boulder, CO, USA, 16–20 October 2018; pp. 400–408.
18. Hu, X. Wenjuanxing Official Website. Available online: https://www.wjx.cn/ (accessed on 19 December 2022).
19. Electro, P. Polar Verity Sense. Available online: https://www.polar.com/us-en/products/accessories/polar-verity-sense (accessed on 21 December 2022).
20. Xinweilai. BMD101 ECG Detection Package. Taobao. Available online: https://item.taobao.com/item.htm?spm=a230r.1.14.22.a4734ab0qwBJQL&id=618036232572&ns=1&abbucket=1&mt= (accessed on 19 June 2023).
21. NeuroSky. MindWave Mobile Setup Kit. Available online: https://mwm2.neurosky.com/ (accessed on 8 August 2022).
22. Mekruksavanich, S.; Hnoohom, N.; Jitpattanakul, A. A Deep Residual-based Model on Multi-Branch Aggregation for Stress and Emotion Recognition through Biosignals. In Proceedings of the 2022 19th International Conference on Electrical Engineering/Electronics, Computer, Telecommunications and Information Technology (ECTI-CON), Prachuap Khiri Khan, Thailand, 24–27 May 2022; IEEE: New York, NY, USA, 2022; pp. 1–4.
23. Fawaz, H.I.; Lucas, B.; Forestier, G.; Pelletier, C.; Schmidt, D.F.; Weber, J.; Webb, G.I.; Idoumghar, L.; Muller, P.-A.; Petitjean, F. Inceptiontime: Finding alexnet for time series classification. *Data Min. Knowl. Discov.* **2020**, *34*, 1936–1962. [CrossRef]
24. Li, R.; Liu, Z. Stress detection using deep neural networks. *BMC Med. Inform. Decis. Mak.* **2020**, *20*, 285. [CrossRef] [PubMed]
25. Arsalan, A.; Majid, M. Human stress classification during public speaking using physiological signals. *Comput. Biol. Med.* **2021**, *133*, 104377. [CrossRef] [PubMed]
26. Behinaein, B.; Bhatti, A.; Rodenburg, D.; Hungler, P.; Etemad, A. A Transformer Architecture for Stress Detection from ECG. In Proceedings of the 2021 ACM International Symposium on Wearable Computers, Virtual, 21–26 September 2021; pp. 132–134. [CrossRef]
27. Egilmez, B.; Poyraz, E.; Wenting, Z.; Memik, G.; Dinda, P.; Alshurafa, N. UStress: Understanding college student subjective stress using wrist-based passive sensing. In Proceedings of the 2017 IEEE International Conference on Pervasive Computing and Communications Workshops (PerCom Workshops), Kona, HI, USA, 13–17 March 2017; pp. 673–678. [CrossRef]
28. Seo, W.; Kim, N.; Kim, S.; Lee, C.; Park, S.-M. Deep ECG-Respiration Network (DeepER Net) for Recognizing Mental Stress. *Sensors* **2019**, *19*, 3021. [CrossRef] [PubMed]
29. Bobade, P.; Vani, M. Stress Detection with Machine Learning and Deep Learning using Multimodal Physiological Data. In Proceedings of the 2020 Second International Conference on Inventive Research in Computing Applications (ICIRCA), Coimbatore, India, 15–17 July 2020; pp. 51–57. [CrossRef]
30. Hwang, B.; You, J.; Vaessen, T.; Myin-Germeys, I.; Park, C.; Zhang, B.-T. Deep ECGNet: An optimal deep learning framework for monitoring mental stress using ultra short-term ECG signals. *Telemed. e-Health* **2018**, *24*, 753–772. [CrossRef] [PubMed]
31. Rastgoo, M.N.; Nakisa, B.; Maire, F.; Rakotonirainy, A.; Chandran, V. Automatic driver stress level classification using multimodal deep learning. *Expert Syst. Appl.* **2019**, *138*, 112793. [CrossRef]

sensors

MDPI

Article

Two-Step Approach for Occupancy Estimation in Intensive Care Units Based on Bayesian Optimization Techniques

José A. González-Nóvoa [1,*], Laura Busto [1], Silvia Campanioni [1], José Fariña [2], Juan J. Rodríguez-Andina [2], Dolores Vila [3] and César Veiga [1]

1 Galicia Sur Health Research Institute (IIS Galicia Sur), Álvaro Cunqueiro Hospital, 36310 Vigo, Spain
2 Department of Electronic Technology, University of Vigo, 36310 Vigo, Spain
3 Intensive Care Unit Department, Complexo Hospitalario Universitario de Vigo (SERGAS), Álvaro Cunqueiro Hospital, 36213 Vigo, Spain
* Correspondence: jose.gonzalez@iisgaliciasur.es

Abstract: Due to the high occupational pressure suffered by intensive care units (ICUs), a correct estimation of the patients' length of stay (LoS) in the ICU is of great interest to predict possible situations of collapse, to help healthcare personnel to select appropriate treatment options and to predict patients' conditions. There has been a high amount of data collected by biomedical sensors during the continuous monitoring process of patients in the ICU, so the use of artificial intelligence techniques in automatic LoS estimation would improve patients' care and facilitate the work of healthcare personnel. In this work, a novel methodology to estimate the LoS using data of the first 24 h in the ICU is presented. To achieve this, XGBoost, one of the most popular and efficient state-of-the-art algorithms, is used as an estimator model, and its performance is optimized both from computational and precision viewpoints using Bayesian techniques. For this optimization, a novel two-step approach is presented. The methodology was carefully designed to execute codes on a high-performance computing system based on graphics processing units, which considerably reduces the execution time. The algorithm scalability is analyzed. With the proposed methodology, the best set of XGBoost hyperparameters are identified, estimating LoS with a MAE of 2.529 days, improving the results reported in the current state of the art and probing the validity and utility of the proposed approach.

Keywords: artificial intelligence; automated machine learning; Bayesian optimization; ICU occupancy; intensive care unit; length of stay; machine learning; MIMIC; XGBoost

1. Introduction

The intensive care units (ICUs) of hospitals have a variety of devices to monitor the patients' health states that generate a large amount of data, allowing healthcare personnel to be aware of the patient's vital sign values and to make the most appropriate decisions to ensure their correct evolution [1]. It is very difficult to analyze all of this volume of information manually, so it is of great interest to use artificial intelligence (AI) tools that automate and help in these tasks [2], especially in extraordinary situations such as the one experienced due to the COVID-19 pandemic, where ICUs were overwhelmed.

The patient's ICU length of stay (LoS) is an important metric from a clinical point of view. Due to the high occupational pressure suffered by the ICU, a correct estimation of the patients' LoS in the ICU is of great interest to predict possible situations of collapse, to help healthcare personnel to select appropriate treatment options and to predict patients' conditions. The design of a reliable estimator system is useful to anticipate future collapse situations and to take the corresponding actions [2], such as conditioning an area similar to the ICU urgently that accommodates future patients. In this field there are several studies. The vast majority of them approach the problem as a binary classification, predicting, for example, which patients will stay in the ICU for more or less than a defined number of days [3–5]. However, other studies approach the problem from the point of view of the exact calculation of LoS [6–8], which allows for a finer prediction of the ICU situation. In this article, it was decided to approach the problem from this second point of view, estimating the ICU stay duration from the monitoring data obtained during the first 24 h since the moment the patient accessed the ICU. In order to obtain more precise results, it was decided to delve into the optimization of the hyperparameters of the model, a task which requires a high computational load.

One of the goals of high-performance computing systems is to reduce the execution time of a given task [9]. Due to the increased computational load of artificial intelligence problems, it is necessary to use these systems to reduce the execution time. In recent years, performance enhancement from one processor generation to another has stagnated, making it necessary to find other ways to optimize the performance of algorithms. One of the main alternatives is the use of GPUs (graphics processing units), which help to continue to reduce execution times.

Occupancy prediction is a frequent topic within the healthcare ecosystem [10,11], as well as in other related fields, such as building energy systems [12], building performance [13] and heating, ventilation and air conditioning (HVAC) system control [14]. In this work, we propose a new methodology based on a novel two-step Bayesian optimization approach to improve LoS estimation methods used to predict ICU occupancy. This method allows one to optimize an LoS XGBoost predictor that uses clinical variables extracted from ICU monitoring, by finding the best set of hyperparameters of the model. The proposed approach improves the results of LoS predictions that would be obtained in the case of carrying out the optimization in the conventional way (regular Bayesian) and also improves the results obtained in other state-of-the-art works [3–8].

Due to the high computational load that this task entails, an important part of the methodology is the parallelization of the problem on a GPU architecture, allowing it to be solved in a computationally efficient way. XGBoost is used as estimator model. All these contributions are supported by experimental results.

The remainder of the article is structured as follows. First, the materials used are detailed. Then, the proposed methodology is explained and the experimental results are presented. Finally, the discussion and conclusions of the work are presented.

2. Materials

2.1. Data Source

In this work, the MIMIC-III (Medical Information Mart for Intensive Care III) [15] ICU database, developed by the MIT (Massachusetts Institute of Technology), was used. It contains data from 61,532 ICU stays at Beth Israel Deaconess Medical Center.

The database collected demographic data, vital sign measurements made at the bedside (1 data point per hour), laboratory test data, procedures, medications, caregiver notes, imaging reports, signals (electrocardiography (ECG), photoplethysmogram (PPG), arterial blood pressure (ABP)), etc. It contains the variable to be estimated in this work (LoS in the ICU) with a resolution of $\pm 10^{-4}$ days ($\pm$8.64 s). Figure 1 shows the different data sources of an ICU. The detailed structure of the database can be found in the MIMIC-III original publication [15].

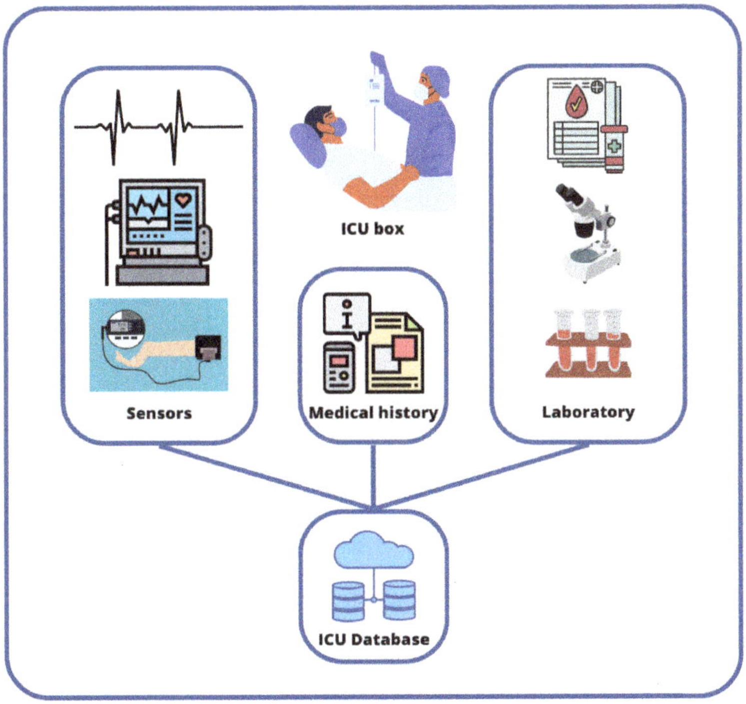

Figure 1. Intensive care unit structure.

2.2. XGBoost

XGBoost [16] is a gradient boosting technique based on ensemble learning. These techniques correct errors made by previous models in successive ones, optimizing a loss function. This function (1) is modified at each iteration t. The successive models are built using the exact greedy algorithm, which analyzes all possible split loss reduction options, until the stop condition is achieved. A more detailed explanation may be found in [16].

$$L^{(t)} = \sum_{i=1}^{n} l(y_i, \hat{y}_i^{(t-1)}) + \Omega(f_t) \quad (1)$$

where l is a differentiable convex loss function that must be transformed into another one in a Euclidean domain using Taylor's Theorem, the pair (y_i, x_i) represents the training set, $\hat{y}_i$ is the final prediction and $\Omega(f_t)$ is the regularization term used to penalize more complex models through both Lasso and Ridge regularization and to prevent overfitting.

XGBoost is one of the most popular algorithms in the state of the art [17], highlighting its computational efficiency and GPU support. It is one of the most used algorithms in recent biomedical works based on tabular data [18–20].

As stated above [21,22], XGBoost uses three types of parameters: general, booster and learning task parameters. General parameters indicate the booster used: tree or linear model. Booster parameters are related with the booster employed and define the internal performance parameters, e.g., learning rate or number of estimators, while learning task parameters indicate the corresponding learning objective. The methodology presented in this work focuses on learning task parameters seeking to optimize the performance of the XGBoost regressor model. These parameters are:

- Learning rate: Step-size shrinkage used in updates to prevent overfitting. After each boost step, the weights of the new features can be obtained directly, and the learning rate reduces the weights of the features to make the boost process more conservative.
- Maximum delta step: the maximum delta step each leaf output is allowed to be.
- Maximum depth of a tree: increasing this value will make the model more complex.
- Maximum leaves: maximum number of nodes to be added.
- Minimum child weight: Minimum sum of instance weight (hessian) needed in a child. If the tree partition step results in a leaf node with the sum of instance weight being less than the minimum child weight, then the building process will give up further partitioning.
- Number of estimators: The number of trees in the ensemble. It is equivalent to the number of boosting rounds.
- Region alpha: L1 regularization term on weights. Increasing this value will make the model more conservative.
- Region lambda: L2 regularization term on weights. Increasing this value will make the model more conservative. It is normalized to number of training examples.
- Scale pos weight: controls the balance of positive and negative weights, useful for unbalanced classes.
- Subsample: Subsample ratio of the training instances. Setting it to 0.5 means that XGBoost would randomly sample half of the training data prior to growing trees. Subsampling will occur once in every boosting iteration.

2.3. *Bayesian Optimization*

One of the relevant parts of a machine learning pipeline is the optimization of the estimator model. This task implies a high computational load, so optimizing it efficiently is a key factor. Given a dataset D, the goal of hyperparameter optimization [23] is to find λ in (2):

$$\lambda^* = \arg\min_{\lambda \epsilon \Lambda} \mathbb{E}_{D_{train}, D_{test} \epsilon D} \mathbf{V}(\mathcal{L}, \mathcal{A}_\lambda, D_{train}, D_{test}) \quad (2)$$

where λ is a vector of hyperparameters from the hyperparameter search space Λ, $\mathcal{A}$ is the predictive model and $\mathbf{V}(\mathcal{L}, \mathcal{A}_\lambda, D_{train}, D_{test})$ measures the loss of the model $\mathcal{A}$.

In this work, Bayesian techniques [24] were used to carry out this task, which stand out for their computational efficiency when performing the search. The surrogate model (a Gaussian stochastic probabilistic model) was used. The optimization was divided into several stages: The first stage was the set-up of the surrogate model. Next, the best hyperparameter combination was sought, and it was applied to the real objective function. Finally, the surrogate model was updated. This process was repeated iteratively until the defined criteria were achieved.

One of the most used objective functions is that of expected improvements (3), because it can be calculated in closed form if the estimator model y with the configuration λ follows a normal distribution (4) [23].

$$\mathbb{E}[\mathbb{I}(\lambda)] = \mathbb{E}[max(f_{min} - y, 0)] \quad (3)$$

$$\mathbb{E}[\mathbb{I}(\lambda)] = (f_{min} - \mu(\lambda))\Phi(\frac{f_{min} - \mu(\lambda)}{\sigma}) + \sigma\phi(\frac{f_{min} - \mu(\lambda)}{\sigma}) \quad (4)$$

where $\phi(\cdot)$ and $\Phi(\cdot)$ are the standard normal density and standard normal distribution, and f_{min} is the best observed value.

In this work, the open-source package Hyperopt [25], which uses Bayesian optimization as the search technique, was used.

3. Methodology

The proposed methodology to optimize the estimator model of the patient's LoS is divided into several stages. Figure 2a shows an outline of these stages, while a more detailed explanation of the hyperparameter search using Bayesian optimization is shown in Figure 2b. The first stage is related to the data preprocessing and the feature extraction. The second one is devoted to building the LoS estimator. In this work, we propose to use the state-of-the-art approach based on gradient boosting, namely XGBoost, due to the good results obtained in estimation tasks and the high level of computational improvement that could be achieved on a GPU architecture during the optimization stage. The third stage is devoted to optimizing the model hyperparameters in order to improve the performance of the estimator. In this work, a novel two-step Bayesian optimization approach is proposed, implemented on a GPU architecture to reduce the execution time. Finally, the estimator model is validated.

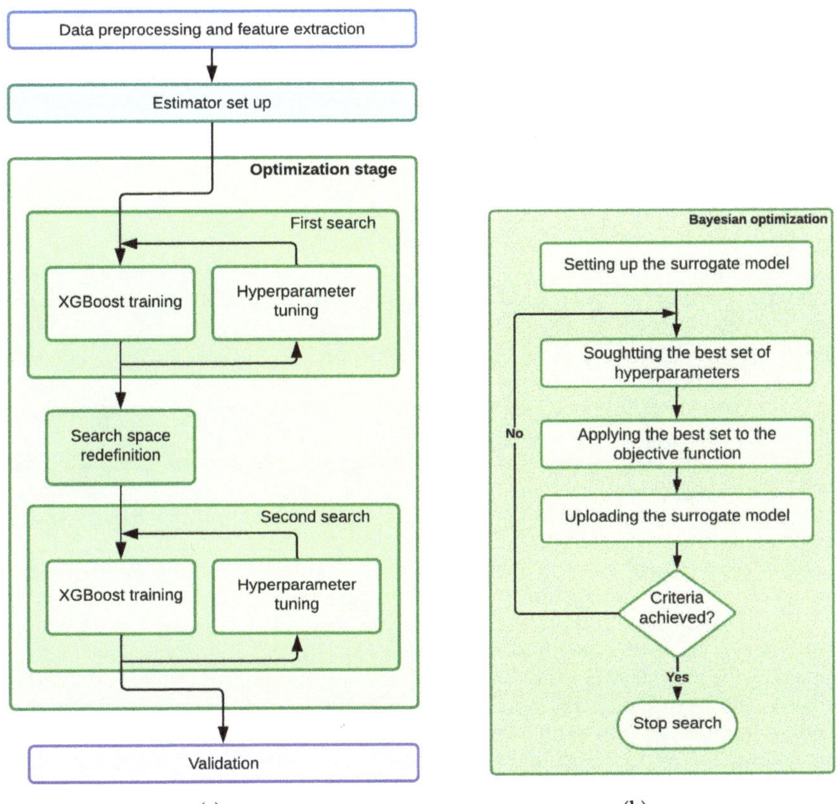

Figure 2. Outline of the proposed methodology. (**a**) Shows a general outline, while a more detailed explanation of the hyperparameter search using Bayesian optimization is shown in (**b**).

3.1. Data Preprocessing and Feature Extraction

The first stage of the pipeline was devoted to data preprocessing to obtain a set of features, derived from the clinical data, used to fit and validate the estimator model. Such a stage required two tasks.

The first task consisted of selecting the variables used to produce the dataset from the original database. In this approach, all available clinical variables were considered, selecting only the ones that were present in at least 80% of the patients for building the model, as in other published works [21,26]. The reason why all of the variables were not available for all patients was because, depending on the pathology and the patient's clinical condition, only a certain set of variables was monitored. Patients who did not have values in at least 2/3 of these variables were subsequently discarded. In order to deal with missing data, a schema for input data was required to fill values on such empty variables. Although there are other imputation methods, such as MissForest [27] or generative adversial networks [28], in this work we have proposed using K–Nearest Neighbors, as it is one of the most widely used in the current state of the art, in addition to its simplicity in implementation and usage.

The next task consisted of the conversion from variables to features, in this case by computing the mean value, standard deviation and maximum and minimum values of the clinical variables gathered during the first 24 h of the patient's ICU stay, except for the volume of urine, for which only the total volume in this time interval was used.

3.2. LoS Estimation Model Building

Once the previous preprocessing stage was completed, it was necessary to proceed with the configuration stage of the ICU LoS estimator model. As mentioned above, the XGBoost model was used, which was fitted using the above described features. In order to train the model, a random split train and test of these data was carried out following a ratio of 80/20. To validate the model, the mean absolute error (MAE) (5) was used, which is one of the most popular metrics in the current state of the art, allowing for the comparison of the results.

$$MAE = \sum_{i=1}^{n} |\hat{y}_i - y_i| \tag{5}$$

where n is the number of patients, $\hat{y}_i$ is the LoS estimated by XGBoost model and y_i is the golden standard for the LoS, obtained from the database.

3.3. XGBoost Optimization: Two-Step Approach

To improve the accuracy of the model, we delved into the automatic optimization of the hyperparameters of XGBoost instead of using the default values.

This stage was divided into two steps. A first search step was executed using the initially defined hyperparameter search space. Then, a second search was performed once the search space was modified after the hyperparameters' evolution during the first search stage was analyzed.

The optimization started with the hyperparameter search space (Λ) definition, which consisted of indicating which parameters should be varied during the different iterations (2) and within which limits, in addition to indicating the type of search space. There are three types of search space distributions: uniform, log uniform and q uniform, which return real values uniformly distributed between defined limits. Log uniform is more suitable for geometric series, whereas uniform and q uniform are more suitable for arithmetic series, with the difference being that q uniform returns round values, so the selection of the search space depends on the hyperparameter type. This part was fundamental, since an incorrect definition of space can cause the process to not work efficiently in computational terms.

In addition to defining the search space, it was necessary to define an objective function to minimize in each iteration, that is, the statistical metric that defined the quality of the estimator model needed to be defined. In this work, it was decided to use the MAE (5) as a metric. Firstly, the maximum number of evaluations (m) was defined as 5000.

Once this first search was completed, the search space was modified to improve the results obtained avoiding the bottlenecks derived from the initial defined limits and refocusing on the area where the best value was found after the first search, but expanding the search limits, as shown in the Results section. Table 1 summarizes the search space distribution used in the first and second phases, and the types of distributions used.

Table 1. Hyperparameter distributions.

Hyperparameter	Distribution Type	First Step Distribution		Second Step Distribution	
		Min	Max	Min	Max
Learning rate	loguniform	−8	0	−7.5	−4.5
Max. depth	quniform	1	15	14.5	25
Min. child weight	quniform	0	10	6	12
Max. delta step	quniform	0	10	0	0.5
Subsample ratio	uniform	0.1	1	0.4	0.7
Lambda region	uniform	0.1	1	0.7	1.2
Alpha region	uniform	0.1	1	0.4	0.8
Scale pos weight	uniform	0.1	1	0.45	0.85
Max. number of leaves	quniform	0	10	0.1	1
Number of estimators	quniform	1	10,000	100	2500

4. Results

In this section, the results obtained using the proposed methodology are presented and analyzed, in terms of the set of hyperparameters identified during the optimization stage, the performance of the model estimator using MAE as the metric and the computational performance. For this analysis, the initial dataset is randomly split into train (27,077 patients) and test (6770 patients) sets, calculating the MAE for the test set.

4.1. Hyperparameter Tuning Stage

Regarding the hyperparameter optimization stage, the method permits one to reduce the MAE, and consequently improve the estimations of the model, by identifying the set of parameters that provides the lower MAE. Table 2 shows the best combination of hyperparameters obtained both after the first search and after the second search, explained and described in Section 3.

Table 2. Hyperparameter optimal values.

Hyperparameter	Best Value	
	First Search	Second Search
Learning rate	0.00135	0.00132
Max. depth	15	24
Min. child weight	9	12
Max. delta step	0	0
Subsample ratio	0.525	0.529
Lambda region	0.891	1.184
Alpha region	0.661	0.559
Scale pos weight	0.674	0.755
Max. number of leaves	0	0
Number of estimators	745	812

The whole evolution of interactions of both optimization steps are plotted in Figure 3, which shows the minimum value of the MAE obtained (grouped into intervals of 50 iterations), that clearly shows the downward trend. The figure also shows how after redefining the search space from the results of the first search step (black vertical line), there is a sharp drop in the value of the MAE, obtaining a minimum value of 2.529 in iteration 6962.

Another important element when analyzing the optimization stage of the model is the evolution in the search, that is, how the values of the hyperparameters vary throughout the different trials. Figure 4 shows the value of one of the model hyperparameters (the maximum depth) in each trial, with its corresponding MAE. The figures corresponding to the rest of the hyperparameters are shown in the Supplementary Materials. From these figures, what was indicated above can also be seen, meaning that there is no direct correlation between the value of the individual hyperparameter and the value of the MAE obtained. However, it shows that the intervals between the optimal MAE value obtained and the trend in the search are found. From these results, a second optimization step was performed by redefining the search space of the Bayesian model, modifying the search space limits. After this second optimization step, a MAE of 2.529 was obtained.

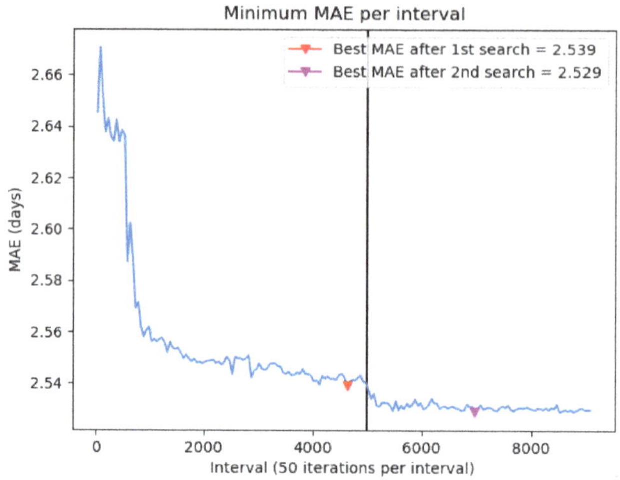

Figure 3. Evolution of minimum MAE across 50 iteration intervals.

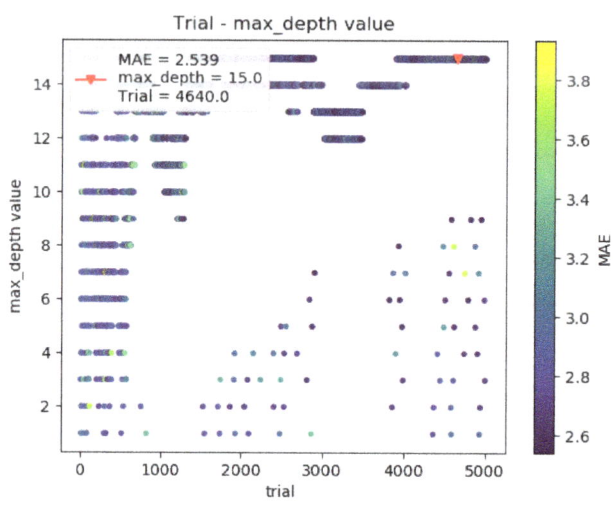

Figure 4. Evolution of maximum depth value and its corresponding MAE across the trials.

4.2. Estimator Validation

Only the first stay of each patient was selected in the database to avoid possible information leakage and to compare the results obtained with related works [8], resulting in a total of 46,476 stays. As indicated in Section 3, a second filtering was subsequently carried out to discard patients for whom more that 1/3 of the variables were unavailable, obtaining a total of 33,847 ICU stays. Table 3 shows the main characteristics of the resulting cohort.

Table 3. Characteristics of the data source.

	MIMIC-III	Resulting Cohort
Number of patients	46,476	33,847
Number of ICU stays	61,532	33,847
Average age	64.93	74.65
Gender	F: 20,380 M: 26,096	F: 19,319 M: 14,528
LoS average	4.92	4.32
LoS standard deviation	9.64	6.21
LoS variance	92.91	38.55

Table 4 shows the clinical ICU variables and features extracted from each patient. It consists of a total of 35 variables as 139 features.

Table 4. Features extracted from each clinical variable.

Clinical Variables	Statistics
Age	Value in admission
Gender	F/M
Urine output	Accumulated value after 24 h
Glasgow coma motor scale	
Glasgow coma verbal scale	
Glasgow coma eyes scale	
Systolic blood pressure	
Heart rate	
Body temperature	
PaO_2	
FiO_2	
Serum urea nitrogen level	
Sodium level	
Potassium level	
Bilirubin level	
Respiratory rate	
Glucose	
Albumin	
Anion gap	Average, maximum, minimum and standard deviation value after 24 h
Chloride	
Creatinine	
Lactate	
Calcium	
Haematocrit	
Hemoglobin	
INR [1]	
Platelets	
Prothrombin time test	
Activated thromboplastin time	
Base excess	
$PaCO_2$	
$FiCO_2$	
PH	
Total CO_2	

[1] International normalized ratio of prothrombin time.

To quantify the precision of the model when estimating the patient's LoS in the ICU, the MAE was used as the metric. As mentioned above, the initial dataset was split into train (27,077 patients) and test (6770 patients) datasets, calculating the MAE for the test set. The LoS was estimated with a resolution of $\pm 10^{-4}$ days, obtaining a MAE of 2.529 days, lower than the rest of the current state-of-the-art works consulted. Rouzbahman et al. [29] obtain a MAE of 5.07 days, while Alghatani et al. [30] obtain a MAE of 2.64 days, both using MIMIC as the database. Moreover, the final MAE obtained is lower than the MAE obtained with the default hyperparameters (3.040) and after the first optimization step (2.539). Although it is true that there are studies with a slightly lower MAE value, they focus on a specific group of patients, instead of addressing the problem from a generic point of view, as we do. Our work improves the results of those who face problems from a generic point of view. Table 5 shows the results obtained.

Table 5. Comparison of the MAE obtained using the default model with respect to that obtained after the first and second steps of the optimization phase.

Default Estimator (Without Optimization)	Optimized Model (After First Step)	Optimized Model (After Second Step)
3.040	2.539	2.529

4.3. Computational Performance

Model fitting and optimization are extremely expensive processes. To carry out this work, a high-performance computing system was required. This architecture consisted of a 24-core CPU, 256 GB of RAM and one NVIDIA A100 GPU unit. The GPU had 40 GB of memory and consisted of 108 multiprocessors, allowing the execution of 2024 threads per multiprocessor, a total of 221,184 threads per GPU [31]. One of the main objectives of the methodology proposed in this article was computational efficiency. For this, it was necessary that the software used could be executed on the GPU.

In order to benchmark the effects that hardware and dataset dimensions have on the performance of the proposed method, several tests were conducted, both on the GPU and on the CPU (using one core and three cores). Each hardware configuration was tested with three different dataset versions, namely the original dataset used in a previous experiment (dataset), this dataset reduced to half of the number of patients (Subdataset 1) and a dataset with half of the features (Subdataset 2). Both the selection of patients and characteristics were performed randomly. Table 6 shows the results obtained in minutes using 100 iterations (m).

Once the drastic difference in time between the CPU and GPU was verified, the number of iterations was increased to 500, only using the GPU to analyze how it influenced the execution time. It was observed that the execution time increases practically in the same proportion as the number of iterations. The variation in this proportion is due to the fact that depending on the value of the hyperparameters in which it is iterating, the execution time differs.

Table 6. Time comparison. m is the number of iterations.

	CPU		GPU	
	1 Core (m = 100)	3 Core (m = 100)	m = 100	m = 500
Dataset	85.86′	39.26′	9.70′	75.23′
Subdataset 1	82.56′	41.86′	11.82′	63.06′
Subdataset 2	44.63′	24.50′	9.07′	44.63′

5. Discussion

The proposed methodology allows one to identify the set of hyperparameters that provides the best performance of the predictor in terms of minimizing the mean absolute error (MAE). With the best combination of hyperparameters, the LoS was estimated with a MAE of 2.529 days, lower than the rest of the current state-of-the-art works consulted. It is also lower than the MAE obtained after the first search stage (MAE = 2.539 days) and lower than using the default XGBoost hyperparameters set without executing the optimization task (MAE = 3.04 days). Although it is true that there are studies with a slightly lower MAE value, they focus on a specific group of patients, instead of addressing the problem from a generic point of view, as was indicated in Section 1.

XGBoost was used as an estimator model due to being one of the ones that obtains the best results in other current state-of-the-art works and its computational efficiency. However, the same methodology could be applied to other models (Random Forest, Support

Vector Machine, etc.). The feature extraction stage could be modified adding data from other ICU sensors, extracting other statistics or considering another time window instead of 24 h.

One of the fundamental reasons for obtaining better results is the optimization phase of the model. Adapting the search methodology to work in the GPU allowed us to intensify the search for the best combination of hyperparameters. In addition, as already indicated above, the search technique used based on Bayesian optimization was characterized by performing the search for hyperparameters more efficiently from a computational point of view. The results also confirm that the two-step approach proposed for the optimization of the estimator model improves the results obtained. This same proposal could be applied to another problem.

A stagnation of the MAE improvement is observed with the passing of the trials, which gives rise to an analysis of the relationship of computational cost–improvement obtained, in which it would be necessary to assess the available hardware and the allowable error in each particular situation. Regarding the relationship between the individual values of the hyperparameters and the value of the MAE, no clear patterns are observed, which justifies that what is really relevant is the combination of the values of the different hyperparameters, not the values of each one separately. This demonstrates the need to perform the optimization of the model automatically, as was the case in this work, instead of performing it manually, which would be unfeasible in terms of work times, and worse results would be obtained.

From the computational results, how the use of the GPU drastically reduces the execution time is observed. It is also interesting to analyze how the variations in the size of the dataset hardly imply variations in execution times in the GPU. This makes sense, since although the size of the dataset is reduced, GPUs are characterized by processing large amounts of data in parallel in a single clock cycle, so the variation in the size of the dataset does not always imply a variation in the timing of execution. If the same methodology was used in the classic architectures based solely on the CPU, the execution time would have been drastically higher, which would limit the ability to analyze the possible improvements to be made to the methodology for a real clinical application.

As for future lines of work, this methodology can be extended in several ways. The first line of research is related to the application domain, regarding using this novel methodology to estimate other time variables that are different to LoS, as well as to estimate other variables aside from time within the ICU, and extending the applicability to other hospital areas and by extension to any other social prediction. Another future line of research could be exploring the configuration domain, namely the modifications in the internal configuration of the methodology (the estimator model used to change the XGBoost predictor model by any other regressor model that could provide a time estimation). Modifications in the optimization stage and in the feature extraction stage are also of great interest, using more data sources or using different feature extraction techniques.

6. Conclusions

This article presents a methodology to estimate the LoS in the ICU using data collected by UCI sensors and other sources (laboratory and medical history) during the first 24 h of a patient's admission, focusing on the optimization stage of the estimator model, both from a computational point of view and at the estimation precision level. To do this, the methodology was adapted so that the training of the model was executed on the GPU and the search for the best combination of hyperparameters was carried out automatically using Bayesian optimization techniques in a novel two-step approach.

The results using 33,847 patients demonstrate the validity of the proposed methodology, obtaining a MAE of 2.529 days, lower than that of other works from the current state of the art consulted. In addition, an improvement in model precision by dividing the model optimization phase into two steps instead of performing it in a single step is also demonstrated.

This work opens several future lines of research for applying the present methodology to predict other variables within the ICU, or in other hospital areas. Another future line of research could be exploring modifications in the internal configuration of the methodology (the estimator model used, modifications in the optimization stage, using different features, etc.).

Supplementary Materials: The following supporting information can be downloaded at: https://www.mdpi.com/article/10.3390/s23031162/s1. Figures similar to Figure 4 are added as Supplementary Material for each hyperparameter.

Author Contributions: J.A.G.-N.: conceptualization, data curation, formal analysis, investigation, methodology, software, validation, visualization, writing—original draft, writing—review editing; L.B.: validation, investigation; S.C.: investigation; J.F.: investigation, supervision; J.J.R.-A.: investigation, supervision; D.V.: investigation, validation; C.V.: conceptualization, investigation, formal analysis, methodology, resources, writing—review editing, supervision, funding acquisition. All authors have reviewed the manuscript. All authors have read and agreed to the published version of the manuscript.

Funding: The research was partially supported by Agencia Gallega de Innovación (GAIN) through "Proxectos de investigación sobre o SARS-CoV-2 e a enfermidade COVID-19 con cargo ao Fondo COVID-19" Program, with code number IN845D-2020/29, and through "Axudas para a consolidación e estruturación de unidades de investigación competitivas e outras accións de fomento nos organismos públicos de investigación de Galicia e noutras entidades do sistema galego de I+D+i - GPC" with code number IN607B-2021/18.

Institutional Review Board Statement: Not applicable.

Informed Consent Statement: Not applicable.

Data Availability Statement: The datasets analyzed in the current study are available in the PhysioNet repository, https://physionet.org/content/mimiciii/1.4/ (accessed on 10 January 2021).

Conflicts of Interest: The authors declare no conflict of interest. The founders had no role in the design of the study; in the collection, analyses or interpretation of data; in the writing of the manuscript or in the decision to publish the results.

Abbreviations

The following abbreviations are used in this manuscript:

CPU	central processing unit
GPU	graphics processing units
ICU	intensive care unit
INR	international normalized ratio
LoS	length of stay
MAE	mean absolute error
MIMIC	Medical Information Mart for Intensive Care

References

1. Gutierrez, G. Artificial Intelligence in the Intensive Care Unit. *Crit. Care* **2020**, *24*, 1–9. [CrossRef] [PubMed]
2. van de Sande, D.; van Genderen, M.E.; Huiskens, J.; Gommers, D.; van Bommel, J. Moving from bytes to bedside: A systematic review on the use of artificial intelligence in the intensive care unit. *Intensive Care Med.* **2021**, *47*, 750–760. [CrossRef] [PubMed]
3. Peres, I.T.; Hamacher, S.; Oliveira, F.L.C.; Thomé, A.M.T.; Bozza, F.A. What factors predict length of stay in the intensive care unit? Systematic review and meta-analysis. *J. Crit. Care* **2020**, *60*, 183–194. [CrossRef] [PubMed]
4. Awad, A.; Bader-El-Den, M.; McNicholas, J. Patient length of stay and mortality prediction: A survey. *Health Serv. Manag. Res.* **2017**, *30*, 105–120. [CrossRef]
5. Su, L.; Xu, Z.; Chang, F.; Ma, Y.; Liu, S.; Jiang, H.; Wang, H.; Li, D.; Chen, H.; Zhou, X.; et al. Early Prediction of Mortality, Severity, and Length of Stay in the Intensive Care Unit of Sepsis Patients Based on Sepsis 3.0 by Machine Learning Models. *Front. Med.* **2021**, *8*, 883. [CrossRef]
6. Alsinglawi, B.; Alnajjar, F.; Mubin, O.; Novoa, M.; Alorjani, M.; Karajeh, O.; Darwish, O. Predicting Length of Stay for Cardiovascular Hospitalizations in the Intensive Care Unit: Machine Learning Approach. In Proceedings of the 2020 42nd Annual

International Conference of the IEEE Engineering in Medicine & Biology Society (EMBC), Montreal, QC, Canada, 20–24 July 2020; pp. 5442–5445. [CrossRef]

7. Houthooft, R.; Ruyssinck, J.; van der Herten, J.; Stijven, S.; Couckuyt, I.; Gadeyne, B.; Ongenae, F.; Colpaert, K.; Decruyenaere, J.; Dhaene, T.; et al. Predictive modelling of survival and length of stay in critically ill patients using sequential organ failure scores. *Artif. Intell. Med.* **2015**, *63*, 191–207. [CrossRef]
8. Purushotham, S.; Meng, C.; Che, Z.; Liu, Y. Benchmarking deep learning models on large healthcare datasets. *J. Biomed. Inf.* **2018**, *83*, 112–134. [CrossRef]
9. Kindratenko, V.; Trancoso, P. Trends in High-Performance Computing. *Comput. Sci. Eng.* **2011**, *13*, 92–95. [CrossRef]
10. Whitt, W.; Zhang, X. Forecasting arrivals and occupancy levels in an emergency department. *Oper. Res. Health Care* **2019**, *21*, 1–18. [CrossRef]
11. Littig, S.J.; Isken, M.W. Short term hospital occupancy prediction. *Health Care Manag. Sci.* **2007**, *10*, 47–66. [CrossRef]
12. Tekler, Z.D.; Chong, A. Occupancy prediction using deep learning approaches across multiple space types: A minimum sensing strategy. *Build. Environ.* **2022**, *226*, 109689. [CrossRef]
13. Tekler, Z.D.; Low, R.; Gunay, B.; Andersen, R.K.; Blessing, L. A scalable Bluetooth Low Energy approach to identify occupancy patterns and profiles in office spaces. *Build. Environ.* **2020**, *171*, 106681. [CrossRef]
14. Dong, J.; Winstead, C.; Nutaro, J.; Kuruganti, T. Occupancy-Based HVAC Control with Short-Term Occupancy Prediction Algorithms for Energy-Efficient Buildings. *Energies* **2018**, *11*, 2427. [CrossRef]
15. Johnson, A.E.W.; Pollard, T.J.; Shen, L.; Lehman, L.w.H.; Feng, M.; Ghassemi, M.; Moody, B.; Szolovits, P.; Anthony Celi, L.; Mark, R.G. MIMIC-III, a freely accessible critical care database. *Sci. Data* **2016**, *3*, 1–9. [CrossRef]
16. Chen, T.; Guestrin, C. XGBoost: A Scalable Tree Boosting System. In *KDD '16: Proceedings of the 22nd ACM SIGKDD International Conference on Knowledge Discovery and Data Mining*; Association for Computing Machinery: New York, NY, USA, 2016; pp. 785–794. [CrossRef]
17. Nielsen, D. Tree Boosting With XGBoost—Why Does XGBoost Win "Every" Machine Learning Competition? Ph.D. Thesis, NTNU, Trondheim, Norway, 2016.
18. González, B.S.; Conceição, R.; Pimenta, M.; Tomé, B.; Guillén, A. Tackling the muon identification in water Cherenkov detectors problem for the future Southern Wide-field Gamma-ray Observatory by means of machine learning. *Neural Comput. Applic.* **2022**, *34*, 5715–5728. [CrossRef]
19. Site, A.; Vasudevan, S.; Afolaranmi, S.O.; Lastra, J.L.M.; Nurmi, J.; Lohan, E.S. A Machine-Learning-Based Analysis of the Relationships between Loneliness Metrics and Mobility Patterns for Elderly. *Sensors* **2022**, *22*, 4946. [CrossRef]
20. Yang, J.; Clifton, D.; Hirst, J.E.; Kavvoura, F.K.; Farah, G.; Mackillop, L.; Lu, H. Machine Learning-Based Risk Stratification for Gestational Diabetes Management. *Sensors* **2022**, *22*, 4805. [CrossRef]
21. González-Nóvoa, J.A.; Busto, L.; Rodríguez-Andina, J.J.; Fariña, J.; Segura, M.; Gómez, V.; Vila, D.; Veiga, C. Using Explainable Machine Learning to Improve Intensive Care Unit Alarm Systems. *Sensors* **2021**, *21*, 7125. [CrossRef]
22. González-Nóvoa, J.A.; Busto, L.; Santana, P.; Fariña, J.; Rodríguez-Andina, J.J.; Juan-Salvadores, P.; Jiménez, V.; Íñiguez, A.; Veiga, C. Using Bayesian Optimization and Wavelet Decomposition in GPU for Arterial Blood Pressure Estimation. In Proceedings of the 2022 44th Annual International Conference of the IEEE Engineering in Medicine & Biology Society (EMBC), Glasgow, UK, 11–15 July 2022; pp. 1012–1015. [CrossRef]
23. Feurer, M.; Hutter, F. Hyperparameter Optimization. In *Automated Machine Learning*; Springer: Cham, Switzerland, 2019; pp. 3–33. [CrossRef]
24. Shahriari, B.; Swersky, K.; Wang, Z.; Adams, R.P.; de Freitas, N. Taking the Human Out of the Loop: A Review of Bayesian Optimization. *Proc. IEEE* **2015**, *104*, 148–175. [CrossRef]
25. Bergstra, J.; Yamins, D.; Cox, D.D. Making a Science of Model Search. *arXiv* **2012**. [CrossRef]
26. Jiang, Z.; Bo, L.; Xu, Z.; Song, Y.; Wang, J.; Wen, P.; Wan, X.; Yang, T.; Deng, X.; Bian, J. An explainable machine learning algorithm for risk factor analysis of in-hospital mortality in sepsis survivors with ICU readmission. *Comput. Methods Programs Biomed.* **2021**, *204*, 106040. [CrossRef] [PubMed]
27. Stekhoven, D.J.; Bühlmann, P. MissForest—non-parametric missing value imputation for mixed-type data. *Bioinformatics* **2012**, *28*, 112–118. [CrossRef] [PubMed]
28. Low, R.; Tekler, Z.D.; Cheah, L. Predicting Commercial Vehicle Parking Duration using Generative Adversarial Multiple Imputation Networks. *Transp. Res. Rec.* **2020**, *2674*, 820–831. [CrossRef]
29. Rouzbahman, M.; Jovicic, A.; Chignell, M. Can Cluster-Boosted Regression Improve Prediction of Death and Length of Stay in the ICU? *IEEE J. Biomed. Health Inf.* **2016**, *21*, 851–858. [CrossRef] [PubMed]
30. Alghatani, K.; Ammar, N.; Rezgui, A.; Shaban-Nejad, A. Predicting Intensive Care Unit Length of Stay and Mortality Using Patient Vital Signs: Machine Learning Model Development and Validation. *JMIR Med. Informat.* **2021**, *9*, e21347. [CrossRef]
31. NVIDIA. NVIDIA A100 GPUs Power the Modern Data Center. 2022. Available online: https://www.nvidia.com/en-us/data-center/a100/ (accessed on 23 May 2022) .

sensors

MDPI

Article

The Trade-Off between Airborne Pandemic Control and Energy Consumption Using Air Ventilation Solutions

Ariel Alexi [1,*,†], Ariel Rosenfeld [1,†] and Teddy Lazebnik [2]

1 Department of Information Science, Bar-Ilan University, Ramat-Gan 5290002, Israel
2 Department of Cancer Biology, Cancer Institute, University College London, London WC1E 6DD, UK
* Correspondence: ariel.alexi@live.biu.ac.il; Tel.: +972-50-845-2481
† These authors contributed equally to this work.

Abstract: Airborne diseases cause high mortality and adverse socioeconomic consequences. Due to urbanization, more people spend more time indoors. According to recent research, air ventilation reduces long-range airborne transmission in indoor settings. However, air ventilation solutions often incur significant energy costs and ecological footprints. The trade-offs between energy consumption and pandemic control indoors have not yet been thoroughly analyzed. In this work, we use advanced sensors to monitor the energy consumption and pandemic control capabilities of an air-conditioning system, a pedestal fan, and an open window in hospital rooms, classrooms, and conference rooms. A simulation of an indoor airborne pandemic spread of Coronavirus (COVID-19) is used to analyze the Pareto front. For the three examined room types, the Pareto front consists of all three air ventilation solutions, with some ventilation configurations demonstrating significant inefficiencies. Specifically, air-conditioning is found to be efficient only at a very high energy cost and fans seem to pose a reasonable alternative. To conclude, a more informed ventilation policy can bring about a more desirable compromise between energy consumption and pandemic spread control.

Keywords: airborne transmission and control; indoors pandemic spread; air ventilation; energy consumption

1. Introduction

Large-scale pandemic outbreaks that occurred in the last several decades such as the Hong Kong flu (Influenza A/H3N2), HIV/AIDS, and the coronavirus (COVID-19) pandemics are associated with wide-reaching economical damage, social disaster, and high mortality [1–3]. Wu et al. [4] suggested that the globalization processes that transpired in the last century have invigorated new pandemic outbreaks [5]. In particular, urbanization in the developing world is bringing more people into denser neighborhoods, which results in a higher infection rate, at which new diseases spread [6,7]. Indeed, the world's population is concentrated in cities, and megalopolises in particular [8], which keep growing over time [9]. In addition, most individuals who live in cities spend most of their time indoors, mainly in building (e.g., home, school and workplace) [10].

In parallel, a global sharp increase in energy and goods consumption, associated with the increase in population size and life quality [11–13], has led to growing recognition for the importance of sustainable consumption in order to minimize costs and, most importantly, protect the planet from degradation so that it can support the needs of the present and future generations [14]. Indeed, in 2015, all UN member countries have adopted the 2030 Agenda for Sustainable Development [15] thus committing to making fundamental changes in the way that our societies produce and consume goods and services. In addition, the UN has highlighted the need to Ensure healthy lives and promote well-being as one of its core 17 sustainable development goals. In the context of pandemic control, minimizing energy consumption and reducing the spread of a pandemic are, in fact, partially conflicting goals [16,17]. Most importantly, in indoors settings, policymakers should consider pandemic control through a multi-objective perspective where both the pandemic spread as well as the energy consumption of the implemented measures are considered together. Indoor settings often provide fruitful grounds for the integration of sensors which can facilitate a real-time system to pursue its desired objective. In our context, sensors are an obvious candidate for measuring key measures associated with a pandemic spread and energy consumption [18].

A large body of work has focused on a broad spectrum of pandemic intervention policies (PIPs) such as lockdown [19,20], mask-wearing [21], social distance [22], inspection units [23], and vaccination [24] for airborne pandemics. Focusing on PIP for indoor settings, Lazebnik and Alexi [25] analyzed the spread of an airborne pandemic indoors in four types of buildings (home, school, office, and mall) while implementing different PIPs, showing significantly different spread patterns between them. Wei and Li [26] review the release, transport, and exposure of expiratory droplets in a close environment such as a room. The authors concluded that droplets or droplet nuclei are transported by airflow and suggested that face masks, as well as air ventilation, can significantly reduce the infection rate. Peng et al. [27] explored the pandemic spread of the COVID-19 pathogen for indoor settings using a spatio-temporal version of the Wells-Riley models [28]. The authors provided a formula for the connection between population density and pandemic spread. They show that mask-wearing is partially effective against air-borne pandemics in a close room, which a decreasing effectiveness over time. These and similar PIPs have shown promising results in controlling airborne pandemic spreads in both theory and practice [29,30]. The main limitation associated with most PIPs are their dependency on the willingness of the population to comply with them [31]. Historical records reveal that, indeed, populations across the globe are only partially complaint with PIPs even if these are formally regulated and enforced [32,33].

Promisingly, implementing air ventilation solutions such as air-conditioning have shown to be an effective PIP in mitigating the spread of pandemics in indoor settings [34]. In addition to the proved reduction in airborne transmission, their effectiveness does not require any active participation from the occupants of the ventilated space. Dai and Zhao [35] estimated the association between the infection probability and changes per hour of air in the room (i.e., ventilation rate), obtaining that to in order to ensure an infection probability of less than one percent, a ventilation rate larger than the regularly practiced in the US is required. Memarzadeh and Xu [36] showed that while the ventilation rate is important in reducing the pandemic spread, the dominant factor that affects the transmission and control of contaminants in closed rooms is the path between the contaminant source and exhaust. D'Orzio et al. [37] investigated the COVID-19 spread in university buildings, extending the analysis from a single room to a larger, more realistic, indoor environment. The authors used an agent-based simulation, relying on historical records, and obtained that standard air ventilation is not able to constrain pandemic outbreaks such as the COVID-19 pandemic without the implementation of additional PIPs, such as masks. However, the authors further suggested that a higher level of ventilation will allow less reliance on masks.

Naturally, controlling a pandemic using any air ventilation solution without explicitly considering its associated costs could bring about an unwarranted economical and ecological burden. On the other hand, considering only the latter would likely entail poor pandemic control as well. To the best of our knowledge, to date, no study has examined this multi-faceted challenge of controlling pandemic spread indoors using air ventilation solutions while accounting for the potentially significant energetic costs associated with them. We present the Pareto front of using three standard air ventilation solutions: open window, air-conditioning and a pedestal fan, which we study across three common room types: inpatient room, classroom, and conference room. The Pareto front, which consists of all Pareto efficient solutions, allows us to explore the trade-off within this set and identify potential inefficient solutions and configurations which rationally should be avoided. Thus, the novelty of the proposed work lies in the investigation of the energy consumption and pandemic spread control of several (well-distributed) air ventilation solution in several types of rooms using sensors and in silico pandemic spread simulation.

The paper is organized as follows: Section 2 describes the study design. Section 3 reports the results obtained in this study which are then discussed in Section 4. Future work directions are highlighted as well.

2. Methods

In this study, we consider three standard air ventilation solutions: open window, an air-conditioning system and a pedestal fan which are studied across three room types: inpatient room, classroom and conference room. Overall, nine air ventilation to room type crossovers are considered in this work. We follow a simple three-phased study design consisted of in-the-field data collection, using a set of sensors, pandemic spread simulation, and epidemiological-energy Pareto-front analysis, as schematically presented in Figure 1. Each phase is discussed in detail below.

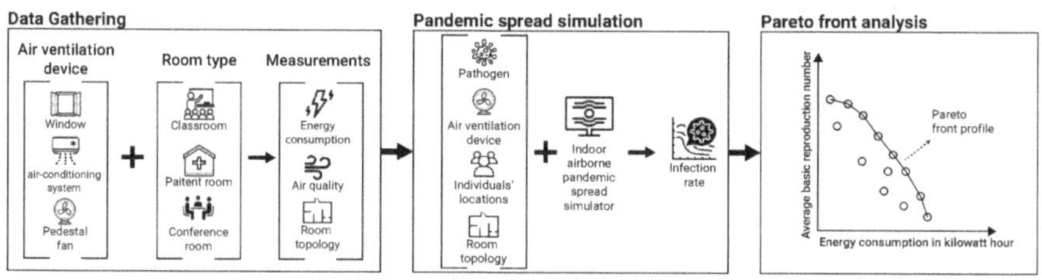

Figure 1. A schematic view of the research's design, divided into data gathering, pandemic spread simulation, and epidemiological-energy Pareto front analysis. During the data gathering phase, we used several air quality and ventilation measurements alongside energy consumption to collect the required data for the pandemic spread analysis. Using the collected data, we estimate the average pandemic spread reduction using each ventilation solution via simulation. Finally, we perform a Pareto front analysis to study the trade-off between pandemic control and energy consumption and identify possible inefficient configurations.

2.1. Data Collection

We first collected key measures for each of the nine crossovers in question. Starting with the room types, in order to ensure the credibility of our measures, three separate yet highly similar and closely located rooms were selected as representative for each type. For inpatient rooms, we used three 3 by 6 Meters inpatient rooms from one of the internal wards of Sheba Hospital (Ramat-Gan, Israel). Per Israeli regulations, all three inpatient rooms were occupied by 2 to 4 patients. Three roughly 8 by 8 Meters elementary school classrooms were further selected (Tel-Aviv, Israel) as representatives, each occupied by 31 to 36 pupils and a teacher. For conference rooms, we used three rectangle-shaped rooms from a modern office building (Ramat-Gan, Israel). The rooms slightly varied in size from 30 m^2 to 40 m^2 and were occupied by 6 to 11 individuals. All rooms were scanned using a 3D LiDAR-based scanning app. Namely, the *RoomScan LiDAR* application on an iPhone 12 device, which captured the room's topology and occupants' locations [38].

Considering the air ventilation solutions, each of the nine rooms had a window to the outdoors with a surface area of 6 by 11 m. The window was either closed (i.e., inactive) or completely open (i.e., active). Each room further had a modern, high quality and well-maintained air-conditioning system which was directed roughly to the center of the room. Since we are not interested in artificially changing the room's temperature (see further discussion in Section 4), when activated, the air-conditioning system was set to the room's current temperature and the blower was set to either low, medium or high speed. Similarly, a high quality, out-of-the-box pedestal fan, roughly pointing to the center of the room, was set to either low, medium or high speed when activated.

The key measures in our data collection, which are essential for simulating a pandemic spread as discussed next, are the room's temperature, humidity, air quality, and associated energy consumption over time. The room temperature, humidity and air quality were monitored using the *Air Quality Monitor/Recorder, model PM-1064SD* device mounted at the center of the room. The energy consumption over time was recorded by a *Poniie-PN2000* device which was directly connected to the power supply.

The data collection in all room types were conducted in parallel in August 2022 in order to avoid external confounds. Before any air ventilation solution was activated, the air quality was sampled over 30 min in order to capture a baseline. Next, one of the systems was activated for 90 min, followed by a 60 min break, thus allowing the room to roughly return to the baseline condition. The process is repeated three time in total, one for each of the examined systems. The order at which the systems were tested was chosen at random. During the measurements, the occupants were asked to behave normally, yet avoid leaving the room or opening doors or windows. During the data recording, little to no movement by the occupants was detected. This may be attributed the functionality of the selected room types and population (i.e., hospitalized patients in a hospital room) which, unlike the occupants of a restaurant or an airport, present more static behavior.

2.2. Pandemic Spread Simulation

Measuring the pandemic spread directly is impossible as it would involve the intentional infection of individuals with a pathogen. Instead, we used a computer simulation to estimate the pandemic spread over time from the measured data collected by our sensors [39]. Specifically, we used the standard model proposed by [34]. This model uses a spatio-temporal simulation approach combining agent-based simulation [40] with breathing dynamics and a Susceptible-Exposed-Infected (SEI) epidemiological model [41] in addition to a Computational Fluid Dynamics (CFD) [42] for the air movement in a room. In particular, the model accepts a spatial configuration in the form of the room's topology, the average locations of the individuals in the room, and the location and average ventilation rate of an air ventilation solution. In addition, the model accepts a temporal configuration in the form of the room's temperature, and the pathogen's basic infection and recovery rates. We provide the model with the room's topology using the aforementioned LiDAR-based scanning. Moreover, the average locations of the individuals in the room were introduced manually based on the sitting/laying locations the individuals spent most of their time in the room. The average ventilation rate of the air ventilation solution is fitted to the airspeed data obtained during the experiments, using the gradient descent optimization algorithm [43] while taking into consideration the relative location of the device in the room and its average air suction force as reported by the manufacturer. For the epidemiological parameters required by the model, we used the COVID-19 pathogen (the *Alpha* variant) as it is the most recorded pandemic in history [44]. Specifically, we follow the values proposed by [34]. The temperature over time is provided to the model as well based on our measurements.

The model adopted in this work (which based on [34]) is constructed around three main components: a population of individuals, an environment (room) representing a three-dimensional space, in which the population is located, and air flow dynamics. The simulation is working in discrete steps over time. Initially, the simulation is used with the space's geometry as obtained from the LiDAR scannings in the form of a dot cloud. In addition, the individuals' locations are set manually, assuming they present little to no movement in the environment (as is the case in our data). Specifically, the location of the individuals' faces are set to be 1.2 m above the floor and oriented approximately according to where they were looking. The simulation is assuming the individuals are breathing normally. We adopted the breathing dynamics as suggested in the original model. Afterwards, in each step in time, five actions take place: First, based on the individuals' stage in the breathing cycle, a force is allocated to the location in which their faces are. In particular, infected individuals generate pathogen particles into the air. In a similar manner, the average air flow of each air ventilation solution is treated as a constant air flow force. Second, an adaptive mesh for the CFD model is computed [45]. Third, the air movement in a single step in time is computed using the RANS method [46] for the CFD model. Fourth, the number of pathogen particles that enter individuals' bodies are updated. Finally, the epidemiological state of each individual in the population is updated based on the number of pathogen particles in the individuals' bodies and their current epidemiological state, following the classic SEI model. This process repeats for $T = 54{,}000$ time steps in order to obtain 10 simulations steps each second for the 90 min of experiment. This value is aimed at balancing the simulation's accuracy, that increases as the number of steps per second increases, and the computational demand that increases as well.

2.3. *Pareto front Analysis*

We computed the Pareto front of the energy consumption and pandemic spread reduction for each solution in question and room type using Matlab's *multiobjective optimization* tool (version 2020a) [47] with the average basic reproduction number (R_0) [48] as the epidemiological signal and the average energy consumption of the implemented solution across the three samples as the energy consumption signal. In particular, we use the "gamultiobj" library that implementing the genetic algorithm [49,50] that can search for a globally optimal solution under given constrains as required to obtain the Pareto frontier. The statistical significance level was set at ($p < 0.05$).

3. Results

Figure 2 displays the pandemic spread and energy consumption associated with each of the examined air ventilation solutions and room types where the x-axis is the energy consumption in kilowatt hour ([10^3 wh]) and the y-axis is the average basic reproduction number ($E[R_0]$). Each point indicates a 90-minutes experiment. The Pareto front is identified and highlighted in orange.

Starting with the open window condition, it significantly reduces the pandemic spread compared to the baseline of no air ventilation at all. In that respect, our simulation replicates a well-known result [51]. Naturally, open window has no associated energetic cost and thus it is depicted in Figure 2 as a single point along the dashed horizontal line. Interestingly, using the pedestal fan with a low fan speed is dominated by the open window condition for all three room types and thus it is not part of the Pareto front. Similarly, for all three room types, using the air-conditioning system with a low blower speed is dominated by the pedestal fan with a high fan speed, thus it is not part of the front either. The four remaining combinations (pedestal fan with either medium or high fan speed and air-conditioning with either a medium or high blower speed) are, in fact, efficient, with the fan-based outcomes presenting significantly lower energy consumption and higher pandemic spread compared to air-conditioning. In other words, for all three room types, the Pareto front consists of all three examined air ventilation solutions but not all their respective configurations.

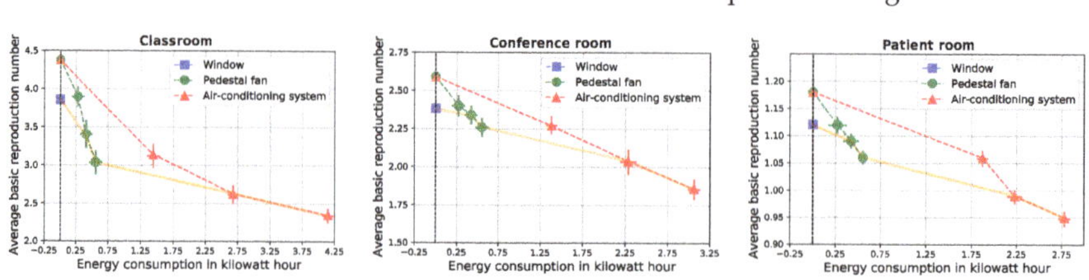

Figure 2. The energetic-epidemiological profiles for open window (blue square), pedestal fan (green dots), and air-conditioning system (red triangles) across three room types. The results shown as average ± standard deviation over 90 min. The x-axis indicates the average energy consumption in kilowatt hour of the air ventilation solution and the y-axis indicates the average basic reproduction number. The dashed black vertical line indicates no energy consumption. The solid orange line is the Pareto front.

4. Discussion

For all examined room types, the Pareto front consists of all three examined air ventilation solutions. By itself, this result means that none of the examined solutions should be a priori preferred or excluded from consideration, as at least one of its configurations is efficient. However, the results further demonstrate several significantly inefficient configurations as well. For example, using an air-conditioning system with low speed blower, one is expected to waste 2.5 to 4 times more energy compared to a pedestal fan at high fan speed, while resulting in even higher pandemic spread. Obviously, inefficient configurations are best to be avoided.

Deciding which of the efficient air ventilation configurations to implement in practice is a personalized task, one which depends on the decision-maker's preference, budget and various other human factors (see [52] for an overview). Nevertheless, air-conditioning systems seem to be efficient only at very high energetic costs. Given that air-conditioning systems are also expensive and operationally challenging to install and modify in existing buildings, fans seem to offer a reasonable alternative. In many cases, and especially ones with limited budget, we believe that fans can strike a more desirable compromise between energy consumption and pandemic control. However, in cases where altering an existing air ventilation solution is challenging or impossible, policymakers can still take advantage of the identified trade-off profile between pandemic control and energy consumption of the given air ventilation solution to make a more informed decision. Taken jointly, the proposed work provides a first analysis of the trade-off between energy consumption and pandemic spread control for well-distributed air ventilation solutions across important room types. A natural conclusion one can draw from our results is that, while implementing some PIPs such as social distancing is laborious and restrictive, implementing air ventilation solutions to control for pandemic spread is relatively accessible, easy to monitor, and widely distributed even before an airborne pandemic begins.

It is important to note that our study does not consider additional factors which may influence the decision to adopt and configure an air ventilation solution in a certain way (see [53,54] for more details). For example, thermal comfort may play an important role as some may be more or less willing than others to reduce their energy consumption at the expense of some thermal discomfort [55]. Since thermal comfort is extremely hard to obtain, and highly personal in nature, we leave its investigation for future work. It is important to note that thermal comfort is highly interwoven with the energy cost and the pandemic spread. Specifically, the environmental conditions in a room such as temperature and humidity, which are accounted for in our simulation (see Section 2.2), have a direct biophysical influence on the pathogen's particles and the pandemic spread [56]. As such, in contrast to the natural negative association between energy consumption and pandemic control identified here, there are presumably more complex and non-trivial connections that link the thermal comfort to the energy consumption and pandemic control which offer a fruitful avenue for future work. In addition, since modern pandemics show high mutation rate [57,58], a natural extension of this work is to evaluate the effectiveness of the examined air ventilation solutions on multi-strain pandemic settings by replacing the SEI model used in the simulation with multi-strain epidemiological models such as [59–61]. Moreover, as room-level air ventilation policy can be implemented in many cases, an additional promising research direction would be to find the optimal air ventilation type, location and configuration for each room given its topology, occupants, use patterns, etc. One can obtain a dynamic and room-type tailored policy using relatively cheap and easy to install sensors, such as those used in this work, thus using the proposed methodology seems to be appropriate [18,62,63].

Author Contributions: Formal analysis and investigation, coding, and experiment conduction were performed by A.A. Formal analysis and investigation, supervision, and manuscript editing were performed by A.R. Conceptualization, formal analysis and investigation, coding, supervision, visualization, and original draft were performed by T.L. All authors have read and agreed to the published version of the manuscript.

Funding: This research received no external funding.

Institutional Review Board Statement: Not applicable.

Informed Consent Statement: Written informed consent has been obtained from the patient(s) to publish this paper.

Data Availability Statement: The data that support the findings of this study are available on request from the corresponding author.

Acknowledgments: The authors wish to thank Guy Masad for helping with the data gathering and organization of the experiments.

Conflicts of Interest: The authors declare no conflict of interest.

References

1. Conti, A.A. Historical and methodological highlights of quarantine measures: From ancient plague epidemics to current coronavirus disease (COVID-19) pandemic. *Acta Bio-Medica Atenei Parm.* **2020**, *91*, 226–229.
2. Snacken, R.; Kendal, A.P.; Haaheim, L.R.; Wood, J.M. The next influenza pandemic: Lessons from Hong Kong, 1997. *Emerg. Infect. Dis.* **1999**, *5*, 195–203. [CrossRef] [PubMed]
3. Piot, P.; Bartos, M.; Ghys, P.D.; Walker, N.; Schwartkander, B. The global impact of HIV/AIDS. *Nature* **2001**, *410*, 968–973. [CrossRef] [PubMed]
4. Wu, T.; Perrings, C.; Kinzig, A.; Collins, J.P.; Minteer, B.A.; Daszak, P. Economic growth, urbanization, globalization, and the risks of emerging infectious diseases in China: A review. *Ambio* **2017**, *46*, 18–29. [CrossRef]
5. Brodeur, A.; Gray, D.; Islam, A.; Bhuiyan, S. A Literature Review of the Economics of COVID-19. IZA Discussion Paper No. 13411. 2020. Available online: https://ssrn.com/abstract=3636640 (accessed on 1 September 2022).
6. Lederberg, J. Medical Science, Infectious Disease, and the Unity of Humankind. *JAMA* **1988**, *260*, 684–685. [CrossRef]
7. Enright, J.; Kao, R.R. Epidemics on dynamic networks. *Epidemics* **2018**, *24*, 88–97. [CrossRef]
8. Kingsley, D. *The Urbanization of the Human Population*; Routledge: London, UK, 2015.
9. Ronald, L. The Outlook for Population Growth. *Science* **2011**, *333*, 569–573.
10. Diffey, B.L. An overview analysis of the time people spend outdoors. *Br. J. Dermatol.* **2010**, *164*, 848–854. [CrossRef]
11. Ahn, K.U.; Park, C.S. Correlation between occupants and energy consumption. *Energy Build.* **2016**, *116*, 420–433. [CrossRef]
12. Fan, J.L.; Zhang, Y.J.; Wang, B. The impact of urbanization on residential energy consumption in China: An aggregated and disaggregated analysis. *Renew. Sustain. Energy Rev.* **2017**, *75*, 220–233. [CrossRef]
13. York, R. Demographic trends and energy consumption in European Union Nations, 1960–2025. *Soc. Sci. Res.* **2007**, *36*, 855–872. [CrossRef]
14. Munier, N. *Introduction to Sustainability*; Springer: New York, NY, USA, 2005.
15. The Sustainable Development Agenda-United Nations Sustainable Development. Available online: https://www.un.org/sustainabledevelopment/development-agenda/ (accessed on 1 September 2022).
16. Wang, X.; Si, C.; Gu, J.; Liu, W.; Qui, J.; Zhao, J. Electricity-consumption data reveals the economic impact and industry recovery during the pandemic. *Sci. Rep.* **2021**, *11*, 19960. [CrossRef] [PubMed]
17. Klemes, J.J.; Fan, Y.V.; Jiang, P. COVID-19 pandemic facilitating energy transition opportunities. *Int. J. Energy Res.* **2020**. [CrossRef]
18. Erisen, S. Real-Time Learning and Monitoring System in Fighting against SARS-CoV-2 in a Private Indoor Environment. *Sensors* **2022**, *22*, 7001. [CrossRef] [PubMed]
19. Acemoglu, D.; Chernozhukov, V.; Werning, I.; Whinston, M.D. *Optimal Targeted Lockdowns in a Multi-Group SIR Model*; Working Paper 27102; National Bureau of Economic Research: Cambridge, MA, USA, 2020.
20. Lazebnik, T.; Blumrosen, G. Advanced Multi-Mutation With Intervention Policies Pandemic Model. *IEEE Access* **2022**, *10*, 22769–22781. [CrossRef]
21. Li, T.; Liu, Y.; Man, L.; Qian, X.; Dai, S.Y. Mask or no mask for COVID-19: A public health and market study. *PLoS ONE* **2020**, *15*, e0237691. [CrossRef]
22. Desvars-Larrive, A.; Dervic, E.; Haug, N.; Niederkrotenthaler, T.; Chen, J.; Di Natale, A.; Lasser, J.; Gliga, D.S.; Roux, A.; Sorger, J.; et al. A structured open dataset of government interventions in response to COVID-19. *Sci. Data* **2020**, *7*, 285. [CrossRef]
23. Alexi, A.; Rosenfeld, A.; Lazebnik, T. A Security Games Inspired Approach for Distributed Controlling of Pandemic Spread. *arXiv* **2022**.

24. King, J.C.; Stoddard, J.J.; Gaglani, M.J.; More, K.A.; Magder, L.; McClure, E.; Rubin, J.D.; Englund, J.A.; Neuzil, K. Effectiveness of School-Based Influenza Vaccination. *N. Engl. J. Med.* **2006**, *355*, 2523–2532. [CrossRef]
25. Lazebnik, T.; Alexi, A. Comparison of pandemic intervention policies in several building types using heterogeneous population model. *Commun. Nonlinear Sci. Numer. Simul.* **2022**, *107*, 106176. [CrossRef]
26. Wei, J.; Li, Y. Airborne spread of infectious agents in the indoor environment. *Am. J. Infect. Control* **2016**, *44*, S102–S108. [CrossRef] [PubMed]
27. Peng, Z.; Rogas, P.A.L.; Kropff, E.; Bahnfleth, W.; Buonanno, G.; Dancer, S.J.; Kurnitski, J.; Li, Y.; Loomans, M.G.L.C.; Marr, L.C.; et al. Practical Indicators for Risk of Airborne Transmission in Shared Indoor Environments and Their Application to COVID-19 Outbreaks. *Environ. Sci. Technol.* **2020**, *56*, 1125–1137. [CrossRef] [PubMed]
28. Riley, E.C.; Murphy, G.; Riley, R.L. Airborne Spread of Measles in a Suburban Elementary School. *Am. J. Epidemiol.* **1978**, *107*, 421–432. [CrossRef] [PubMed]
29. Fumanelli, L.; Ajelli, M.; Manfredi, P.; Vespignani, A.; Merler, S. Inferring the structure of social contacts from demographic data in the analysis of infectious diseases spread. *PLoS Comput. Biol.* **2012**, *8*, e1002673. [CrossRef] [PubMed]
30. Kozyreff, G. Hospitalization dynamics during the first COVID-19 pandemic wave: SIR modelling compared to Belgium, France, Italy, Switzerland and New York City data. *Infect. Dis. Model.* **2021**, *6*, 398–404. [CrossRef]
31. Zhao, S.; Stone, L.; Gao, D.; Musa, S.S.; Chong, M.K.C.; He, D.; Wang, M.H. Imitation dynamics in the mitigation of the novel coronavirus disease (COVID-19) outbreak in Wuhan, China from 2019 to 2020. *Ann. Transnatl. Med.* **2020**, *8*, 448. [CrossRef]
32. Didier, B.; Elspeth, G.; Elif, K.M. Obedience in times of COVID-19 pandemics: A renewed governmentality of unease? *Interdiscip. J. Curr. Aff.* **2021**, *11*, 471–489. [CrossRef]
33. Nivette, A.; Ribeaud, D.; Murray, A.; Steihoff, A.; Bechtiger, L.; Hepp, U.; Shanahan, L.; Eisner, M. Non-compliance with COVID-19-related public health measures among young adults in Switzerland: Insights from a longitudinal cohort study. *Soc. Sci. Med.* **2021**, *268*, 113370. [CrossRef]
34. Lazebnik, T.; Alexi, A. High Resolution Spatio-Temporal Model for Room-Level Airborne Pandemic Spread. *arXiv* **2022**, arXiv:2210.03431.
35. Dai, H.; Zhao, B. Association of the infection probability of COVID-19 with ventilation rates in confined spaces. *Build. Simul.* **2020**, *13*, 1321–1327. [CrossRef]
36. Memarzadeh, F.; Xu, W. Role of air changes per hour (ACH) in possible transmission of airborne infections. *Build. Simul.* **2012**, *5*, 15–28. [CrossRef] [PubMed]
37. D'Orazio, M.; Bernardini, G.; Quagliarini, E. A probabilistic model to evaluate the effectiveness of main solutions to COVID-19 spreading in university buildings according to proximity and time-based consolidated criteria. *Build. Simul.* **2021**, *14*, 1795–1809. [CrossRef] [PubMed]
38. Jie, S.; Sampath, A. Urban DEM Generation from Raw Lidar Data. *Photogramm. Eng. Remote Sens.* **2005**, *71*, 217–226.
39. Ahn, S.Y.; Kim, M.; Bae, J.E.; Bang, I.S.; Lee, S.W. Reliability of the In Silico Prediction Approach to In Vitro Evaluation of Bacterial Toxicity. *Sensors* **2022**, *22*, 6557. [CrossRef]
40. Raberto, M.; Cincotti, S.; Focardi, S.M.; Marchesi, M. Agent-based simulation of a financial market. *Phys. A Stat. Mech. Its Appl.* **2001**, *299*, 319–327. [CrossRef]
41. Paeng, S.H.; Lee, J. Continuous and discrete SIR-models with spatial distributions. *J. Math. Biol.* **2017**, *74*, 1709–1727. [CrossRef]
42. Yu, Y.; Megri, A.C.; Jiang, S. A review of the development of airflow models used in building load calculation and energy simulation. *Build. Simul.* **2019**, *12*, 347–363. [CrossRef]
43. Haskell, B.C. The method of steepest descent for non-linear minimization problems. *Quart. Appl. Math.* **1944**, *2*, 258–261.
44. Piret, J.; Boivin, G. Pandemics Throughout History. *Front. Microbiol.* **2021**, *11*, 631736. [CrossRef]
45. Zheng, J.; Wu, X.; Fang, F.; Li, J.; Wang, Z.; Xiao, H.; Zhu, J.; Pain, C.; Linden, P.; Xiang, B. Numerical study of COVID-19 spatial–temporal spreading in London. *Phys. Fluids* **2021**, *33*, E10. [CrossRef]
46. Su, F.; Kinnas, S.A.; Jukola, H. Application of a BEM/RANS Interactive Method to Contra-Rotating Propellers. In Proceedings of the Fifth International Symposium on Marine Propulsion, Espoo, Finland, 12–15 June 2017.
47. Tian, Y.; Cheng, R.; Zhang, X.; Jin, Y. PlatEMO: A MATLAB Platform for Evolutionary Multi-Objective Optimization [Educational Forum]. *IEEE Comput. Intell. Mag.* **2017**, *12*, 73–87. [CrossRef]
48. Breda, D.; Florian, F.; Ripoll, J.; Vermiglio, R. Efficient numerical computation of the basic reproduction number for structured populations. *J. Comput. Appl. Math.* **2021**, *384*, 113165. [CrossRef] [PubMed]
49. Holland, J.H. Genetic Algorithms. *Sci. Am.* **1992**, *267*, 66–73. [CrossRef]
50. Bo, L.; Rein, L. Comparison of the Luus–Jaakola optimization procedure and the genetic algorithm. *Eng. Optim.* **2005**, *37*, 381–396.
51. Ahmadzadeh, M.; Shams, M. Passenger exposure to respiratory aerosols in a train cabin: Effects of window, injection source, output flow location. *Sustain. Cities Soc.* **2021**, *75*, 103280. [CrossRef]
52. Rosenfeld, A.; Kraus, S. Predicting human decision-making: From prediction to action. *Synth. Lect. Artif. Intell. Mach. Learn.* **2018**, *12*, 1–150.
53. Shang, W.; Deng, L.; Liu, J. A Novel Air-Door Opening and Closing Identification Algorithm Using a Single Wind-Velocity Sensor. *Sensors* **2022**, *22*, 6837. [CrossRef]
54. Hattori, S.; Iwamatsu, T.; Miura, T.; Tsutsumi, F.; Tanaka, N. Investigation of Indoor Air Quality in Residential Buildings by Measuring CO_2 Concentration and a Questionnaire Survey. *Sensors* **2022**, *22*, 7331. [CrossRef]

55. Azaria, A.; Rosenfeld, A.; Kraus, S.; Goldman, C.V.; Tsimhoni, O. Advice provision for energy saving in automobile climate-control system. *AI Mag.* **2015**, *36*, 61–72. [CrossRef]
56. Yao, M.; Zhang, L.; Ma, J.; Zhou, L. On airborne transmission and control of SARS-CoV-2. *Sci. Total Environ.* **2020**, *731*, 139178. [CrossRef]
57. Marquioni, V.M.; de Aguiar, M.A.M. Modeling neutral viral mutations in the spread of SARS-CoV-2 epidemics. *PLoS ONE* **2021**, *16*, e0255438. [CrossRef] [PubMed]
58. Khyar, O.; Allali, K. Global dynamics of a multi-strain SEIR epidemic model with general incidence rates: Application to COVID-19 pandemic. *Nonlinear Dyn.* **2020**, *102*, 489–509. [CrossRef] [PubMed]
59. Lazebnik, T.; Bunimovich-Mendrazitsky, S. Generic approach for mathematical model of multi-strain pandemics. *PLoS ONE* **2022**, *17*, e0260683. [CrossRef] [PubMed]
60. Dang, Y.X.; Li, X.Z.; Martcheva, M. Competitive exclusion in a multi-strain immuno-epidemiological influenza model with environmental transmission. *J. Biol. Dyn.* **2016**, *10*, 416-4-56. [CrossRef]
61. Minayev, P.; Ferguson, N. Improving the realism of deterministic multi-strain models: Implications for modelling influenza A. *J. R. Soc. Interface* **2008**, *6*, 509–518. [CrossRef]
62. Aguilar, A.J.; de la Hoz-Torres, M.L.; Martínez-Aires, M.D.; Ruiz, D.P. Monitoring and Assessment of Indoor Environmental Conditions after the Implementation of COVID-19-Based Ventilation Strategies in an Educational Building in Southern Spain. *Sensors* **2021**, *21*, 7223. [CrossRef]
63. Perez, A.O.; Bierer, B.; Scholz, L.; Wöllenstein, J.; Palzer, S. A Wireless Gas Sensor Network to Monitor Indoor Environmental Quality in Schools. *Sensors* **2018**, *18*, 4345. [CrossRef]

sensors

MDPI

Article

A Sensor and Machine Learning-Based Sensory Management Recommendation System for Children with Autism Spectrum Disorders †

Lingling Deng [1,*] and Prapa Rattadilok [2]

1 School of Computer Science, University of Nottingham Ningbo China, Ningbo 315100, China
2 University of Hertfordshire, Hatfield AL10 9AB, UK; p.rattadilok@herts.ac.uk
* Correspondence: lingling.deng2@nottingham.edu.cn

† This paper is an extended version of our paper published in Deng, L.; Rattadilok, P. A Machine Learning-Based Monitoring System for Attention and Stress Detection for Children with Autism Spectrum Disorders. In Proceedings of the ICIMH 2021: 2021 the 3rd International Conference on Intelligent Medicine and Health, Macau, China, 13–15 August 2021.

Abstract: Sensory processing issues are one of the most common issues observed in autism spectrum disorders (ASD). Technologies that could address the issue serve a more and more important role in interventions for ASD individuals nowadays. In this study, a sensory management recommendation system was developed and tested to help ASD children deal with atypical sensory responses in class. The system employed sensor fusion and machine learning techniques to identify distractions, anxious situations, and the potential causes of these in the surroundings. Another novelty of the system included a sensory management strategy making a module based on fuzzy logic, which generated alerts to inform teachers and caregivers about children's states and risky environmental factors. Sensory management strategies were recommended to help improve children's attention or calm children down. The evaluation results suggested that the use of the system had a positive impact on children's performance and its design was user-friendly. The sensory management recommendation system could work as an intelligent companion for ASD children that helps with their in-class performance by recommending management strategies in relation to the real-time information about the children's environment.

Keywords: assistive technology; autism spectrum disorders; sensors; wearables; sensory management; machine learning; fuzzy logic

1. Introduction

Autism spectrum disorders (ASDs) refer to a group of neurodevelopmental disabilities that affect an individual's social interactions, communication, interests, and behavior [1]. Sensory processing issue is one of the most common issues observed in individuals with ASD [2]. As many as 90% of ASD individuals may have experienced atypical sensory responses to auditory, visual, touching, tasting, and smelling stimuli [3]. According to the Diagnostic and Statistical Manual of Mental Disorders [2], atypical sensory responses in ASD involve hyper- or hypo-sensitivity to sensory input. ASD individuals who are hypo-sensitive may fail to notice sensory stimuli which typically developing (TD) people can easily detect, resulting in behavioral outcomes such as having difficulty paying attention. Conversely, those who are hyper-sensitive are likely to experience distress to sensory stimuli [4]. Therefore, for ASD individuals with atypical sensory responses, any of the senses at any random time could become hyper-sensitive or hypo-sensitive, which could further trigger discomfort, stress, or distraction [5,6]. ASD individuals could have more problems in sensory management, such as attention and stress management, than their TD peers when processing everyday sensory information [4,7,8]. The sensory issue presents a significant challenge for ASD children and their caregivers, especially within a classroom setting, as it affects their learning efficiency and behavioral control.

The rapid development of information technology has encouraged many researchers to explore technology-based interventions that can be used in daily life for ASD children with atypical sensory responses. Some causes of the above-mentioned discomfort, stress, or distraction in ASD individuals could be found as being part of an 'unfriendly environment' [9] (page 92), such as noisy or sudden sounds, bright or dark lights, and warm or cold weather [10,11]. The identification of a triggering event for atypical sensory responses in ASD by technologies could be an important aspect to optimize the surrounding causes related to explosive behaviors in ASD. Traditionally, to measure sensory responses, various manual methods, including rating scales and questionnaires, were used in some systems for predicting the comfort level based on sensory inputs or triggers [8,12]. For instance, Mauro et al. [13] proposed a personalized recommendation system to predict points of interest for individuals with ASD using a Top-N model. Their system used a self-defined sensory profiling questionnaire to obtain information about sensory aversion and preference of people with ASD, generating suggestions for places that are expected to be comfortable and safe for the users. However, even in a safe place, unpleasant sensory factors around the child could appear. To provide effective intervention, exact and real-time identification of sensory causes could be required, which was difficult to achieve merely by manual methods [9].

Recent advancements in sensing technologies have made the sensors work similarly to human senses [9]. Developments in sensor fusion and artificial intelligence have driven the integration of multiple sensors and machine learning for capturing ASD individuals' real-time sensation in different environments. For instance, a stress-monitoring system for people with ASD was created by Tomczak et al. [14] using low-power wearable sensors. The detection module of the system was built using a heuristic rule-based model. Heart rate, skin conductivity, body temperature, and hand movements were the main indicators to identify a person's stress responses. Coronato et al. [15] used wearable accelerometers to create a situation-aware system that can identify abnormal motor behaviors. The employment of a neural network model led to an accuracy close to 92% on anomalous gestures of an individual with ASD. Additionally, the internet-connected smart devices had also demonstrated their potential by reducing the reliance on dimensions for monitoring systems [9]. Sula et al. [16] designed a system based on the Internet-of-Things (IoT) that could monitor the body movement and the sensory environment of children with ASD. It could send information about the children's state in real time to therapists and caregivers using Peer to Peer (P2P) technology. Khullar et al. [9] designed another IoT-based system to detect the environmental information and process the information using fuzzy logic. The system was also able to generate alerts to caregivers of children with ASD and provide auditory-video feedback to calm down children with ASD. Schmidt et al. [17] designed a spherical video-based virtual reality (SVVR) mobile application to facilitate social skill training for individuals with ASD. Their study evaluated user experience, usability of application, and utilization feasibility, and found an overall positive user experience. On the other hand, robot-assisted therapies were commonly seen in emerging ASD studies. Robots were always built with various sensors, through which robots could get the necessary signals and interact as a human-like friend with ASD children [18]. For example, an NAO robot was a programmable robot equipped with cameras, microphones, and tactile sensors [19]. In a recent study conducted by Ali et al. [20], the NAO robot was programmed to provide visual, auditory, and tactile stimuli to engage children with ASD. Its camera monitored and recorded children's joint attention. KASPAR is another child-sized robot equipped with tactile sensors. Costa et al. [21] used the KASPAR robot to help teach children with ASD appropriate tactile interactions. However, these robots did not have body-worn sensors, so children's atypical sensory responses were manually analyzed based on observations in previous robotic studies.

It has been found that previous studies have used a range of methods, including a sensory profiling questionnaire, sensors, machine learning techniques, or IoT, to measure sensory responses and associated behavioral outputs, implying the feasibility of combining several off-the-shelf technologies to obtain comprehensive information for the development of a real-time monitoring system for children with ASD. Therefore, the goal of this study was to design a sensory management recommendation system (hereinafter referred as SMRS) specifically for children with ASD, using phone devices and sensors. The SMRS could understand a user's sensory processing pattern, record and analyze sensory input, and act as a 'specialist' companion to recommend sensory management strategies in real time. The SMRS was implemented on a mobile phone because the growing usage of phones and sensors provided low-cost platforms for people to track information and access support [22]. Furthermore, it was suggested that the stigma associated with the disability could be reduced by using mobile phones, given that mobile phones could successfully address the challenges of public exposure [22,23]. The affordability and portability of the SMRS may make it easier for families of ASD children to access, in particular for those from relatively low-income countries. The SMRS in this study was evolved from the prototype which was tested successfully in researchers' previous pilot study [4]. This study extended the previous paper by describing the full infrastructure of the SMRS and evaluating its effectiveness.

2. Materials and Methods

2.1. System Infrastructure

The proposed sensory management recommendation system (SMRS) was capable of collecting a user's (child with ASD) sensory profile and allowing real-time data acquisition from the surroundings of the user through sensors. Firstly, the caregiver of the user should sign up for an account and complete a Sensory Profile of Children Three to Ten Years Caregiver Questionnaire [24], which is a standard sensory profiling tool that assesses the sensory processing pattern of a child. This questionnaire could elicit children's sensory preferences or limitations, and classify their sensory pattern under four quadrants based on Dunn's model of sensory processing [25]. Table 1 described the characteristics of individuals under the four quadrants. The measuring module of the SMRS collected physiological data such as skin conductivity, heart rate, and hand movements, and environmental data such as temperature, humidity, light exposure, and noise. The recording frequency was 1 hertz. Moreover, on the basis of acquired sensory information and machine learning models, the SMRS predicted the level of stress and attention of the child. The real-time data and predicted stress and attention level were further transmitted to the cloud server embedded with fuzzy logic controllers for sensory management strategy making. If the analyzed outcome was 'High Risk' (atypical sensory responses situation) and a sensory management recommendation was given, the alert could be initiated automatically by the SMRS. The message subscription allowed an alert to be sent to the caregivers or teachers via Short Message Service (SMS) regarding the occurrence of atypical situation (which was identified as 'High Risk' by fuzzy logic controllers) so that they could provide an intervention as suggested by the SMRS timely. The overall system infrastructure and working flow were presented in Figure 1.

Table 1. Characteristics of four sensory patterns [4].

Sensory Pattern	Characteristics
Low registration	Less likely to notice sensory input, may behave as passive or easy going.
Sensory seeking	Prone to add sensory events to daily life, may be very active or keep busy.
Sensory sensitivity	More likely to get distracted by sensory inputs, often show discomfort and sensitivity towards daily events.
Sensory avoiding	Prone to withdraw from overwhelming sensory stimulation, may be very ritualistic and rule-bound.

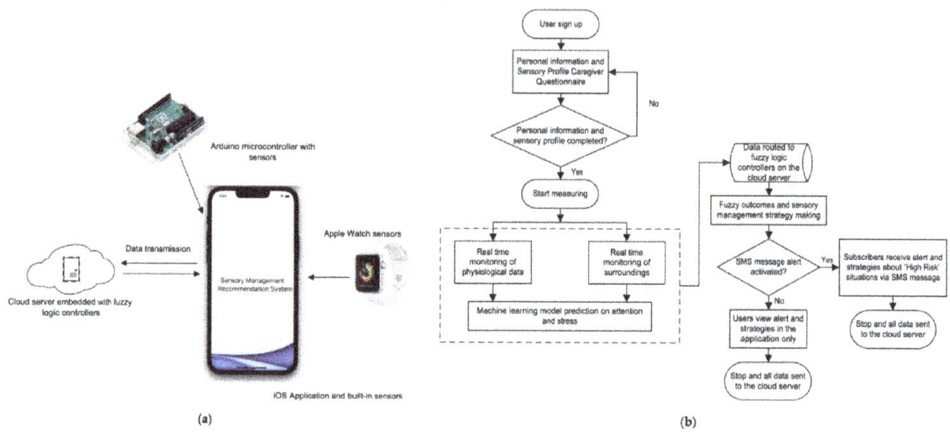

Figure 1. System infrastructure. (**a**) Data transmission network of the SMRS; (**b**) working flow of the SMRS.

2.2. Sensor Fusion and Data Management

To achieve proposed functions, the design of SMRS consisted of sensing devices, an iOS-based software application, and a cloud server for data management. iPhone [26], Apple Watch [27], and Arduino sensors [28] were the sensing devices used to record environmental and physiological data. The sensors and microcontrollers used in the SMRS are listed in Table 2, along with the purpose of each component. Figure 2a presents the circuit diagram of the Arduino board and Figure 2b is a prototype of the Arduino board. More technical details are presented as guidelines for sensor fusion and data management in the Supplementary Materials.

Table 2. Sensors and microcontrollers.

Sensor/Microcontroller	Unit	Purpose
Arduino UNO Rev3	N/A *	To fetch and transmit signal from sensors.
Apple Watch three-axis accelerometer	Sensor value	To identify the hand movements.
Apple Watch heart rate sensor	Beats per minute (bpm)	To measure heart rate.
DHT11 temperature and humidity sensor	Celsius (°C) for temperature, percentage (%) for humidity	To measure temperature and humidity level.
iPhone microphone	Decibel (dB)	To measure noise level.
iPhone barometer	Kilopascal (kPa)	To measure air pressure.
Light sensor (photoresistor)	Lux (lx)	To measure brightness level.
SEEED Grove Galvanic Skin Response (GSR) sensor	Sensor value	To detect skin conductivity.

* N/A: Not applicable.

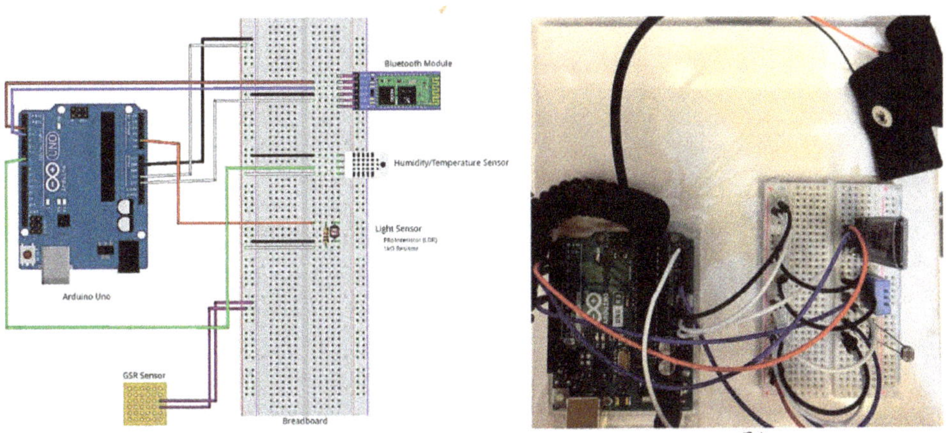

(a) **(b)**

Figure 2. Arduino UNO circuit connection. (**a**) Circuit diagram; (**b**) prototype of the Arduino board.

An iOS-based application was created to connect the Arduino UNO and gain access to the built-in sensors on the iPhone and Apple Watch. The application allowed the caregivers to select a language they preferred (Figure 3a), to complete the sensory profile caregiver questionnaire (Figure 3b), and to view the environmental/physiological changes around the ASD user (Figure 3c) and sensory management strategies.

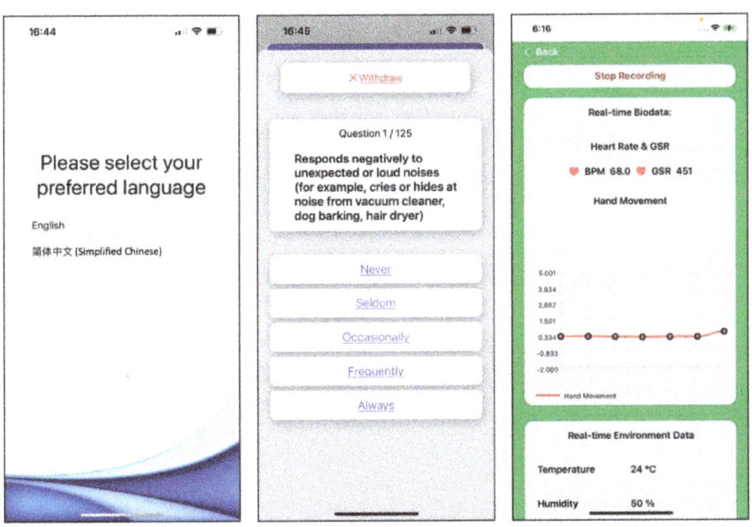

(a) **(b)** **(c)**

Figure 3. SMRS application interface. (**a**) Language selection; (**b**) sensory profile questionnaire; (**c**) sensor data visualization.

The cloud server for the SMRS was equipped with an Intel Xeon Processor, 4 GB RAM, and a Linux Ubuntu Operating System, which could handle multiple threads in parallel to receive data from different devices. The fuzzy logic controllers deployed on the cloud server could process the data received and return outputs to the SMRS. When the SMRS completed a recording session, sensor data were packed up with personal information (e.g., gender, date of birth, sensory profile) and sent to the cloud server so that researchers could access and manage the data accordingly.

2.3. Data Acquisition and Machine Learning Model Training

To acquire a training set of data with indicative labels for attention and stress detection, this study firstly used the SMRS prototype to record data in some children with ASD at a testing room of a rehabilitation center for several months.

Data acquisition was conducted with 35 children with ASD (mean age: 5.3; 29 boys, 6 girls, gender ratio: 4.83:1). Each child was told to complete 15 data collection sessions in different settings where environmental influences (i.e., temperature, noise, and brightness) were controlled during the session. In each session, three classical attention tasks (i.e., counting, picture matching, and drawing) for ASD children were given with audio instructions. Children's performances on tasks were rated by an accuracy score from 0 (0% correct) to 1 (100% correct). Detailed descriptions of these procedures can be found in [4]. Finally, 521 valid datasets were obtained. Each dataset recorded sensor data during one session for approximately 15 min at a frequency of 1 hertz. Two healthcare professionals (qualified ASD specialists) were invited as assessors to classify children's attention and stress levels at the end of tasks. They labelled the attention level of children as low or normal. Stress levels were labelled as low, moderate, and high. Since the data acquisition was long-term work, two healthcare professionals were not always available during that period. They successfully labelled 222 datasets. By combining task accuracy scores with classifications from professionals, researchers calculated the entropy and information gain [29] of the labelled sample to determine the best split points on attention and stress for all the 521 datasets. The calculation results showed that when employing the following classification, the data had the highest information gain values, indicating the best quality of the splits. A task accuracy score higher than 0.6 was considered to be within a normal level of attention, indicating that the child could generally pay attention to the task. A task accuracy score less than or equal to 0.6 was classified as the inverse of normal, indicating a low level of attention. Children relaxing in moderate environmental conditions were classified as having low stress, whereas performing tasks in moderate and extreme environmental conditions were classified as creating moderate stress and high stress, respectively. Features extracted as predictors for attention and stress detection were provided in Table 3.

Table 3. Extracted data features.

Category	Included Features
Environmental features	Temperature, noise, humidity, brightness, air pressure
Sensory profile features	Low registration, sensory seeking, sensory sensitivity, sensory avoiding
Physiological features	GSR, heart rate, watch accelerometer (mean absolute value of three axis)
Personal characteristics	Gender, age

Several machine learning models were implemented and compared for attention and stress detection. The investigated models included Logic Regression (LR), K Nearest Neighbor (KNN), Random Forest (RF), Artificial Neural Network (ANN), and Gradient Boosting Decision Tree (GBDT). The pre-processed data were split into the training and testing datasets by adopting 80:20 as the ratio of training:testing dataset. Five-fold cross-validation was performed with Grid Search on the training dataset to prevent overfitting and to obtain the best parameters for LR, KNN, RF, and GBDT. The best parameters evaluated by the mean cross-validation scores were used for the final model training. For ANN, loss functions with L2 regularization were used to optimize the model [30]. The performance of each model on the testing dataset was evaluated by accuracy and F1-score, which were computed by the following equations:

$$Accuracy = \frac{\text{number of correct predictions}}{\text{total number of predictions made}}, \quad (1)$$

$$F1 - score = 2 \times \frac{\text{precision} \times \text{recall}}{\text{precision} + \text{recall}}, \quad (2)$$

$$Precision = \frac{\text{true positive}}{\text{true positive} + \text{false positive}}, \quad (3)$$

$$Recall = \frac{\text{true positive}}{\text{true positive} + \text{false negative}}. \quad (4)$$

For attention detection, which was a binary classification, a weighted F1-score was calculated, while a macro F1-score was computed for stress detection, which was a multi-class problem [31]. Considering the machine learning model needed to be implemented on the mobile phone devices, the response time of a model was another factor that researchers investigated. All the models were processed on a laptop CPU and the inference time of each model was calculated and compared:

$$Inference\ time = \frac{\text{total time taken to calculate the outputs}}{\text{number of samples}}. \quad (5)$$

The accuracy, F1-score, and inference time on the testing data set of LR, KNN, RF, ANN, and GBDT models with optimal hyperparameters for stress and attention detection are presented in Table 4. It could be noticed that GBDT significantly outperformed the other models on attention detection with the highest accuracy (86.67%) and F1-score (0.8772). Machine learning models had overall better performance on stress detection than attention detection. The accuracies of RF, ANN, and GBDT models on stress detection were over 95%. RF had a higher accuracy (98.82%) and F1-score (0.9851) than ANN and GBDT. Most models could process an input within 0.1 millisecond (ms). Finally, two ensemble learning models, GBDT and RF, with the highest accuracy and generally short inference time, were chosen to be embedded into the SMRS for attention and stress detection, respectively [4].

Table 4. Model performance on attention and stress detection.

	Attention Detection			Stress Detection		
Model	**Accuracy (%)**	**Weighted F1**	**Inference Time (ms)**	**Accuracy (%)**	**Macro F1**	**Inference Time (ms)**
LR	65.71	0.6949	0.0052	65.30	0.5712	0.0013
KNN	81.90	0.8319	0.0291	93.92	0.9251	0.1041
RF	79.05	0.8000	0.0958	98.82	0.9851	0.0182
ANN	80.95	0.8246	0.0040	96.89	0.9592	0.0021
GBDT	86.67	0.8772	0.0046	98.50	0.9812	0.0366

2.4. Sensory Management Strategy Making

Real-time data collected by sensor fusion and predicted outcomes were further processed through the sensory management strategy making module where fuzzy logic controllers were implemented. The study employed fuzzy logic because it is a very classical and easily implemented method that imitates the human strategy-making mechanism [32]. It took the best decision for the given conditions based on some set of rules [33]. Before the fuzzy logic controllers could be used for the SMRS, all the inputs and fuzzy logic rules must be predefined. Therefore, researchers firstly gathered information about sensory management strategies through focus group consultations with 10 ASD specialists, combining with knowledge from ASD sensory toolkits [34,35]. Key environmental conditions that may trigger sensory management strategies included loud noises, bright or dark lights, and warm or cold room temperatures. As suggested by the focus group, the length of time that atypical sensory responses lasted was a factor that should be considered. The strategy for long-term and short-term atypical sensory responses could be different given that children with ASD have some degree of self-management ability. Secondly, in order to obtain more validated combinations of fuzzy logic rules and outcomes, researchers conducted a survey with 242 ASD specialists. Scenarios with risky conditions were shortlisted in the survey. Sensory management strategies obtained from focus group consultations and toolkits were listed as options. Survey results were finally interpreted into 63 fuzzy logic rules. Three independent fuzzy logic controllers were designed to process brightness, temperature, and noise stimuli in parallel. Each fuzzy logic controller contained 21 rules. A full list of rules is attached in Appendix A. Inputs to the fuzzy logic controllers included sensory stimuli (temperature/noise/brightness), duration of atypical sensory responses, and attention and stress levels. Trapezoidal and Gaussian membership functions were used to fuzzify the inputs. Figure 4 depicted the membership functions of the inputs and output.

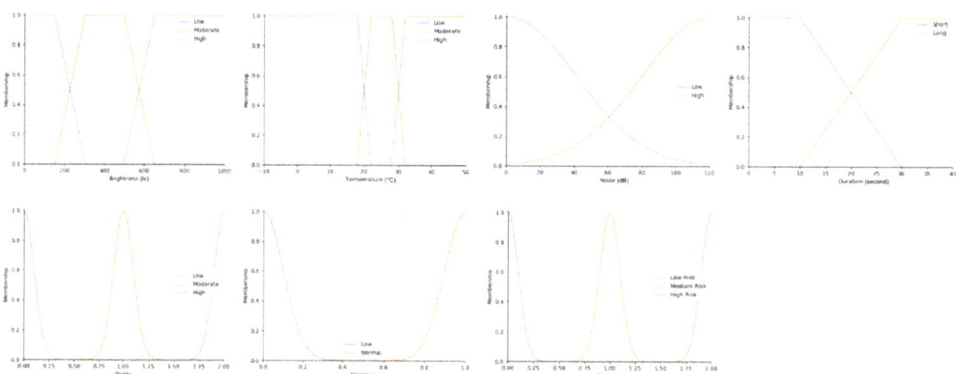

Figure 4. The graphs of the membership functions for the inputs and outcome.

The selection of the defuzzification method usually influenced overall performance of the fuzzy logic controllers [36]. This study opted for Largest of Maximum (LOM) method because it was more suitable for the general design of the fuzzy logic controllers. Researchers compared it with another most popular defuzzification method, which was Centroid method [37]. Two methods were tested on 21 different combinations of inputs. Centroid method only returned 79.4% of outcomes as expected in the testing because it usually led to a reasonable control action. To simplify, if there were two rules: 'IF Temperature is High AND Duration is Short, THEN Outcome is Low Risk', 'IF Temperature is High AND Duration is Long, THEN Outcome is High Risk', when the temperature was high, and the duration of atypical sensory responses was approaching long, the Centroid method averaged the two possible outputs to get the unwanted result 'Medium Risk'. In this study, it was more important to detect 'High Risk' accurately. Therefore, the LOM method, which selected the largest output value whose membership value reached the maximum [38], was used in this study. Expectedly, the LOM method yielded superior results by returning all outcomes accurately in the testing. Table 5 provides some examples of the tested inputs, outcomes based on the LOM method, and strategies recommended by the SMRS.

Table 5. Outcome responses of fuzzy logic controllers and strategy making module on different combinations of inputs.

Inputs				Outcome Responses	
Sensory Stimuli	**Attention**	**Stress**	**Duration (s)**	**Fuzzy Outcome**	**Recommended Strategy**
Brightness = 100 lx	Low	High	25	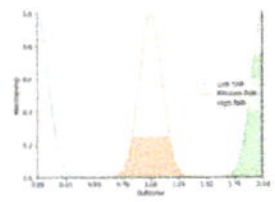	Brightness level is low. Enhance indoor brightness (e.g., draw the curtains open), use a phone to show pictures or videos that the child likes for comfort and attention.
Brightness = 400 lx	Normal	Low	40	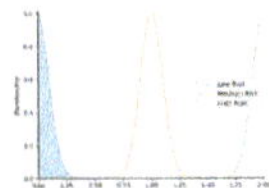	Brightness level is moderate. No impact.
Brightness = 750 lx	Normal	High	10	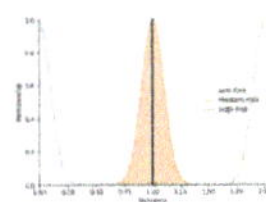	Brightness level is high. Reduce indoor brightness (e.g., draw the curtains). Keep observing.
Temperature = 15 °C	Normal	High	10	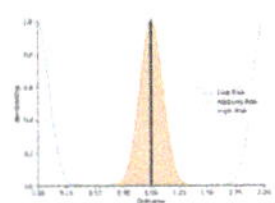	Temperature level is low. Enhance temperature level (e.g., turn up the air-conditioner). Keep observing.
Temperature = 26 °C	Normal	Moderate	25	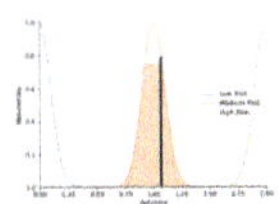	Temperature level is moderate. Keep observing.
Temperature = 32 °C	Low	High	40	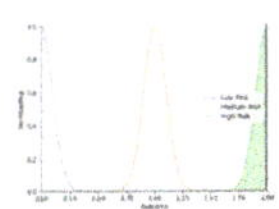	Temperature level is high. Reduce temperature level (e.g., turn on the fan). Provide some deep pressure (e.g., hugs or massage) input to child for comfort and attention.
Noise = 60 dB	Low	Low	25	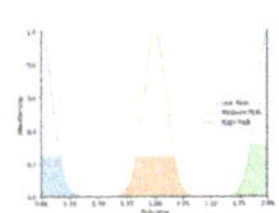	Noise level is moderate. Check other factors that may distract your child.
Noise = 70 dB	Normal	High	10	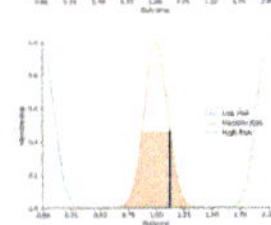	Noise level is moderate-high. Keep observing.
Noise = 80 dB	Low	Moderate	40	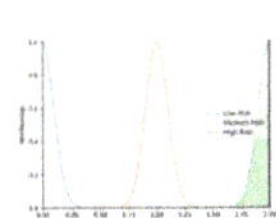	Noise level is high. Try to reduce loud (e.g., use noise-cancelling headphones or play calming music). Provide a fidget toy with texture that child likes for comfort and attention.

Fuzzy logic controllers were coded by the Python language and deployed on the cloud server. The SMRS application on the phone was able to receive the outcomes of fuzzy logic controllers. When the outcome was 'Low Risk', the SMRS would show 'No impact' (Figure 5a). When the outcome approached 'High Risk', then the SMRS identified the triggering sensory input and recommended a proper strategy. The SMRS allowed an automatic message alert of a 'High Risk' situation. The recommended strategy could be sent to a corresponding caregiver or teacher via SMS message if they activate the automatic message alert by entering their phone number (Figure 5b).

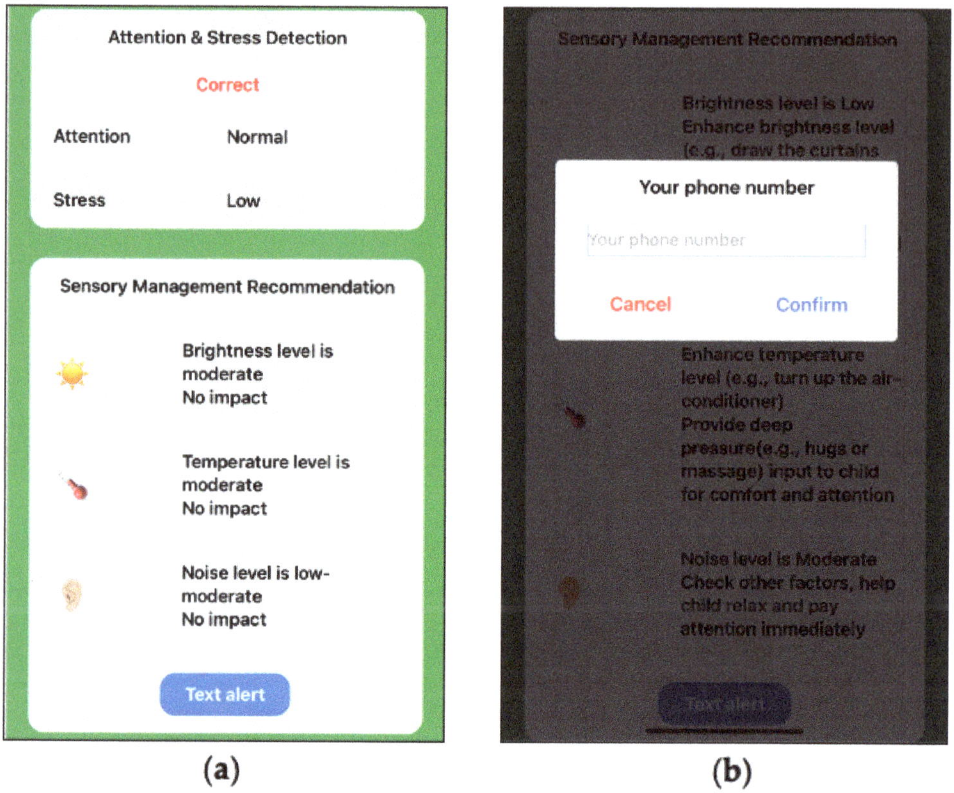

Figure 5. Interface of sensory management strategy recommendation. (**a**) Feedback interface; (**b**) message alert subscription.

2.5. Real-Life System Evaluation

The SMRS Beta App has been released on TestFlight for real-life system evaluation since March 2022. TestFlight is Apple's beta testing service with which developers can invite testers simply by sharing a public link [39]. In this study, the evaluation of SMRS measured how accurately the SMRS identified the abnormal attention and stress levels of the children with ASD in real-life cases of different conditions of sound, light, and temperature. Separately, the study investigated the effectiveness of management strategies recommended by the SMRS on children's performance improvement. Furthermore, the caregivers' level of satisfaction in terms of system utilization, such as user interface, intention of long-term use, were assessed as well. 'Real-life' here referred to the regular environment in daily life in which there was no preliminary control on the generation of sound, visual, or tactile-related stimuli unless the SMRS recommended them to make an adjustment.

The evaluation was performed on 30 preschool children formally diagnosed as ASD. Another 30 gender- and age-matched TD children were involved as well for comparison. Table 6 highlights the demographic characteristics of participants. As shown in Table 6, the testing sites included several childcare centers in Zhejiang Province of China. The SMRS were tested across sites to demonstrate that the application and results of the SMRS intervention were not limited to a specific place.

Table 6. Demographic information of participants of evaluation study.

Condition	Testing Site	Number of Participants	Average Age	Gender Ratio (Male:Female)
ASD	An ASD Rehabilitation Center in Wenzhou	15	4.3	12:3
	An ASD Rehabilitation Center in Ningbo	15	4.0	12:3
TD	A Public Kindergarten in Wenzhou	15	4.4	12:3
	A Private Childcare Center in Ningbo	15	4.3	12:3

The evaluation sessions were conducted in normal classrooms equipped with a desk, chairs, and necessary facilities (such as Figure 6a). The Arduino board was placed near the participant. An Apple Watch and GSR sensor were worn by the participant (Figure 6b). As shown in Figure 6c, in each testing room, there were one teacher and one participant at a time, with the caregiver using the SMRS and observing around the corner.

(a) (b) (c)

Figure 6. The classroom setup in the evaluation sessions. (**a**) A real-life classroom setting; (**b**) child worn sensors in the evaluation; (**c**) the setup for a SMRS session.

Before the formal evaluation sessions, caregivers were invited to install the SMRS Beta App and sign up an account for their children. If the caregivers did not own an iPhone or Apple Watch, test iOS devices that belonged to the research team were lent to them. All the participants and corresponding stakeholders (i.e., caregivers and teachers) were given coaching sessions about testing settings in advance. Each participant should undergo three sessions: no-SMRS session, SMRS session 1, and SMRS session 2 as described in Sections 2.5.1–2.5.3. Duration of each individual session was 30 min. Any two sessions for each participant should not be scheduled in one day to avoid a possible short-term effect such as fatigue and stress.

2.5.1. No-SMRS Session

Prior to the first testing session with SMRS, the classroom teacher and caregiver provided a baseline rating regarding the child's attention and stress in a no-SMRS condition. The child took a class as normal. After the session, the teacher and caregiver rated the child's performance by using a report form adapted from the Caregiver-Teacher Report Form for Ages 1½-5 (C-TRF) [40]. C-TRF is a well-validated instrument which evaluates behavior problems that occur in the classroom across multiple domains including anxiety, stress, attention, and social interaction [41]. Each item on the problem section of the C-TRF contains a statement about a child's behavior. Response choices include: 'Not True' (scored as 0), 'Somewhat or Sometimes True' (scored as 1), and 'Very True or Often True' (scored as 2). The adapted C-TRF in this study includes items listed in 'Anxious/Depressed' and

'Attention Problem' subcategories. The adapted version of C-TRF was attached in the Appendix B.

2.5.2. SMRS Session 1

In the first SMRS session, children and caregivers used the SMRS in the same class as the No-SMRS session. The researchers helped the child put the watch and GSR sensor on before the class started. The caregiver held the phone and used the SMRS on the phone in the classroom. If the detection results regarding attention or stress were wrong, the caregiver should make a real-time correction on the anomaly detection by clicking the 'correct' button and providing true labels on the phone (Figure 7). In this session, the caregiver or teacher would not follow immediate strategies recommended by the system for attracting the child's attention or calming down the child. After the session, the caregiver and teacher completed the adapted C-TRF. Results from this session were used to evaluate how accurately the SMRS identifies the abnormal attention and stress levels of the children with ASD. By comparing the reported scores between the no-SMRS session and SMRS session 1, the researchers could discuss whether the implementation of wearable devices would influence children's attention and stress in the classroom or not.

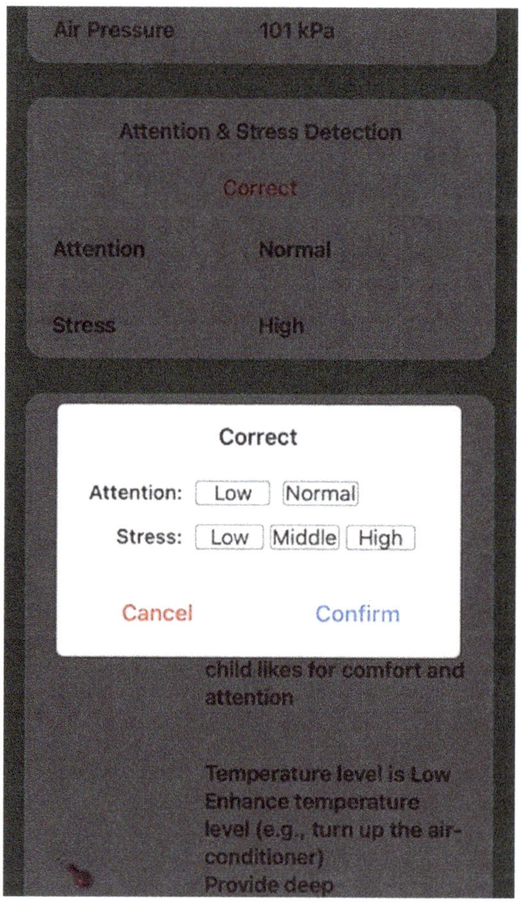

Figure 7. Caregiver feedback interface.

2.5.3. SMRS Session 2

The procedures described for the testing preparation in the SMRS session 1 were identical for the SMRS session 2. However, during this session, when the SMRS identified an abnormal situation and the system generated a recommended strategy, the teacher in the classroom would simultaneously receive a text message of the recommended strategy. The teacher should take actions quickly as instructed by the strategy, such as using deep pressure (Figure 8a), fidget toys (Figure 8b), or playing a calming video (Figure 8c), to help the child pay attention or calm down during the class. Similarly, after the session, the caregiver and teacher completed the adapted C-TRF. By comparing the reported scores between SMRS session 1 and SMRS session 2, the researchers could investigate whether the management strategies recommended by the SMRS were helpful on children performance improvement. If no alerts happened when conducting sessions with a child from ASD group, the child would be asked to go through SMRS sessions again with the caregiver's consent.

(a) (b) (c)

Figure 8. Teachers applied strategies recommended by the SMRS. (**a**) Deep pressure strategy; (**b**) fidget toy strategy; (**c**) calming video strategy.

2.5.4. Post-Session Evaluation

Following the completion of the testing, caregivers of the participants were invited to evaluate the overall functionality of SMRS by completing a System Usability Scale (SUS) questionnaire [42]. The SUS is a 10-item standardized questionnaire designed to measure users' perceived usability and satisfaction of a system. As shown in the Appendix C, statements arranged as odd numbers are positively expressed and statements with even numbers are negatively expressed. Responses of each statement range from 'Strongly Disagree' to 'Strongly Agree' on a 5-point Likert scale.

2.6. Ethics Statement and Material Interpretation

The research protocol of this study was reviewed and approved by the Research Ethics Committee of the University of Nottingham Ningbo China. Informed consent was obtained from all participants involved in the study. For children, consent must have been given by their caregivers before participation. Caregivers had the right to withdraw their children from the study at any time.

The materials used in this study, including software application, questionnaires, research protocol, and participant information sheet and consent form, were prepared in English initially. However, participants in this study were living in China and using Chinese. The validated Chinese versions of sensory profile questionnaire, C-TRF and SUS questionnaire were used in this study. Other materials which did not have official Chinese versions were interpreted into Chinese by the first author who comes from China. The results and materials were presented in English in this study.

3. Results

This study utilized affordable sensors and off-the-shelf mobile devices to achieve proposed measurement, analysis, and strategy making. In all the testing cases, users' sensory profile, physiological data, and information about user environment were successfully recorded by the SMRS. Machine learning models integrated in the SMRS could compute real-time user attention and stress using the collected sensory information. Fuzzy logic controllers that were deployed on the cloud server could generate fuzzy outcomes, which were responsible for activating the alert and making sensory management recommendations.

In the evaluation study, all the ASD and TD participants completed the required three test sessions. Each caregiver observed their child's performance in each of the three sessions. The duration of every individual session was set to 30 min by researchers. During each session, a caregiver's real-time reports on incorrect predictions through an interface in Figure 7 were interpreted into the number of wrong prediction cases. The wrong prediction cases were averaged for both groups of participants to give an indication of, overall, how many false predictions were made by machine learning algorithms in 30 min. Average wrong prediction cases, average ratings from C-TRF of ASD and TD groups, and standard deviations (SDs) are presented in Table 7.

Table 7. ASD and TD group data of different measures.

Group	Sample	Session	Rating Parameters: Mean (SD)					
			Wrong Prediction Cases—Attention	Wrong Prediction Cases—Stress	C-TRF Attention Score—Caregiver	C-TRF Attention Score—Teacher	C-TRF Stress Score—Caregiver	C-TRF Stress Score—Teacher
ASD	30	No-SMRS	/	/	8.1 (3.5)	8.6 (2.8)	4.3 (3.5)	4.9 (4.0)
		SMRS #1	21.9 (15.7)	6.7 (5.0)	8.3 (3.2)	8.7 (3.3)	4.4 (3.6)	4.9 (3.9)
		SMRS #2	11.0 (7.7)	4.2 (2.5)	6.5 (2.8)	7.0 (2.9)	3.4 (3.3)	3.7 (3.4)
TD	30	No-SMRS	/	/	1.6 (1.6)	2.0 (2.0)	1.6 (1.9)	1.5 (1.9)
		SMRS #1	5.2 (4.2)	18.6 (8.4)	1.8 (1.8)	2.2 (2.2)	1.7 (2.1)	1.7 (1.9)
		SMRS #2	4.8 (3.4)	13.3 (7.2)	1.5 (1.5)	1.8 (2.0)	1.2 (1.8)	1.3 (1.6)

3.1. Detection Accuracy

When examining wrong prediction cases, researchers identified that the ASD group reported more wrong prediction cases on attention than stress in the real-life situation. This meant that the accuracy of the attention model was not as satisfactory as the stress model in real-life practice, consistent with the results of model training in Section 2.3. The TD group obtained lower average ratings of C-TRF on both attention and stress domains, indicating that TD children might have a better ability of attention and stress management than the ASD group. TD children were more likely to make their attention or stress stable at a moderate level, with a higher tolerance to an unfriendly environment. The attention model could detect most of the state of TD children correctly as well. However, caregivers of TD children reported more wrong prediction cases of stress than those of ASD children. One reason for this could be that the data used for machine learning training were all from ASD children. Inputs corresponding to 'uncomfortable level' for an ASD child might be still within a TD child's 'comfort zone', making the machine learning model generate wrong predictions for TD children.

By comparing the detection accuracy between SMRS session 1 and session 2, the number of wrong prediction cases dropped in session 2 where teachers implemented strategies to adjust the environment and help the children. It was suggested that machine learning models embedded in the SMRS were more usable for ASD children and had better performance in a comfortable environment. Admittedly, prediction accuracy of the attention model needs further improvement.

3.2. Effectiveness of the SMRS Intervention

To investigate the effectiveness of the SMRS intervention on children's performance improvement, the *t*-test was employed to check if differences existed between the no-SMRS session and the SMRS session 2 for two groups. The magnitude of the differences between the no-SMRS session and SMRS session 2 was examined by calculating the effect size (d). The effect size was computed by dividing the mean change by standard deviation between two sessions. Cohen [43] labeled an effect size 'small' if $d \geq 0.2$ and < 0.5, 'moderate' if $d \geq 0.5$ and < 0.8, or 'large' if $d \geq 0.8$.

A summary of the *t*-test and effect sizes are presented in Table 8. The analyses for each rating score given by caregivers and teachers on the C-TRF revealed significant performance differences in the ASD group between the no-SMRS session and the SMRS session 2 ($p < 0.001$). It indicated that the use of SMRS and the application of strategies recommended by the SMRS could help improve ASD children's attention and reduce stress. Although differences in the TD group were not significant on the C-TRF rating of attention, caregivers also observed reduced stress in TD children ($p < 0.05$). Overall, *t*-test results suggested the positive impact of the SMRS intervention on sensory management in children with ASD. However, another index—effect size—was found only to be moderate for attention improvement and small for stress relief.

Table 8. Summary of the *t*-test and effect sizes for both groups.

Measures	No-SMRS—SMRS Session 2					
	ASD			TD		
	t	Sig.* (2-Tailed)	d	t	Sig. (2-Tailed)	d
C-TRF Attention Score—Caregivers	4.732	<0.001	0.505	0.769	0.448	0.065
C-TRF Attention Score—Teachers	4.533	<0.001	0.561	1.229	0.229	0.100
C-TRF Stress Score—Caregivers	4.160	<0.001	0.265	3.340	0.002	0.216
C-TRF Stress Score—Teachers	5.288	<0.001	0.323	1.649	0.110	0.114

* Sig.: Significance.

3.3. Level of Satisfaction in Terms of System Utilization

Caregivers' perceived satisfaction in terms of system utilization was measured using the SUS. The average mean score for the 10 SUS items are presented in Table 9. Caregivers' rating for the 6th statement was all scored below 2, suggesting that they generally disagreed that there was too much inconsistency in the system. They also did not perceive the system to be cumbersome to use or unnecessarily complex, given that the scores of the 2nd and 8th statement were low. Mean scores relating to the statement 'I found the various functions in this system were well integrated' were particularly high with similar standard deviations for both ASD and TD groups. However, it was noticeable that the statement 'I think that I would need the support of a technical person to be able to use this system' also obtained high scores, indicating that some instruction and assistance were required by the user group before they were able to use the system themselves.

The overall SUS scores in Table 9 were calculated with reference to the practical guidance developed by Sauro [44]. The scores for each statement were interpreted to a new number on a normalized scale of 0–4, summed, and then multiplied by 2.5. Generally, a SUS score above 68 was considered to have above-average usability. The mean SUS score of the ASD group and the TD group was 70.5 and 68.3, respectively, over the average SUS rating of 68. According to the practical guideline on the interpretation of a SUS score [44], a score above 70 suggested that the user-friendliness was good.

Table 9. Caregivers' SUS rating of the SMRS.

Statement	Mean Score (SD) and Range	
	ASD Group	TD Group
1. I think that I would like to use this system frequently.	3.87 (0.64); range: 3–5	3.67 (0.72); range: 3–5
2. I found the system unnecessarily complex.	1.67 (0.72); range: 1–3	2.07 (0.80); range: 1–3
3. I thought the system was easy to use.	3.67 (0.72); range: 3–5	3.73 (0.80); range: 3–5
4. I think that I would need the support of a technical person to be able to use this system.	3.73 (0.80); range: 3–5	3.93 (0.70); range: 3–5
5. I found the various functions in this system were well integrated.	4.20 (0.68); range: 3–5	4.33 (0.62); range: 3–5
6. I thought there was too much inconsistency in this system.	1.07 (0.26); range: 1–2	1.07 (0.26); range: 1–2
7. I would imagine that most people would learn to use this system very quickly.	3.53 (0.83); range: 2–5	3.40 (0.51); range: 3–4
8. I found the system very cumbersome to use.	1.13 (0.35); range: 1–2	1.20 (0.56); range: 1–3
9. I felt very confident using the system.	3.33 (0.90); range: 2–5	3.40 (0.74); range: 2–5
10. I needed to learn a lot of things before I could get going with this system.	2.80 (0.68); range: 2–4	2.93 (0.46); range: 2–4
Overall SUS Score (calculated as per [44])	70.5 (3.92); range: 62.5–80	68.3 (3.62); range: 62.5–77.5

4. Discussion

The evaluation results suggested the benefit of the SMRS for preschool ASD children in real-life classroom settings. The SMRS could provide overall correct predictions on the attention and stress levels of children with ASD, identifying distractions and anxious situations. Statistical analysis revealed that the application of strategies recommended by the SMRS had a positive impact on ASD children's sensory management in class, improving their attention level and reducing stress. The results of SUS survey suggested that caregivers of ASD children contended that the SMRS was user-friendly and various functions, such as real-time monitoring, detection, alerts, and strategy making, were well integrated.

The utilization of the SMRS in this study revealed that such a sensor and machine learning-based system could work as an efficient 'specialist' companion in real-life classroom settings for ASD children. The SMRS sensed ASD children's environment, detected their attention and stress, and provided sensory management strategies to help mediate the negative affect of an unfriendly environment.

4.1. Comparisons with Existing Sensory-Based Technologies

The researchers compared the SMRS with some existing sensory-based technologies for ASD children. Related work referred to in the Introduction section targeting sensory issues of ASD individual were compared with the SMRS. As there were no previous studies that utilized the exact same materials and methods, comparisons mainly focused on the system features and methodological quality.

As shown in Table 10, it can be found that most technologies take advantage of sensors for monitoring. However, many of them fail to profile users' sensory processing patterns to make recommendations on management strategies to help users. There is evidence to suggest that sensory processing patterns are idiosyncratic in individuals with ASD [25]. Researchers' previous work compared prediction accuracy of machine learning models on attention detection before and after removing sensory profile features [4]. The results showed that the prediction accuracy of all models dropped dramatically after the sensory profile features were removed, which indicated that the inclusion of a sensory profile could help ensure the correctness of detection. Although many sensory profiling tools have been widely used in healthcare services, very few technologies have involved these tools in their design.

Table 10. Comparison with other sensory-based technologies for ASD individuals.

Reference	Technology Features					Methodology Quality	
	Sensory Profiling	**Physiological Monitoring**	**Environmental Monitoring**	**Data Analysis**	**Strategy Making**	**Evaluation**	**ASD Sample in the Evaluation**
This study	Yes	Yes	Yes	Yes	Yes	Yes	30
[13]	Yes	No	Yes	Yes	Yes	Yes	20
[14]	No	Yes	No	Yes	No	Yes	20
[15]	No	Yes	No	Yes	No	Not reported	Not reported
[16]	No	Yes	Yes	No	No	Yes	1
[9]	No	No	Yes	Yes	Yes	Yes	10
[20]	No	Yes	No	No	No	Yes	12
[21]	No	Yes	No	No	No	Yes	8

Regarding data analysis, references [16,20,21] still depended on an ASD expert's manual analysis, which would require the continuous involvement of expert assistance. It complicated the use of technologies in daily life and increased the cost for ASD families. Five out of eight studies in Table 10 have used machine learning or cloud computing to enable on-device data analysis. Most of them were published in or after 2019, suggesting an emerging trend of studies executing computation directly in the system.

To date, to the researchers' best knowledge, this is the first work that has combined a standardized sensory profiling tool, sensor monitoring, data analysis, and sensory management strategies in one low-cost system for supporting ASD families to deal with sensory issues. These features highlighted the novelty of the SMRS. Moreover, one challenge for many previous studies was to evaluate the intervention with a large sample of ASD individuals. Some studies only involved a few ASD participants in the evaluation or did not report evaluation. This study conducted an evaluation with a larger sample than previous studies, following a well-defined protocol aiming to make the results more generalizable.

4.2. Limitations and Recommendations for Further Study

The development of the SMRS complemented other existing research by providing a comprehensive sensor and machine learning-based monitoring system. The SMRS was available on TestFlight which could be installed after being invited. This allowed continuous test of the SMRS and data collection in children with ASD. With increased data size, machine learning models will be further trained to improve prediction accuracy. Researchers in this study further used data from 30 ASD participants in the evaluation study to train the GBDT model for attention detection. Experiments using an increased training sample witnessed an improvement of the performance of GBDT model with the detection accuracy increased from 86.67% to 92.61%. This implies that after a period of beta test involving more ASD users, the SMRS can have better performance in identifying the risk situations for children with ASD.

Considering the moderate effect size of the SMRS intervention in an ASD group whose sample size was still limited, further efforts could be made to expand the strategy knowledge base by adding more effective intervention strategies, and to involve more ASD participants in the evaluation.

Another limitation of this study related to the restricted age range and condition (ASD) of the users. The sensory profile questionnaire in the SMRS was for children aged between three and ten only. The age of children involved in this study ranged from three to five. However, sensory issues in ASD may last throughout an individual's life. Using an adolescent/adult sensory profile and the replication of the study with other age groups would increase the utility of the SMRS. Sensory issues were not only observed in ASD but also in disorders such as Attention Deficit Hyperactive Disorder or Sensory Processing Disorder. The SMRS could also be tested for other sensory-related disorders in the future.

Moreover, researchers noticed a slight increase of C-TRF scores in SMRS session 1 compared to the no-SMRS session, suggesting that sensors that touched the body of the child might make the child stressed or more easily distracted. A further system update may try to minimize the influence of tactile defensiveness due to the touch of sensor. For example, non-invasive sensors could be used, such as Kinect or apparel-based sensors. The IoT nowadays creates opportunities for human-machine or machine-to-machine communications, in which way the 'things' can not only identify sensory events in their surroundings but also react autonomously without human intervention [16]. Future work could make use of IoT to create a smart environment to implement the environmental control and calming strategy automatically so that the SMRS will not require the continuous involvement of a human assistant.

Supplementary Materials: The following supporting information can be downloaded at: https://www.mdpi.com/article/10.3390/s22155803/s1. Figure S1: Arduino UNO circuit diagram; Figure S2: SMRS application interface.

Author Contributions: Conceptualization, L.D. and P.R.; methodology, L.D. and P.R.; software, L.D.; validation, L.D. and P.R.; formal analysis, L.D. and P.R.; investigation, L.D.; resources, L.D.; data curation, L.D.; writing—original draft preparation, L.D.; writing—review and editing, L.D. and P.R.; visualization, L.D.; supervision, P.R.; project administration, L.D.; funding acquisition, L.D. and P.R. All authors have read and agreed to the published version of the manuscript.

Funding: This study was funded in part by the University of Nottingham Ningbo China Faculty of Science and Engineering Innovation Lab Project Grant.

Institutional Review Board Statement: The study was conducted in accordance with the Declaration of Helsinki, and the research protocol of this study was approved by the Research Ethics Committee of University of Nottingham Ningbo China (Date of approval: 23 February 2022).

Informed Consent Statement: Informed consent was obtained from all subjects involved in the study.

Data Availability Statement: The data presented in this study are available on request from the corresponding author. The data are not publicly available due to ethical restrictions.

Acknowledgments: We would like to thank all the participants involved in this study. Thanks also go to the healthcare professionals at collaborative institutes for supporting us to complete this research.

Conflicts of Interest: The authors declare no conflict of interest. The funders had no role in the design of the study; in the collection, analyses, or interpretation of data; in the writing of the manuscript, or in the decision to publish the results.

Appendix A

Fuzzy Logic Rules

```
Temperature_rule1 = ctrl.Rule(antecedent = (((Temperature['Low'] | Temperature['High'])
& Duration['Short'] & Attention['Normal'] & Stress['Moderate']) | ((Temperature['Low'] |
Temperature['High']) & Duration['Short'] & Attention['Low'] & Stress['Low']) |
(Attention['Normal'] & Stress['Low'])), consequent = Outcome['Low Risk'], label = 'Low
Risk')

Temperature_rule2 = ctrl.Rule(antecedent = (((Temperature['Low'] | Temperature['High'])
& Duration['Short'] &Attention['Normal'] & Stress['High']) | ((Temperature['Low'] |
Temperature['High']) & Duration['Short'] & Attention['Low'] & Stress['Moderate']) |
(Temperature['Moderate'] & Duration['Short'] & Attention['Normal'] & Stress['Moderate'])
| (Temperature['Moderate'] & Duration['Short'] & Attention['Normal'] & Stress['High']) |
(Temperature['Moderate'] & Duration['Short'] & Attention['Low'] & Stress['Low']) |
(Temperature['Moderate'] & Duration['Short'] & Attention['Low'] & Stress['Moderate']) |
(Temperature['Moderate'] & Duration['Short'] & Attention['Low'] & Stress['High']) |
(Temperature['Moderate'] & Duration['Long'] & Attention['Normal'] &
Stress['Moderate'])), consequent = Outcome['Medium Risk'], label = 'Medium Risk')
```

```
Temperature_rule3 = ctrl.Rule(antecedent = (((Temperature['Low'] | Temperature['High']) & Duration['Short'] & Attention['Low'] & Stress['High']) | ((Temperature['Low'] | Temperature['High']) & Duration['Long'] & Attention['Normal'] & Stress['Moderate']) | ((Temperature['Low'] | Temperature['High']) & Duration['Long'] & Attention['Normal'] & Stress['High']) | ((Temperature['Low'] | Temperature['High']) & Duration['Long'] & Attention['Low'] & Stress['Low']) | ((Temperature['Low'] | Temperature['High']) & Duration['Long'] & Attention['Low'] & Stress['Moderate']) | ((Temperature['Low'] | Temperature['High']) & Duration['Long'] & Attention['Low'] & Stress['High']) | (Temperature['Moderate'] & Duration['Long'] & Attention['Normal'] & Stress['High']) | (Temperature['Moderate'] & Duration['Long'] & Attention['Low'] & Stress['Low']) | (Temperature['Moderate'] & Duration['Long'] & Attention['Low'] & Stress['Moderate']) | (Temperature['Moderate'] & Duration['Long'] & Attention['Low'] & Stress['High'])), consequent = Outcome['High Risk'], label = 'High Risk')

Noise_rule1 = ctrl.Rule(antecedent = ((Noise['High'] & Duration['Short'] & Attention['Normal'] & Stress['Moderate']) | (Noise['High'] & Duration['Short'] & Attention['Low'] & Stress['Low']) | (Noise['High'] & Duration['Short'] & Attention['Low'] & Stress['Moderate']) | (Attention['Normal'] & Stress['Low'])), consequent = Outcome['Low Risk'], label = 'Low Risk')

Noise_rule2 = ctrl.Rule(antecedent = ((Noise['High'] & Duration['Short'] & Attention['Normal'] & Stress['High']) | (Noise['Low'] & Duration['Short'] & Attention['Normal'] & Stress['Moderate']) | (Noise['Low'] & Duration['Short'] & Attention['Normal'] & Stress['High']) | (Noise['Low'] & Duration['Short'] & Attention['Low'] & Stress['Low']) | (Noise['Low'] & Duration['Short'] & Attention['Low'] & Stress['Moderate']) | (Noise['Low'] & Duration['Short'] & Attention['Low'] & Stress['High']) | (Noise['Low'] & Duration['Long'] & Attention['Normal'] & Stress['Moderate'])), consequent = Outcome['Medium Risk'], label = 'Medium')

Noise_rule3 = ctrl.Rule(antecedent = ((Noise['High'] & Duration['Short'] & Attention['Low'] & Stress['High']) | (Noise['High'] & Duration['Long'] & Attention['Normal'] & Stress['Moderate']) | (Noise['High'] & Duration['Long'] & Attention['Normal'] & Stress['High']) | (Noise['High'] & Duration['Long'] & Attention['Low'] & Stress['Low']) | (Noise['High'] & Duration['Long'] & Attention['Low'] & Stress['Moderate']) | (Noise['High'] & Duration['Long'] & Attention['Low'] & Stress['Moderate']) | (Noise['High'] & Duration['Long'] & Attention['Low'] & Stress['High'])), consequent = Outcome['High Risk'], label = 'High Risk')

Brightness_rule1 = ctrl.Rule(antecedent=(((Brightness['Low'] | Brightness['High']) & Duration['Short'] & Attention['Normal'] & Stress['Moderate']) | ((Brightness['Low'] | Brightness['High']) & Duration['Short'] & Attention['Low'] & Stress['Low']) | ((Brightness['Low'] | Brightness['High']) & Duration['Short'] & Attention['Low'] & Stress['Moderate']) | (Attention['Normal'] & Stress['Low'])), consequent=Outcome['Low Risk'], label='Low Risk')
Brightness_rule2 = ctrl.Rule(antecedent=(((Brightness['Low'] | Brightness['High']) & Duration['Short'] & Attention['Normal'] & Stress['High']) | ((Brightness['Low'] | Brightness['High']) & Duration['Short'] & Attention['Low'] & Stress['High']) | ((Brightness['Low'] | Brightness['High']) & Duration['Long'] & Attention['Normal'] & Stress['Moderate']) | ((Brightness['Low'] | Brightness['High']) & Duration['Long'] & Attention['Low'] & Stress['Low']) | ((Brightness['Low'] | Brightness['High']) & Duration['Long'] & Attention['Low'] & Stress['Moderate']) | (Brightness['Moderate'] & Duration['Short'] & Attention['Normal'] & Stress['Moderate']) | (Brightness['Moderate'] & Duration['Short'] & Attention['Normal'] & Stress['High']) | (Brightness['Moderate'] & Duration['Short'] & Attention['Low'] & Stress['Low']) | (Brightness['Moderate'] & Duration['Short'] & Attention['Low'] & Stress['Moderate']) | (Brightness['Moderate'] & Duration['Short'] & Attention['Low'] & Stress['High']) | (Brightness['Moderate'] & Duration['Short'] & Attention['Low'] & Stress['High']) | (Brightness['Moderate'] & Duration['Long'] & Attention['Normal'] & Stress['Moderate'])), consequent=Outcome['Medium Risk'], label='Medium Risk')
Brightness_rule3 = ctrl.Rule(antecedent=(((Brightness['Low'] | Brightness['High']) & Duration['Long'] & Attention['Normal'] & Stress['High']) | ((Brightness['Low'] | Brightness['High']) & Duration['Long'] & Attention['Low'] & Stress['High']) | (Brightness['Moderate'] & Duration['Long'] & Attention['Normal'] & Stress['High']) | (Brightness['Moderate'] & Duration['Long'] & Attention['Low'] & Stress['Low']) | (Brightness['Moderate'] & Duration['Long'] & Attention['Low'] & Stress['Moderate']) | (Brightness['Moderate'] & Duration['Long'] & Attention['Low'] & Stress['High'])), consequent=Outcome['High Risk'], label='High Risk')
```

Appendix B

Adapted Caregiver-Teacher Report Form

TODAY'S DATE
Mo. ____ Day ____ Year _____

Your role: ◆ teacher ◆ caregiver

Below is a list of items that describe children. For each item that describes the child ***over the past 30 min***, please circle the **2** if the item is ***very true or often true*** of the child. Circle the ***1*** if the item is ***somewhat or sometimes true*** of the child. If the item is ***not true*** of the child, circle the ***0***. Please answer all items as well as you can, even if some do not seem to apply to the child.

0 = Not True 1 = Somewhat or Sometime True 2 = Very True or Often True

Anxious/Depressed	0	1	2	1. Clings to adults or too dependent
	0	1	2	2. Feelings are easily hurt
	0	1	2	3. Gets too upset when separated from parents
	0	1	2	4. Looks unhappy without good reason
	0	1	2	5. Nervous, high-strung, or tense
	0	1	2	6. Self-conscious or easily embarrassed
	0	1	2	7. Too fearful or anxious
	0	1	2	8. Unhappy, sad, or depressed
Attention Problem	0	1	2	9. Cannot concentrate, cannot pay attention for long
	0	1	2	10. Cannot sit still, restless, or hyperactive
	0	1	2	11. Difficulty following directions
	0	1	2	12. Fails to carry out assigned tasks
	0	1	2	13. Fidgets
	0	1	2	14. Poorly coordinated or clumsy
	0	1	2	15. Quickly shifts from one activity to another
	0	1	2	16. Inattentive, easily distracted
	0	1	2	17. Wanders away

Appendix C

System Usability Scale

	Strongly Disagree 1	2	3	Strongly Agree 4	5
1. I think that I would like to use this system frequently.					
2. I found the system unnecessarily complex.					
3. I thought the system was easy to use.					
4. I think that I would need the support of a technical person to be able to use this system.					
5. I found the various functions in this system were well integrated.					
6. I thought there was too much inconsistency in this system.					
7. I would imagine that most people would learn to use this system very quickly.					
8. I found the system very cumbersome to use.					
9. I felt very confident using the system.					
10. I needed to learn a lot of things before I could get going with this system.					

References

1. Ousley, O.; Cermak, T. Autism Spectrum Disorder: Defining dimensions and subgroups. *Curr. Dev. Disord. Rep.* **2014**, *1*, 20–28. [CrossRef] [PubMed]
2. American Psychiatric Association. *Diagnostic and Statistical Manual of Mental Disorders*, 5th ed.; American Psychiatric Association: Washington, DC, USA, 2013.
3. Leekam, S.R.; Nieto, C.; Libby, S.J.; Wing, L.; Gould, J. Describing the sensory abnormalities of children and adults with autism. *J. Autism Dev. Disord.* **2007**, *37*, 894–910. [CrossRef] [PubMed]
4. Deng, L.; Rattadilok, P.; Xiong, R. A Machine Learning-Based Monitoring System for Attention and Stress Detection for Children with Autism Spectrum Disorders. In Proceedings of the 3rd International Conference on Intelligent Medicine and Health, Macau, China, 13–15 August 2021; pp. 23–29. [CrossRef]

5. Gomes, E.; Rotta, N.T.; Pedroso, F.S.; Sleifer, P.; Danesi, M.C. Auditory hypersensitivity in children and teenagers with autistic spectrum disorder. *Arq. Neuropsiquiatr.* **2004**, *62*, 797–801. [CrossRef] [PubMed]
6. Talay-Ongan, A.; Wood, K. Unusual sensory sensitivities in autism: A possible crossroads. *Int. J. Disabil. Dev. Educ.* **2007**, *47*, 201–212. [CrossRef]
7. Baranek, G.T.; Foster, L.G.; Berkson, G. Tactile defensiveness and stereotyped behaviors. *Am. J. Occup. Ther.* **1997**, *51*, 91–95. [CrossRef] [PubMed]
8. Brown, C.; Dunn, W. *Adolescent-Adult Sensory Profile: User's Manual*; Therapy Skill Builders: San Antonio, TX, USA, 2002.
9. Khullar, V.; Singh, H.P.; Bala, M. IoT based assistive companion for hypersensitive individuals (ACHI) with autism spectrum disorder. *Asian J. Psychiatr.* **2019**, *46*, 92–102. [CrossRef] [PubMed]
10. Schoen, S.A.; Miller, L.J.; Sullivan, J.C. Measurement in sensory modulation: The sensory processing scale assessment. *Am. J. Occup. Ther.* **2014**, *68*, 522–530. [CrossRef] [PubMed]
11. Shabha, G. An assessment of the impact of the sensory environment on individuals' behavior in special needs schools. *Facilities* **2006**, *24*, 31–42. [CrossRef]
12. Riederer, M.; Schoenauer, C.; Kaufmann, H.; Soechting, E.; Lamm, C. Development of Tests to Evaluate the Sensory Abilities of Children with Autism Spectrum Disorder Using Touch and Force Sensors. In Proceedings of the 4th International Conference on Wireless Mobile Communication and Healthcare, Athens, Greece, 3–5 November 2014; pp. 160–163. [CrossRef]
13. Mauro, N.; Ardissono, L.; Cena, F. Personalized Recommendation of PoIs to People with Autism. In Proceedings of the 28th ACM Conference on User Modeling, Adaptation and Personalization, Genoa, Italy, 14–17 July 2020; pp. 163–172. [CrossRef]
14. Tomczak, M.T.; Wójcikowski, M.; Pankiewicz, B.; Łubiński, J.; Majchrowicz, J.; Majchrowicz, D.; Walasiewicz, A.; Kiliński, T.; Szczerska, M. Stress monitoring system for individuals with autism spectrum disorders. *IEEE Access* **2020**, *8*, 228236–228244. [CrossRef]
15. Coronato, A.; De Pietro, G.; Paragliola, G. A situation-aware system for the detection of motion disorders of patients with Autism Spectrum Disorders. *Expert Syst. Appl.* **2014**, *41*, 7868–7877. [CrossRef]
16. Sula, A.; Spaho, E.; Matsuo, K.; Barolli, L.; Miho, R.; Xhafa, F. An IoT-Based System for Supporting Children with Autism Spectrum Disorder. In Proceedings of the 8th International Conference on Broadband and Wireless Computing, Communication and Applications, Compiegne, France, 28–30 October 2013; pp. 282–289. [CrossRef]
17. Schmidt, M.; Schmidt, C.; Glaser, N.; Beck, D.; Lim, M.; Palmer, H. Evaluation of a spherical video-based virtual reality intervention designed to teach adaptive skills for adults with autism: A preliminary report. *Interact. Learn. Environ.* **2019**, *29*, 345–364. [CrossRef]
18. Cabibihan, J.-J.; Javed, H.; Aldosari, M.; Frazier, T.; Bashir, H. Sensing technologies for Autism Spectrum Disorder screening and intervention. *Sensors* **2017**, *17*, 46. [CrossRef] [PubMed]
19. NAO. Available online: https://www.softbankrobotics.com/emea/en/nao (accessed on 2 March 2022).
20. Ali, S.; Mehmood, F.; Ayaz, Y.; Khan, M.A.; Sadia, H.; Nawaz, R. Comparing the effectiveness of different reinforcement stimuli in a robotic therapy for children with ASD. *IEEE Access* **2020**, *8*, 13128–13137. [CrossRef]
21. Costa, S.; Lehmann, H.; Dautenhahn, K.; Robins, B.; Soares, F. Using a humanoid robot to elicit body awareness and appropriate physical interaction in children with Autism. *Int. J. Soc. Robot* **2014**, *7*, 265–278. [CrossRef]
22. Deng, L.; Rattadilok, P. The need for and barriers to using assistive technologies among individuals with Autism Spectrum Disorders in China. *Assist. Technol.* **2022**, *34*, 242–253. [CrossRef] [PubMed]
23. Morris, M.E.; Aguilera, A. Mobile, social, and wearable computing and the evolution of psychological practice. *Prof. Psychol. Res. Pract.* **2012**, *43*, 622–626. [CrossRef] [PubMed]
24. Dunn, W. *Sensory Profile: User's Manual*; Psychological Corporation: San Antonio, TX, USA, 1999.
25. Dunn, W. The sensations of everyday life: Empirical, theoretical, and pragmatic considerations. *Am. J. Occup. Ther.* **2001**, *55*, 608–620. [CrossRef] [PubMed]
26. iPhone. Available online: https://www.apple.com/iphone/ (accessed on 5 April 2022).
27. Monitor your Heart Rate with Apple Watch. Available online: https://support.apple.com/en-us/HT204666 (accessed on 5 April 2022).
28. Arduino Hardware. Available online: https://www.arduino.cc/en/Main/Products (accessed on 5 April 2022).
29. Shannon, C.E. A mathematical theory of communication. *Bell Syst. Tech. J.* **1948**, *27*, 379–423. [CrossRef]
30. Goodfellow, I.; Bengio, Y.; Courville, A. *Deep Learning*; The MIT Press: Cambridge, MA, USA, 2016.
31. Harrington, P. *Machine Learning in Action*; Manning Publications: Shelter Island, NY, USA, 2012.
32. Description of Fuzzy Logic. Available online: https://ww2.mathworks.cn/help/fuzzy/what-is-fuzzy-logic.html (accessed on 5 April 2022).
33. Foundations of Fuzzy Logic. Available online: https://ww2.mathworks.cn/help/fuzzy/foundations-of-fuzzy-logic.html?searchHighlight=fuzzy%20rules&s_tid=srchtitle_fuzzy%20rules_3 (accessed on 5 April 2022).
34. ASSIST: Autism Sensory Strategies, Information, and Toolkit. Available online: https://paautism.org/resource/assist-toolkit/ (accessed on 25 March 2021).
35. Bundy, A.; Lane, S.; Murray, E. *Sensory Integration: Theory and Practice*, 2nd ed.; F.A. Davis Company: Philadelphia, PA, USA, 2002.
36. Perumal, L.; Nagi, F.H. Switching control system based on Largest of Maximum (LOM) defuzzification-theory and application. In *Fuzzy Logic-Controls, Concepts, Theories and Applications*; Dadios, E.P., Ed.; InTechOpen: London, UK, 2012; pp. 301–324. [CrossRef]

37. Ross, T.J. *Fuzzy Logic with Engineering Applications*, 4th ed.; Wiley: Hoboken, NJ, USA, 2016.
38. Mogharreban, N.; DiLalla, L.F. Comparison of Defuzzification Techniques for Analysis of Non-interval Data. In Proceedings of the 2006 Annual Meeting of the North American Fuzzy Information Processing Society, Montreal, QC, Canada, 3–6 June 2006; pp. 257–260. [CrossRef]
39. Beta Testing Made Simple with TestFlight. Available online: https://developer.apple.com/testflight/ (accessed on 25 April 2022).
40. Achenbach, T.M.; Rescorla, L.A. *Manual for the ASEBA Preschool Forms & Profiles*; University of Vermont: Burlington, VT, USA, 2000.
41. Preschool (Ages $1\frac{1}{2}$-5) Assessments. Available online: https://aseba.org/preschool/ (accessed on 3 January 2022).
42. Brooke, J. SUS: A "quick and dirty" usability scale. In *Usability Evaluation in Industry*; Jordan, P.W., Thomas, B., McClelland, I.L., Weerdmeester, B., Eds.; Taylor & Francis: London, UK, 1996; pp. 189–194.
43. Cohen, J. *Statistical Power for the Behavioral Sciences*; Academic Press: New York, NY, USA, 1977.
44. Sauro, J. *A Practical Guide to the System Usability Scale: Background, Benchmarks & Best Practices*; Measuring Usability LLC: Denver, CO, USA, 2011.

sensors

MDPI

Review

A Comprehensive Review of Hardware Acceleration Techniques and Convolutional Neural Networks for EEG Signals

Yu Xie [1] and Stefan Oniga [1,2,*]

[1] Faculty of Informatics, University of Debrecen, 4032 Debrecen, Hungary; yu.xie@inf.unideb.hu
[2] North University Center of Baia Mare, Technical University of Cluj-Napoca, 400114 Cluj-Napoca, Romania
* Correspondence: oniga.istvan@inf.unideb.hu

Abstract: This paper comprehensively reviews hardware acceleration techniques and the deployment of convolutional neural networks (CNNs) for analyzing electroencephalogram (EEG) signals across various application areas, including emotion classification, motor imagery, epilepsy detection, and sleep monitoring. Previous reviews on EEG have mainly focused on software solutions. However, these reviews often overlook key challenges associated with hardware implementation, such as scenarios that require a small size, low power, high security, and high accuracy. This paper discusses the challenges and opportunities of hardware acceleration for wearable EEG devices by focusing on these aspects. Specifically, this review classifies EEG signal features into five groups and discusses hardware implementation solutions for each category in detail, providing insights into the most suitable hardware acceleration strategies for various application scenarios. In addition, it explores the complexity of efficient CNN architectures for EEG signals, including techniques such as pruning, quantization, tensor decomposition, knowledge distillation, and neural architecture search. To the best of our knowledge, this is the first systematic review that combines CNN hardware solutions with EEG signal processing. By providing a comprehensive analysis of current challenges and a roadmap for future research, this paper provides a new perspective on the ongoing development of hardware-accelerated EEG systems.

Keywords: EEG signal analysis; convolutional neural networks; feature extraction; hardware acceleration

1. Introduction

Electroencephalography (EEG) has become a fundamental area of neuroscience research, clinical diagnostics, and brain–computer interface (BCI) applications. EEG signals can record the activity of the brain's electrical waves and are a key area of interest for practitioners studying neural activity in the human brain and the BCI. Applications of EEG include emotion classification, motor imagery, epilepsy detection, sleep monitoring, attention monitoring, cognitive load assessment, BCI control systems, neurofeedback training, detection of pathological brain activity, and diagnosis of mental disorders. These applications include but are not limited to medical diagnosis, neurorehabilitation, mental health monitoring, education and learning assessment, augmented and virtual reality interaction, user status monitoring, and many other fields. As the relevance of EEG extends across various domains, there is a growing demand for proficient and efficient signal-processing techniques to fully harness the latent capabilities of EEG data. Previous reviews related to EEG have mostly focused on software development. For example, these reviews covered general deep learning methods based on EEG [1], analyzed transfer learning mechanisms for common EEG tasks [1], and listed existing software solutions [2]. However, beyond these software-focused discussions, it is also crucial to consider application scenarios that demand small size, low power consumption, high security, and high accuracy, such as motor imagery and emotion detection. The authors of [3] briefly discussed hardware solutions for EEG-based emotion recognition and introduced classifiers like support vector machines (SVMs), k-nearest neighbor (KNN), and artificial neural networks (ANNs), providing a performance comparison. Unfortunately, they did not extensively cover the hardware design of deep learning methods. Despite some similarities, our work has the following distinct features:

(1) This paper focuses on the challenges of hardware acceleration in EEG-based systems, which we believe is a critical area in the development of wearable medical solutions.
(2) We conduct a systematic analysis of EEG signal characteristics and categorize existing feature extraction methods, which helps identify the most suitable hardware acceleration strategies for different application scenarios and provides a clear direction for further research.
(3) To the best of our knowledge, this paper is the first to systematically review the combination of CNN hardware solutions with EEG signal processing, filling a gap in this specialized field.
(4) This paper comprehensively considers the development of EEG and CNN technologies, identifies existing challenges, and provides insights into future development directions, offering a new perspective for research in this area.

1.1. EEG Signal-Processing Challenges

The characteristics of EEG signals include high temporal resolution and a direct reflection of neural activity [4]. However, they present unique challenges due to their low spatial resolution, sensitivity to noise, and high dimensionality. Extracting meaningful information from EEG data necessitates complex signal-processing techniques. EEG signal processing generally includes preprocessing, parameter determination, and feature extraction. There is currently no established set of paradigms for EEG signal processing.

Non-deep learning methods commonly used in EEG signal processing include techniques such as SVM, kNN, and Linear Discriminant Analysis (LDA), as well as traditional feature extraction approaches like Independent Component Analysis (ICA) and Principal Component Analysis (PCA). While these non-deep learning methods are indeed simpler, faster, and often easier to implement on wearable devices, they may struggle to achieve high accuracy due to the inherent complexity and continuous nature of EEG signals. These traditional methods typically rely on handcrafted features, the selection of which is highly dependent on the researcher's expertise and can be time-consuming. Additionally, these models often have limited analytical power, which may not be sufficient for handling complex tasks. In contrast, deep learning greatly simplifies the processing workflow by automatically extracting features while simultaneously performing decoding. Although deep learning models do involve a large number of parameters, resulting in significant computational demands, this is precisely where hardware acceleration becomes particularly valuable. Hardware acceleration enables the efficient execution of these models, making it feasible to deploy deep learning approaches in real-time EEG signal-processing applications, even in resource-constrained environments. Recent advances in deep learning have demonstrated remarkable capabilities in various fields, such as image processing and natural language understanding [5]. These advancements have spurred interest in utilizing deep learning for EEG signal analysis. Convolutional neural networks (CNNs) are one of the most common applications of deep learning, particularly in image processing. Images, as a form of feature representation, allow for the visualization of EEG signals' temporal and frequency domain characteristics through additional spatial information. This enables CNNs to fully exploit the characteristics of EEG data contained in the input. In addition, CNNs can autonomously learn discriminative features from EEG data, potentially reducing reliance on manual feature engineering and enhancing classification performance. It is, therefore, natural for EEG researchers to attempt to replicate this approach.

1.2. Hardware Acceleration and Efficiency

While CNNs offer promising solutions for EEG signal analysis, they come with substantial computational demands, particularly in real-time applications like BCIs. One possible approach is to deploy processing tasks to cloud servers with more powerful resources [6,7]. But in many cases, applications such as real-time emotion monitoring and motor imagery need to respond immediately to changes in the user's state. Relying on cloud servers introduces network latency. If there is network fluctuation or no network, the real-time requirements may not be met. In addition, EEG data usually contain sensitive information, and transmitting these data to the cloud for processing may raise privacy and security concerns. Therefore, for many wearable EEG devices, hardware acceleration is necessary for achieving key functions such as low power consumption, real-time processing, data security, and independent operation. These factors make hardware acceleration particularly important for wearable devices. Hardware acceleration, especially on Field-Programmable Gate Arrays (FPGAs) and Graphics Processing Units (GPUs), can significantly enhance the speed and efficiency of EEG processing based on CNNs [8]. FPGAs, in particular, provide a reconfigurable hardware platform that allows researchers to optimize CNN architectures for specific EEG applications. FPGA-based implementations are renowned for their parallel processing capabilities, low latency, and energy efficiency. These characteristics make FPGAs an excellent choice for real-time EEG processing [9].

1.3. Research Objectives and Structure

The primary objective of this paper is to provide a comprehensive review of the integration of CNNs and hardware acceleration techniques for EEG signal processing. To achieve this goal, we have structured this paper into several sections. Section 2 reviews EEG-based hardware solutions, including preprocessing, feature extraction, and various features used in EEG analysis. Section 3 discusses efficient CNN architectures, including pruning, quantization, tensor decomposition, knowledge distillation, and neural architecture search. Section 4 explores the challenges in this field and provides recommendations for future research directions. Section 5 summarizes the essential findings and insights of this paper. The purpose of this review is to explore the latest achievements in the field of EEG and CNNs and to help more researchers gain a deeper understanding of the hardware solutions, potentials, and challenges in this field.

2. EEG-Based Hardware Solutions

EEG signals are variable and contain a lot of noise. As researchers study EEG, they have developed many algorithms to extract raw signals and convert them into usable information. This process comprises several critical stages, each employing a unique set of techniques and technologies. In this section, we present the state-of-the-art hardware implementations of various algorithms used in the preprocessing and feature extraction stages.

2.1. Preprocessing

Before exploring the wealth of information hidden within EEG signals, it is typically necessary to preprocess the data to address the substantial noise and artifacts that often accompany them [10,11]. While some datasets offer EEG data with certain artifacts already removed, the hardware implementation of preprocessing blocks provides a more comprehensive and practical evaluation of the hardware resources required for EEG-based solutions.

These artifacts can be both physiological and non-physiological. Physiological artifacts are associated with activities such as head movements (EMG), blinking (EOG), muscle contractions, heart activity (ECG), sweating, and respiration (RS) [12,13]. Non-physiological artifacts occur due to external factors like electrode malfunctions (Ep), power sources (AC), ventilation, magnetic fields, cable movements (Cable), and more [14,15]. As summarized in Table 1, the spectrum of EEG signals spans from 0 to 100 Hz and intertwines with various artifacts across this range.

The frequency range of EEG bands is defined slightly differently in different studies. It is worth noting that in Table 1, the frequency bands of Mu waves and Alpha waves overlap because the Mu band is a special form of the Alpha band. Alpha waves are among the most easily recognizable in EEG and are usually found in the posterior region of the head (occipital lobe) when the subject is relaxed or has their eyes closed. Mu waves usually appear in the sensory-motor area of the human brain. When the subject is at rest, Mu waves can be clearly observed; however, when the subject is moving or exercising, the amplitude of the Mu wave decreases or disappears, which is called motor inhibition. Although the Alpha and Mu bands are very similar, they come from different brain regions and have distinct associated functions. The authors of [16] showed that ICA can help distinguish activities from different brain regions, thereby differentiating Alpha waves from Mu waves.

Table 1. Precision and recall scores for the proposed model.

EEG Band	Frequency (Hz)
Alpha	8–13
Beta	12–30
Delta	0.5–3.5
Theta	4–7
Mu	8–12
Gamma	26–100
Artifacts	**Frequency (Hz)**
ECG	0.5–40
EOG	0.1–17
EMG	0.1–2.5/20–100
RS	0.1–2.5
Ep	0.1–35
Cable	2–15
AC	50

One method for reducing noise in EEG signals is the application of digital filters. In [17], a bandpass filter ranging from 8 to 45 Hz was used. While this configuration successfully eliminated some low-frequency artifacts, it came at the cost of attenuating EEG activity in the Theta and Delta brainwaves. In contrast, [18,19] employed a low cutoff frequency of 2 Hz. This lenient setting allowed for the coexistence of slow eye-movement artifacts with the EEG recordings. Ref. [20] used an 81-order digital filter, which provided a better graphical representation for the analysis of Beta and Mu brainwaves and was successfully implemented on the DE0-Nano-Soc development card. It is worth noting that using only bandpass filtering may not cover complex interference scenarios caused by actual artifacts. Another method was described in [21], which involves three key steps: downsampling, bandpass filtering, and integrating a Mean Average Reference (MAR) filtering block. Downsampling, performed at twice the rate without altering the frequency content, reduces the power spectral amplitude by half and enhances signal clarity. The bandpass filter operates within the range of 2 Hz to 60 Hz, ensuring that EEG signals remain within the desired spectrum. The MAR filter involves re-referencing electrode signals relative to their collective average, effectively reducing the signal's dependence on specific electrode locations [4]. This is achieved by subtracting the average signal value from each electrode signal. With this technique, artifacts associated with the physical reference electrode are effectively removed, while averaging acts as noise suppression, preventing the introduction of new artifacts into the EEG signal.

On the other hand, in [22], the authors initiated an adaptation strategy employing ICA based on kurtosis. This technique can extract artifacts co-occurring within the same frequency band as EEG signals. However, the application of this method is entirely offline, and the feasibility of its hardware implementation remains unconfirmed. In [23], fast ICA was employed for epileptic seizure detection from multi-channel signals. This rapid

ICA assists in artifact removal and reduces energy dissipation. Eigenvalue decomposition (EVD), used in conjunction with the Jacobi algorithm, resulted in a 77.2% reduction in area. The practical performance of this architecture was implemented on an FPGA and validated through appropriate databases. These blind source separation algorithms have become popular methods for hardware implementations of EEG preprocessing [24,25].

Nevertheless, the efficacy of these methods is related to the number of independent components obtained, a quantity dependent on the number of signal sources or electrodes. Consequently, these methods exhibit limitations in situations with a small number of sources. This realization has prompted the exploration of cost-effective alternatives [26,27] for integrating the ICA model, where the authors developed a performance evaluation method that fused matched filtering and threshold cross-checking based on the number of electrodes.

In the field of EEG-based classification, effective methods must encompass a set of techniques to ensure signal quality, including but not limited to detrending, re-referencing, outlier detection, interpolation, detection and correction, and elimination of artifacts.

2.2. Feature Extraction

Preprocessed EEG data include a wealth of information. However, to use this information, it must first be extracted and converted into meaningful features. This section reviews feature extraction techniques that transform EEG signals into quantifiable attributes. We examine five types of feature extraction used in this domain: frequency-based features, time-based features, time-frequency features, spatial features, and raw signals.

2.2.1. Frequency-Based Features

Frequency-based features analyze the rhythmic patterns of brain activity. Given that a large number of studies have identified correlations between spectral components and brainwave rhythms, researchers have a clear tendency to focus on frequency characteristics [5,28].

The complexity of related algorithms varies greatly across software platforms, and researchers rarely consider the issue of computing resource consumption at the software level. Consequently, their high complexity levels make many of them unsuitable for direct hardware implementation. However, methods such as Power Spectral Density (PSD), Fourier Transform, Discrete Cosine Transform (DCT), Singular Spectrum Analysis (SSA), and combinations of filter banks still hold promise for integration into hardware solutions, as the hardware design of these traditional signal-processing modules is well established. Hardware-accelerated Fast Fourier Transform (FFT) modules, with their moderate resource requirements, can efficiently compute PSD. In contrast, higher-order spectral (HOS) analysis, Linear-Frequency Cepstral Coefficients (LFCCs), Mel-Frequency Cepstral Coefficients (MFCCs), the Burg method, and Auto-Regressive Reflection Coefficients (ARRCs), while effective, might require more resource allocation and design work for successful integration into hardware solutions.

Cross-frequency coupling (CFC) functions are promising approaches that leverage synchronized variations across different frequency bands [29,30]. These functions manifest as connectivity graphs or indices. They not only ensure robustness but also have the potential to reduce EEG data rates and memory requirements. However, the computational complexity of CFC algorithms and the hardware demands, such as Coordinate Rotation Digital Computer (CORDIC) processors for computing cross-frequency Hilbert transforms, pose significant challenges for ultra-low-power hardware designs [31].

It is foreseeable that PSD has become one of the most popular methods in the EEG classification domain. In [18,21], PSD approximations were used, with only one power value per channel calculated based on the average activity within the beta band (12 to 30 Hz). While this approach does not necessitate FFT accelerators, making it more hardware-friendly, it introduces information loss within the frequency bands. In contrast, [22] employed an FFT with 128 frequency bins, ensuring sufficient resolution to capture fluctuations both within and between the EEG wavebands.

2.2.2. Time-Based Features

In addition to frequency-based features, time-based features are also a key research direction in EEG signal processing because they can reflect the dynamic changes in brain activity over time and provide indicators, such as maximum and average amplitudes, slopes, and skewness, that reflect transient brain activity.

Notably, Higher-Order Crossings (HOCs) have become a common approach. HOCs are based on tracking zero crossings and their consecutive differences, offering valuable insights into the temporal dynamics of EEG signals. Higuchi's Fractal Dimension (HFD) is well known for its ability to capture the non-linear aspects of data, making it particularly suitable for feature extraction in various EEG events. Other time-based feature extraction methods include Hjorth parameters, Grassberger–Procaccia (GP), Hybrid Adaptive Filtering (HAF), the Hurst exponent, Local Binary Patterns (LBPs), and Event-Related Late Positive Potentials (LPPs).

Compared to frequency-domain approaches, the advantage of time-based methods is their relatively lower hardware resource requirements, making them an attractive choice for hardware implementations. However, it must be acknowledged that methods involving non-linear transformations, such as HFD or Hjorth parameters, often increase the complexity of hardware design. In [19], a time-based approach was employed for feature extraction. The method focuses on extracting HOCs and skewness (SK) information from time series to construct feature vectors. To facilitate hardware-friendly implementation, [19] introduced the Approximate SK Indicator (ASKI), significantly reducing gate counts by a factor of 86. For HOCs, the authors employed a straightforward implementation based on comparators, using 3-bit SR triggers to monitor the previous values of a given sample. Identifying sign changes between consecutive sample values triggers a simple counter to accumulate zero crossings. Both of these metrics are standardized and fed into a dense neural network. Hjorth parameters indicate statistical characteristics in EEG signal analysis. Calculating them requires variance computation for the "activity" parameter, the square root of the ratio for the "mobility" parameter, and division operations for the "complexity" parameter with first-order derivatives [32].

Sample Entropy (SE) is another time-based method [33] that has proven to be relatively hardware-friendly due to its simplicity. SE is a measure of time-series complexity defined as the logarithm of a ratio, making its hardware implementation relatively straightforward. Ref. [34] proposed that SE architecture runs ten times faster on FPGAs than on CPUs. Methods with minimal hardware resource utilization and power consumption include statistical mean and other time-based methods like Root Mean Square (RMS), Variance (Var), and Standard Deviation (SD), as observed in the articles we surveyed, both on software and hardware platforms, these methods all have an average accuracy of no more than 60%. Therefore, they are the least ideal options.

2.2.3. Time-Frequency Features

The fusion of time-domain and frequency-domain techniques gives rise to time-frequency features, which are adept at handling the non-stationary nature of EEG data. Wavelet Transforms (WTs) and their various forms (such as Discrete Wavelet Transform (DWT), Continuous Wavelet Transform (CWT), and Adaptive Tunable Q-Wavelet Transform) play a significant role. The Short-Time Fourier Transform (STFT) and the Hilbert–Huang Transform (HHT), along with its Independent Mode Function (IMF) and Empirical Mode Decomposition (EMD), have also garnered attention. Unlike Fourier and Wavelet Transforms, the HHT employs a unique approach using EMD to decompose signals, extracting quasi-harmonics starting from the highest frequency. This method demonstrates robustness in analyzing non-stationary and non-linear time series, a common characteristic of EEG signals [35]. However, it is worth noting that the calculation of the IMF in the HHT involves an iterative algorithm. The parameters of each layer of this algorithm are highly correlated, making it difficult to take advantage of hardware parallel processing, especially in some real-time processing application scenarios.

Wavelet Transforms (WTs) and fundamental forms of the STFT are two feasible choices for hardware implementation. Ref. [36] proposed an STFT design based on the Xilinx Zedboard and extracted 20 frequency feature components for classification. These 20 features were divided into five groups, each corresponding to five different EEG wave frequency bands. Features were extracted from 128 channels of EEG signals at a clock rate of 50 MHz. Ref. [37] employed the STFT for time-frequency analysis of EEG segments and extracted the regions of interest in terms of frequency bands and features, implementing a deep learning model and an STFT module on an FPGA to enhance seizure detection. Ref. [38] introduced a generalized spectrogram method. The STFT was applied to signals captured from six electrodes, which were subsequently mixed based on asymmetric indices. Ref. [39] implemented a DWT with a two-class LDA classification on the Basys 3 Artix-7 FPGA Board. Ref. [40] presented a design and implementation of the Morlet CWT for EEG signals on the Spartan 3AN FPGA. By optimizing the trade-off between speed and silicon area, it could generate all-scale wavelet coefficients for a 1024-point EEG signal in approximately one millisecond when designed to run at a maximum clock speed of 125 MHz. In the STFT, the signal to be processed is divided into multiple shorter signals according to the size of the window, and a separate FFT is then applied to each small signal. In contrast, the WT processes the complete signal. Both methods yield similar results, with the Wavelet Transform being particularly effective in handling non-periodic and fast-changing features. In contrast, the STFT is favored for real-time processing due to its shorter processing time windows. Moreover, advances in FFT hardware acceleration and various space-efficient techniques for compressed FFT accelerators make them highly attractive for hardware-based solutions.

2.2.4. Spatial Features

Unlike independent features, spatial features depend on the availability of time-based, frequency-based, or time-frequency EEG data for computation. This section introduces a well-known spatial feature method: Common Spatial Pattern (CSP). In a two-classification problem, CSP aims to find spatial filters that maximize the variance of EEG signals for one class while minimizing it for the other class. This results in spatial filters that enhance discriminative features for the two classes, making classification more effective. Here, a simplified explanation of the formula application of CSP is presented. The spatial filter w is calculated by simultaneous diagonalization of the sample covariance matrices from both classes, as follows:

$$J(w) = \frac{w^T \overline{C_1} w}{w^T \overline{C_2} w} \tag{1}$$

where T represents the transpose. $\overline{C_1}$ and $\overline{C_2}$ denote the average covariance matrices of the two different conditions or classes, respectively, and are defined as follows:

$$\overline{C_k} = \frac{1}{N} \sum_{n=1}^{N_k} \frac{D_{(k,n)} D_{(k,n)}^T}{Trace\left(D_{(k,n)} D_{(k,n)}^T\right)},\ k = 1,\ 2 \tag{2}$$

where $Trace\left(D_{(k,n)} D_{(k,n)}^T\right)$ represents the solution of the matrix trace. N_k is the number of samples for the kth class, that is, the number of single-trial data. $D_{(k,n)}$ represents the nth trial data of the kth class. The CSP algorithm frames this as a generalized eigenvalue problem:

$$\overline{C_2}^{-1} \overline{C_1} w = \lambda w \tag{3}$$

The conventional CSP feature extraction method derives features by computing the logarithm of variances from spatially filtered signals, as follows:

$$f_p = \log\left\{\frac{var(Z_p)}{\sum_{i=1}^{2m} var(Z_i)}\right\},\ p = 1,\ 2,\ \cdots,\ 2m \tag{4}$$

where the first and last m rows of the EEG signals are usually taken as the final spatial filter. Here, Z is given by $Z = w^T D$.

In [41], the CSP model was implemented on an Altera Stratix-IV FPGA and verified on the BCI competition dataset. In [42], CSP was used as the feature extraction algorithm with Mahalanobis Distance (MD) as the classifier, and the entire signal-processing task was executed on software embedded in a Nios II processor within a Stratix IV FPGA. This approach reduced FPGA resource consumption but introduced a time delay of 0.5 s. An improvement on the approach in [41] can be found in [42], where the authors accelerated some functions in hardware but still performed most processing on the Nios II, reducing the delay to 0.399 s. In [43], the authors proposed a complete hardware implementation of a BCI embedded on an FPGA using Scalp-Coupled Common Spatial Spectral Patterns (SCSSPs) as the feature extraction algorithm and support vector machines (SVMs) as the classifier.

The previous section mentioned the use of asymmetric indices as a subset of spatial features. These indices are used to compare the relationship between the two hemispheres and thus indicate asymmetric neural activity [44]. This method depends on the calculation of asymmetric indices, which are usually computed between electrode pairs and can be effectively used alongside frequency, time, or time-frequency features. Hardware designs, such as those in [18,21], combine asymmetric indices with channel-based PSD averages, a concept discussed earlier in the frequency-based feature category. In these implementations, utilizing eight channels, the asymmetric indices are calculated based on the PSD averages of electrode pairs.

In addition to CSP and asymmetric indices, another subset of spatial-based features (specific to EEG-based emotion detection systems implemented in software) includes connectivity features, particularly Directed Transfer Function (DTF) features. DTF features serve as causal measurements used to determine brain connectivity patterns [45]. The DTF quantifies the causal influence of one EEG channel on another at specific frequencies. While theoretically, the DTF can be implemented in hardware, it presents more complex challenges compared to asymmetric indices. This complexity arises from the need to calculate the transfer matrices of multivariate autoregressive models and to perform square root and division operations.

2.2.5. Raw Signals

This section introduces some works on EEG classification using raw signals. Due to the complexity of EEG signals, it is difficult for traditional EEG classifiers to obtain good results from raw signals. The shift in researchers' views is mainly due to the popularity of deep neural network (DNN) classifiers. In these new DNN approaches, raw EEG signals can be directly input into the network for feature extraction and classification. Currently, there are no hardware-based implementations that follow this paradigm. This is mainly due to the need to compensate for the lack of specific features in the complexity of DNNs [46,47]. Moreover, most DNN hardware implementations for EEG tend to adopt shallow DNNs to achieve designs that save more computing resources and space. These networks lack the direct capability to process raw EEG data. However, some new material technologies may change this status, such as near/super-threshold techniques in Fully Depleted Silicon on Insulator (FDSOI) technology [48] or adaptive body biasing in advanced semiconductor technologies [49]. These technologies have the potential to facilitate hardware implementations of DNNs that can directly operate on raw EEG signals in wearable classifiers.

Another approach is to use a DNN only as a feature extractor instead of as a classifier. An example can be found in [50], where overlapping filters, as well as Differential Entropy (DE) and PSD analysis, were employed to derive suitable kernels that guide a CNN in automatically extracting features from raw EEG data. The advantage of this approach lies in implicitly merging spectral and entropy information when dealing with raw EEG data for learning and classification.

Table 2 shows the methods discussed in this section.

Table 2. Feature extraction methods.

Feature Domain	Authors	Algorithm	Work Description
Frequency-based	[18,21]	PSD	A power spectrum approximation per channel Hardware friendly, no FFT accelerator required Easily loses band information
	[22]	PSD	Dataset: Own FEXD Hardware: Digilent Atlys with Spartan-6 An FFT accelerator with 128 frequency bins Captures fluctuations within and between frequency bands
	[31]	CFC	Uses connection diagrams to represent EEG Reduces the need to store and compute EEG CORDIC required Hilbert transform of each frequency band
Time-based	[19]	HOC, SK	Hardware: TSMC 0.18 um 1P6M CMOS chip Databases: DEAP and SEED Uses ASKI to reduce gate counts, 3-bit SR triggers
	[32]	Hjorth	Hardware: Zynq-7000 FPGA Dataset: Bonn University EEG database Classifier: KNN
	[33,34]	SE	For statistics, complexity, and irregularity systems For analyzing short and noisy datasets Relatively hardware-friendly
Time-frequency	[36]	STFT	Hardware: Xilinx Zedboard Extracts 20 frequency-band features from 128 channels
	[37]	STFT	Used for epilepsy detection Only simulation; not applied to real FPGA chips
	[38]	STFT	Hardware: ADVANTEST V93000 PS1600 chip Dataset: DEAP STFT features mixed with asymmetric indices
	[39]	DWT	Hardware: Basys 3 Artix-7 FPGA Board Classifier: LDA Suitable for developing low-cost, marketable products
	[40]	CWT	Hardware: Spartan 3AN FPGA Efficient architecture in Fourier space via optimized speed and silicon area

Table 2. *Cont.*

Feature Domain	Authors	Algorithm	Work Description
Spatial	[41]	CSP	Hardware: Altera FPGA platform (Stratix-IV) Dataset: motor imagery tasks in a BCI competition Classifier: MD Time delay of 0.5 s
	[42]	CSP	Improvement on method in [41]; reduces the delay to 0.399 s
	[43]	SCSSP	Hardware: Virtex-6 FPGA Dataset: BCI Competition IV-dataset 2a Classifier: SVM
	[18]	Ai, PSD	Based on a scaled version of IHPR with customized LUT
	[21]	Ai, PSD	Improvement on method in [18]; reduces the gate and power consumption
	[45]	DTF	Describes incidental effect between two channels Difficult to implement in hardware
Raw	[50]	PSDCNN	Dataset: Own For emotion recognition and fatigue-driving

"Ai" denotes asymmetric indices.

3. Efficient Convolutional Neural Networks

In Section 2, we mentioned that DNNs have been widely used in the study of EEG signals. CNNs have showcased remarkable capabilities in various domains, from image recognition to natural language understanding. Their innate ability to autonomously learn hierarchical features has positioned them as the predominant force in software-driven DNNs. The application of CNNs presents an opportunity to fundamentally reshape the approach to neural data analysis, interpretation, and classification. It is well recognized that the surprising progress of CNNs has been accompanied by a large number of parameters that need to be calculated. In particular, the real-time calculation of CNNs is challenging in the field of BCIs. This section introduces five highly efficient CNN hardware techniques specifically used for EEG signal processing: pruning, quantization, tensor decomposition, knowledge distillation, and neural architecture search. The applications we reviewed are summarized at the end of this section.

3.1. Pruning

One approach to reducing the computational complexity of CNNs is pruning, a technique dating back to the early 1990s [51]. Researchers have found that there are some redundant or less informative neural nodes in CNNs. Therefore, deleting these unimportant weights, filters, channels, and even layers during the inference process has little effect on accuracy. The core idea of pruning is to identify and eliminate redundancy in the network to reduce computational and memory requirements. This process can be categorized into two main strategies: weight pruning and structural pruning.

3.1.1. Weight Pruning

Weight pruning is a core technique in the design of efficient neural networks. Removing some unimportant connections does not affect the performance of the network because not all neural nodes are equally important. Several strategies have been developed to effectively execute weight pruning.

Early work in this field, such as Optimal Brain Damage (OBD) [51] and Optimal Brain Surgeon (OBS) [52], introduced the concept of utilizing second-order derivatives (Hessian matrix) of the loss function to identify unnecessary weights. These methods laid the foundation for subsequent weight-pruning techniques.

A widely adopted three-step deep learning method was proposed in [53,54]. First, the network is initially trained to identify essential connections. Subsequently, less important connections are pruned. Finally, the network undergoes retraining to fine-tune the remaining connections. This iterative process gradually leads to a more compressed network structure. Furthermore, dead neurons, with all their connected weight pruned, can be safely discarded. Some research has suggested that pruned larger models can outperform smaller dense models in terms of accuracy under the same memory constraints [55,56]. While traditional weight pruning methods operate in the spatial domain, [57] proposed a frequency-domain approach that transforms spatial weight into frequency-domain coefficients. Pruning is performed dynamically in different frequency bands during each iteration, leading to higher compression ratios.

In the previously mentioned methods, a global magnitude threshold is applied to the entire DNN, which can affect network performance since applying the same compression rate to each layer is not always optimal. Some studies have used variable pruning rates to target layers of different importance. Ref. [51] combined selective learning, identifying weight importance concerning the loss, and discarding other weights by cutting off the gradient flow. Ref. [58] employed a Gaussian Mixture Model (GMM) for each layer's weight distribution to determine layer-wise compression rates. Pruning layers are selected based on the number of weights, with small amplitudes estimated using the GMM.

Additionally, amplitude-aware pruning methods like [59] introduce considerations for energy reduction along with compression ratios and accuracy loss when determining pruning strategies.

Weight-pruning techniques remove less important weights or neurons, regardless of their position within the network. This process results in so-called "unstructured" networks. Such networks may require specialized software or hardware adaptations, thus limiting the universality of the weight-pruning algorithm on hardware.

3.1.2. Structural Pruning

On the other hand, structural pruning focuses on eliminating entire filters, channels, or even whole layers. This approach not only reduces computational requirements but also leads to more compact network architectures, which is especially valuable in resource-constrained environments. Unlike weight pruning, structural pruning does not require specialized hardware or software support. Because it only changes the size of the input and output, it does not change the parameters within the layer. This method maintains strong compatibility. Mainstream structural pruning strategies primarily aim to identify and remove unimportant filters. They are typically categorized into three main branches:

- Criterion-Based Pruning: In this method, filters are ranked based on specific pruning criteria. Methods that consider both L1 and L2 norm criteria, such as L1-norm-based pruning [60] and Soft Filter Pruning (SFP) [61], have been explored. Additionally, activation neuron criteria, such as the Average Percentage of Zeros (APoZ) [62], help identify and remove filters with low APoZ values.
- Ranking-Based Pruning: HRank [63] employs feature map ranking as the metric for filter pruning. Lower-ranked feature maps are considered to contain less information, making them one of the primary candidates for pruning.
- Optimization-Based Pruning: Methods like ThiNet [64] and NISP [65] formulate pruning as an optimization problem. ThiNet prunes filters in the current layer while minimizing the reconstruction error in the next layer. On the other hand, NISP focuses on reducing the reconstruction error in the "Final Response Layer" (FRL).

In addition to filter and channel pruning, structural pruning extends to whole-layer pruning methods [66–68]. These methods selectively remove layers from the network using various criteria, resulting in more compact models. Similarly, these pruned models not only reduce the amount of computation and time but also maintain good accuracy.

The above algorithms all follow the conventional format of training models, pruning, and fine-tuning, except for the method in [69]. In structural pruning, fine-tuning pruned models may produce performance comparable to or even worse than training the same model from scratch with randomly initialized weights. Innovative techniques like network slimming [70] and group LASSO loss [71] automatically identify and prune unimportant channels or neurons during the training process, effectively achieving model compression without the need for post-pruning fine-tuning.

3.2. Quantization

Quantization is another technique aimed at improving the efficiency of CNNs. It reduces the precision of network weights and activations, typically from high-precision formats (e.g., FP32) to lower-bit fixed-point representations. Precision reduction helps lower memory requirements and computational complexity, making CNNs more hardware-friendly. A commonly used CNN hardware implementation strategy for EEG signal processing is to use quantization and pruning together to achieve better computational efficiency.

In the quantization process, neural network data are constrained to a set of discrete levels, offering various distribution options, including uniform or non-uniform distributions. Uniform distribution [72] (characterized by even step sizes) and non-uniform distribution [73] (typically represented by logarithmic distributions) are commonly used quantization schemes. Quantization can be achieved using deterministic methods and stochastic methods. Deterministic methods (such as projecting data onto the nearest discrete levels [71]) ensure precision, while stochastic methods involve probabilistically determining which of the two neighboring discrete levels the data are projected onto.

The roots of quantizing neural network data can be traced back to the early 1990s [74]. This technology offers two primary forms: post-training quantization [75,76] and quantization-aware training [77–79]. In post-training quantization, FP32 weights and activations are converted to lower-precision formats after the model's full training. Conversely, quantization-aware training incorporates the quantization error as part of the training loss, typically resulting in improved model accuracy.

The field of quantization has achieved impressive results. Research has indicated that FP32 precision parameters can be effectively reduced to INT8 without significant loss of accuracy [75]. Some researchers have even developed 4-bit quantization approaches, eliminating the need for fine-tuning post-quantization [76]. Additionally, experiments using INT8 precision for training and inference have shown competitive results, such as a 1.5% accuracy loss in the case of the ResNet-50 model [77]. Furthermore, advancements like generalized bit precision [78] provide greater flexibility by allowing different bit depths for storing weights and activations, offering more flexibility than traditional INT8 quantization. For example, [78] used 1-bit weights and 2-bit activations to quantize AlexNet, achieving a top-1 accuracy of 51%.

Quantization-aware INT8 training methods [79] enhance the practicality and performance of quantization in deep neural networks by cleverly optimizing calculations in both forward and backward passes through the thoughtful incorporation of loss-aware compensation and parameterized range clipping.

Within the scope of quantization strategies, binarization represents the most extreme form, reducing data to only two possible values: (0, 1) or $(-1, 1)$. This radical reduction allows for resource-efficient XNOR and bit-count operations instead of resource-intensive matrix multiplications. Therefore, binarization is particularly suitable for deploying deep neural networks on resource-constrained devices. However, this extreme quantization inevitably results in significant information loss. Additionally, the discontinuous nature of binary values presents optimization challenges for binary neural networks.

In recent years, to address these challenges, researchers have developed several promising algorithms. Innovative methods like BinaryConnect [80], binary neural networks (BNNs) [81], and XNOR-Net [82] have garnered attention in the field of binary neural networks. BinaryConnect introduces random techniques for binarizing neural networks, applying binarization during both forward and backward propagation. BNNs extend BinaryConnect by binarizing activations, marking a significant milestone in binary neural network research. Both BinaryConnect and BNNs employ combinations of deterministic and stochastic binarization functions to simplify hardware implementations. In contrast, XNOR-Net employs a different approach, using scaling factors for binary parameters to approximate floating-point parameters. This allows the weight quantization in XNOR-Net to be represented as $w \approx \alpha * bw$, where α represents the floating-point scaling factor of the binary weight bw.

Recent advancements include optimization-based binarization techniques. Methods like XNOR-Net and DoReFa-Net [72] concentrate on mitigating quantization errors during the training process. For instance, DoReFa-Net quantizes gradients to expedite training. In addition to local layer considerations, other binarization techniques, such as loss-aware binarization [83] and incremental network quantization [84], directly minimize the overall binary weight loss across the entire network. Through the application of an adaptive error decay estimator, IR-Net [85] represents a pioneering solution for preserving information flow during forward and backward propagation.

Different quantization strategies provide researchers with more options, requiring them to strike a balance between computational efficiency and model accuracy based on specific tasks in EEG signal processing.

3.3. Tensor Decomposition

Tensor decomposition is a technique aimed at breaking down weight tensors into smaller tensors or matrices. This process significantly reduces the number of parameters in CNNs while preserving accuracy. DNN parameters within convolutional layer weight tensors and fully connected layer weight matrices typically exhibit low-rank characteristics [86]. Based on this observation, the directions of algorithm design by researchers can be categorized into two main classes: low-rank matrix decomposition and tensor decomposition.

3.3.1. Low-Rank Matrix Decomposition

The core of low-rank matrix decomposition techniques lies in approximating the weight matrices of DNN layers by multiplying several low-rank matrices. Singular Value Decomposition (SVD) [87] is the most popular low-rank approximation method, emphasizing the description of a matrix through its singular values.

SVD decomposes a matrix $A \in R$ into three key components:

$$A = UDV \tag{5}$$

where $U \in R$ and $V \in R$ are orthogonal matrices, and $D \in R$ is a diagonal matrix containing singular values. A compact and efficient network model is achieved by retaining only the most critical components in the decomposition matrices.

In practice, SVD-based techniques have played a crucial role in various applications. For instance, in [88], SVD was utilized to decompose the product of weight matrices and input data, further enhancing model efficiency. The authors of [89] introduced sparsity into low-rank factorized matrices by maintaining a lower rank for less critical neurons, thus improving compression rates. Additionally, in [90], channel-wise SVD decomposition of convolutional layers was employed to enhance efficiency by subdividing kernels into two consecutive layers of different sizes.

Using SVD at every training node results in high computational costs. Therefore, it becomes important to determine the importance of each DNN layer during training. Innovative approaches have emerged to address these issues. In [91], an SVD training technique was proposed to explicitly achieve low-rank approximation without invoking SVD at every training step. Furthermore, [92] introduced a joint matrix factorization scheme that simultaneously decomposes layers with similar structures and optimizes based on SVD.

3.3.2. Tensorized Decomposition

Tensors, multi-way data arrays, provide a universal foundation for representing complex data. In the context of deep learning, these tensors encompass various orders, with two-dimensional matrices representing second-order tensors commonly used in fully connected layers. Meanwhile, the weight tensors of convolutional layers are represented as fourth-order tensors. Higher-rank tensors require higher compression ratios. Tensor decomposition extends the idea of low-rank matrix decomposition to high-order tensors, breaking down weight tensors into smaller tensors and thus reducing the network's complexity. Prominent methods in this field include Tucker decomposition [93], CANDECOMP/PARAFAC (CP) decomposition [94,95], tensor train (TT) decomposition [96–98], and tensor ring (TR) decomposition [99]. In [93], Tucker decomposition was used to compress convolutional weight kernels with predefined ranks. Simultaneously, the CP decomposition strategically breaks down convolutional kernels into a collection of first-order tensors, reducing parameters and training time. This method requires an iterative process that cleverly combines decomposition with fine-tuning for each convolutional layer. Refs. [94,95] differ in whether they decompose all convolutional layers. The former only decomposes a few convolutional layers, while the latter decomposes all convolutional layers. Ref. [100] introduced a robust low-rank decomposition technique that merges Tucker decomposition with CP decomposition. In this approach, the core tensors extracted through Tucker decomposition are further decomposed using CP decomposition. Other tensor decomposition techniques, such as TT decomposition, appear to be more favorable for recurrent neural networks (RNNs) [96], as their application can significantly reduce RNN model parameters by up to 1000 times. However, they face precision–tradeoff issues in CNN compression. Nevertheless, the authors of [97] proposed a TT format suitable for CNN models. Additionally, TT decomposition was found to be practical for compressing three-dimensional CNNs (3DCNNs) in [98], where an innovative approach for selecting TT ranks to maximize compression gains was outlined. Another pioneering approach involves utilizing TR decomposition, as shown in [99], which employs a progressive genetic algorithm for optimizing rank selection.

In summary, CP decomposition represents tensors as the sum of first-order tensors, offering a unique perspective on compression. On the other hand, Tucker decomposition breaks down tensors into groups of matrices and a compact core tensor. TT decomposition constructs complicated three-dimensional tensor structures, particularly adept at handling high-order tensors. TR decomposition is an extension of TT decomposition, emerging as a linear combination of TT decompositions. These tensor decomposition techniques continue to advance the efficiency of EEG signal processing.

3.4. *Knowledge Distillation*

Knowledge distillation is a technique that transfers knowledge from large, complex CNNs (teachers) to smaller, more efficient CNNs (students). The student network is trained to mimic the behavior of the teacher, resulting in a compact model with comparable performance. This concept can be traced back to the pioneering work in [101] and has since been extended to the context of deep learning [102]. The core challenge of knowledge distillation revolves around the techniques used to transfer knowledge from the teacher model to the student model, which involves three fundamental components—knowledge, distillation algorithms—and the architecture defining the relationship between the teacher and student models. In this context, knowledge manifests in various forms, including logits, activations, or features extracted from intermediate layers of the teacher model. The distillation algorithms can be categorized as offline, online, or self-distillation.

Offline distillation [101,103–105] extracts knowledge from a pre-trained teacher model and uses the soft label output of the teacher model to train the student network. The authors of [103] used data augmentation to exploit the output distributions of multiple teacher networks. The authors of [101] introduced a tailored distillation approach for quantized models, demonstrating that quantized student networks can closely match the accuracy of full-precision teacher networks while achieving high compression rates and inference acceleration. In contrast, [104] pioneered a data-free technique, training the student network using synthetic data responses from the complex teacher network. Online distillation [106–108] occurs during the simultaneous training of both the teacher and student models. It employs online knowledge distillation using the soft-label outputs of the teacher network. Ref. [106] proposed training the student model at different checkpoints of the teacher model until convergence is achieved. Meanwhile, Collaborative Learning Knowledge Distillation (KDCL) [107] dynamically generates high-quality soft targets through an ensemble approach for one-stage online training.

Furthermore, as observed in [105,108], knowledge distillation extends its influence to the intermediate layers of the teacher network. Layer Selective Learning (LSL) [105] provides a two-layer selection scheme, known as the inter-layer Gram matrix and layered inter-class Gram matrix, enabling the selection of intermediate layers in both the student and teacher networks for knowledge distillation. In [108], an updated ensemble-based knowledge distillation method was used to leverage features from intermediate layers, facilitating the simultaneous training of the distilled student system and the ensemble of teachers without the need for pre-trained teacher models.

On the other hand, an efficient neural network architecture design for self-knowledge distillation without training a large teacher network has been developed [109,110]. The authors of [109] introduced an auxiliary self-learning network that transfers refined knowledge to the classifier network using soft labels and intermediate feature maps. In [110], effective augmentation strategies were employed to extract knowledge using the best-performing student network from past epochs while simultaneously training the current student network, thereby enhancing network generalization.

In knowledge distillation, hardware constraints can be incorporated during the teacher model's training of the student model to guide it toward meeting specific hardware performance goals. The authors of [111] used knowledge distillation to develop NEMOKD. This system balances latency, accuracy, training time, and hardware costs. It was evaluated on the Movidius Myriad X VPU and Xilinx's programmable Z7020 FPGA. This hardware-aware distillation process minimizes memory usage and computational load, making it possible for the student model to run efficiently on wearable EEG devices.

3.5. Neural Architecture Search

Neural Architecture Search (NAS) is an automated approach for discovering the optimal neural network architecture. NAS defines a search space that includes various CNN models and performance evaluation algorithms. It generally does not require human intervention to discover the most efficient CNN model for specific needs. Unlike methods such as pruning, quantization, and tensor decomposition, as discussed above, NAS requires a specific model. Early NAS methods [112] initially generated a multitude of candidate network architectures based on predefined search space criteria. Each architecture undergoes rigorous training until convergence and is then ranked based on its accuracy on a validation dataset. These rankings provide feedback for fine-tuning the search strategy and generating new neural architectures. This iterative process continues until specified termination conditions are met. Ultimately, the most promising architectures are meticulously selected and evaluated using a test dataset. Since the search space contains thousands of neural networks, training and evaluating each network incurs significant computational and time costs.

Researchers need to find ways to identify efficient network architectures and reduce the computational costs required for training and evaluation. Current NAS methods typically involve three primary stages: training a super network, training and evaluating sampled networks, and finally, improving the discovered networks.

3.5.1. Search Space

The complexity increases exponentially when searching for networks and their interconnected structures in the NAS space [112]. Therefore, the search space needs to be limited. A limited number of network structures are used to form iterative and repeated units. Examples of this approach include NASNet [113], ENAS [114], and AutoDispNet [115]. For instance, AutoDispNet focuses on finding two different cell types, namely normal cells and reduction cells, within a predefined overall network framework. In contrast, ENAS involves an approach where all sub-models share weights collectively, significantly increasing GPU processing time but achieving over a thousandfold improvement compared to traditional NAS [112].

Depending on the specific task, we hope to find a balance between latency (inference time) and memory usage while maintaining satisfactory accuracy. MnasNet [116] is a platform-aware NAS tailored for mobile devices. It studies complex network architectures and analyzes the optimal trade-off between latency and accuracy.

3.5.2. Search Algorithm

In the complex field of NAS, search strategies shape the exploration of a specified search space. Typically, NAS involves a generator that generates sample architectures and an evaluator responsible for assessing their post-training performance. Various search algorithms, such as reinforcement learning (RL) and neural evolution algorithms, have emerged to address this essential issue. RL has achieved significant success, with NAS-RL [112] and MetaQnn [117] demonstrating classification accuracy in image classification tasks. Neuroevolution algorithms originally covered the optimization of neural structures and weights [118], but recently, the focus has shifted primarily toward optimizing neural structures [119]. Ref. [119] showed that both reinforcement learning and neural evolution consistently outperform random search (RS) in terms of final test accuracy. Specifically, neuroevolution consistently manages to strike a balance between high performance and compactness.

3.5.3. Performance Evaluation Strategy

The strategy for evaluating neural architecture performance is a fundamental aspect of the NAS field. In the early days of NAS, the approach was to assess the quality of each sampled network through exhaustive training from scratch [112]. However, this process proved to be highly costly in terms of computational resources and time. Recent advancements have introduced innovative techniques to streamline the performance evaluation step while maintaining high-quality results. Progressive NAS [120] and ReNAS [121] have ingeniously addressed this challenge by constructing accuracy predictors. These predictors are trained using data collected from a subset of sampled network candidates within the search space. During the search process, the trained predictor can evaluate the searched network at no additional cost. Another approach is known as one-shot NAS [122]. It reduces the evaluation cost by training a super network. Each sampled network inherits weights from the super network so that no additional training cost is required. This weight-sharing mechanism is at the core of one-shot NAS and plays a crucial role in reducing the computational burden associated with NAS, thereby significantly improving efficiency.

Automatically designed networks are very powerful but are not suitable for deployment on wearable EEG devices. Recognizing this challenge, hardware-aware NAS methods [123,124] have recently emerged. Unlike previous search strategies, they incorporate hardware factors, such as inference time, power consumption, and memory consumption, into the NAS search process. Ref. [125] proposed a two-stage automated hardware-aware search algorithm, which was evaluated under identical latency constraints across CPU, DPU, and VPU platforms. Compared to manually designed search spaces, this approach achieved higher accuracy. Ref. [126] introduced MemNAS, a search method that takes memory requirements into account. By progressively growing and pruning the network, this algorithm maintains a balance between accuracy and memory consumption across devices with varying resource constraints. To further reduce evaluation costs, hardware-aware NAS focuses on training a single network that can adapt to different hardware architectures, achieving a balance between high performance and cost-effectiveness.

3.5.4. EEG-Based Convolutional Neural Network Hardware Acceleration

In this subsection, we discuss the available hardware architectures suitable for CNNs. Next, we briefly review some cases showing representative CNN accelerators for EEG signal processing. Over the past decade, CNN hardware acceleration technology has developed significantly, mainly due to the emergence of high-performance GPUs, which are able to cope with the increasing memory requirements and computational complexity of CNNs. Although early applications of deep learning mostly focused on large computing platforms, there is now an increasing need to deploy CNNs on edge devices with limited hardware resources and energy, such as portable or wearable EEG signal-processing devices. To this end, hardware solutions for CNNs have expanded from general-purpose architectures (such as CPUs and GPUs [127]) to spatial architectures (such as FPGAs [22,128] and ASICs [129,130]).

The authors of [128] described the implementation of the VGG model on a PYNQ Z1 FPGA development board for motor imagery tasks. Figure 1 shows a CNN design based on a Xilinx FPGA. The authors found that quantizing from a 32-bit floating-point format to a 16-bit fixed-point format reduced BRAM and DSP resource usage by 25%, with no significant loss in accuracy. In contrast, quantizing to an 8-bit fixed-point format resulted in a 7% decrease in accuracy. The authors of [130] implemented the STFT and CNN models on a TSMC 28nm chip and tested them using the DEAP dataset for emotion recognition during movie-watching tasks.

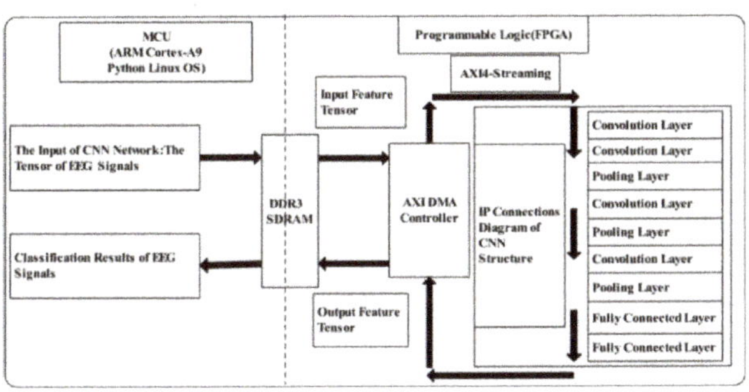

Figure 1. FPGA block diagram from [128].

The authors of [129] used a 180 nm 1P6M CMOS as a coprocessor to accelerate the CNN inference. The accelerator achieved 97.8% accuracy in floating-point operations and 93.5% accuracy in fixed-point operations. As shown in Figure 2, the CNN is transferred to the edge device after training. The authors also defined a coprocessor interface (CCI) to facilitate communication between the RISC-V core and the accelerator.

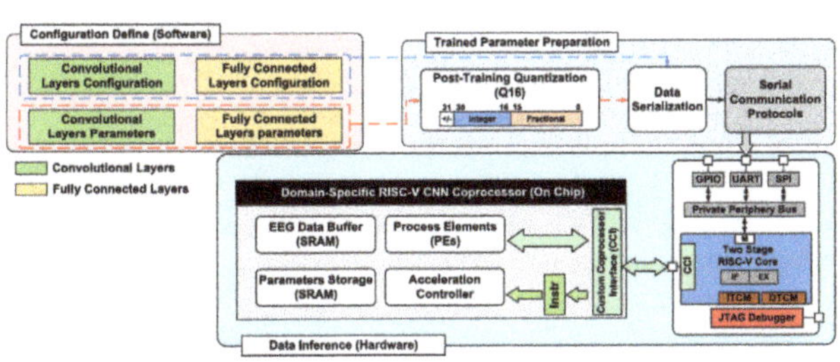

Figure 2. ASIC block diagram from [129].

In the context of portable EEG signal processing, FPGAs have become a mainstream choice for research. The reason is that the architectural design of FPGAs is very suitable for this purpose, and their unique flexibility and efficiency provide high computing performance with low power consumption. In [127], the authors compared three platforms—an Intel Core i7-7500U CPU, an NVIDIA Jetson Nano GPU, and a Xilinx Ultrascale+ ZU7EV FPGA—for motor imagination tasks. As shown in Figure 3, in terms of FOM and energy consumption rankings, the FPGA may be the best solution, followed by the Jetson Nano.

	CPU	FPGA	Jetson Nano	
Power consumption (W)	13.22	1.47	5	10
Inference time (ms)	121.57	19.97	15.85	12.21
FOM (W × s)	1.61	29.36×10^{-3}	79.25×10^{-3}	122.1×10^{-3}
Memory footprint (Mb)	476	6.37	415	
Test accuracy	81%	77%	81%	

Figure 3. Results on different platforms from [127].

In [9], the authors proposed a software and hardware co-design solution for a lightweight CNN. As shown in Figure 4, they proposed another FPGA acceleration implementation. Unlike [128], the calculation results of each layer are transmitted back to the ARM processor through the AXI bus. This implementation method can flexibly configure various parameters of the CNN to adapt to rapidly evolving CNN technology but at the cost of increased data pressure. The results of the eye-state recognition task were evaluated on three platforms. As shown in Figure 5, FPGA has obvious advantages in weight, cost, and power consumption. In contrast, it is slightly less efficient than a high-performance PC in terms of energy consumption.

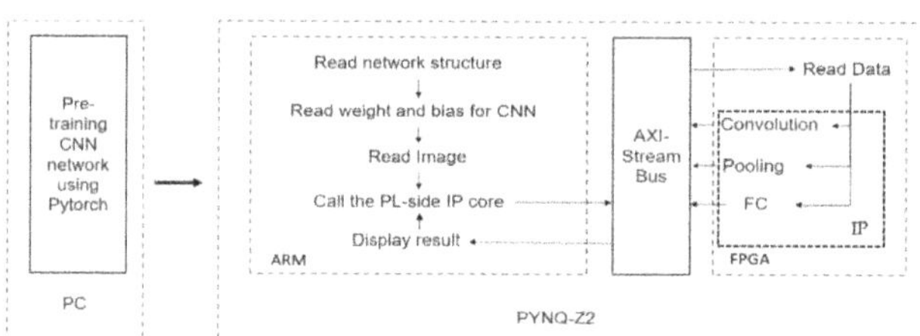

Figure 4. FPGA block diagram from [9].

Hardware Platform	CPU	Accelerator	Memory	Inference Time	Accuracy	Power	Power Efficiency	Price	Weight
Laptop	Intel Core i5-7300HQ @ 2.50 GHz	GTX 1050Ti @ 1620 Mhz	8 GB DDR4 @2666 MHz	0.003 s	89.92%	131.1 W	2.5	~$750	3.61 kg
PYNQ-Z2	ARM Dual-core Cortex A9 @ 650 MHz	-	512 MB DDR3 @525 MHz	5.1 s	89.85%	1.9 W	0.1	~$177	0.22 kg
PYNQ-Z2	ARM Dual-core Cortex A9 @ 650 MHz	FPGA @ 100 Mhz	512 MB DDR3 @525 MHz	0.22 s	89.69%	3.1 W	1.46	~$177	0.22 kg

Figure 5. Results on different platforms from [9].

The authors of [22] introduced BioCNN, which was designed for emotion classification tasks. BioCNN was implemented on a Digilent Atlys board with a low-cost Spartan-6 FPGA and tested using the FEXD and DEAP datasets. It achieved an energy efficiency of 11 GOPS/W and a latency of less than 1 ms, significantly below the standard 150 ms threshold for human–computer interaction and approaching the performance of software-based solutions. Compared with the GPU, the FPGA not only accelerates the CNN inference process by processing a large number of multiply-accumulate (MAC) operations in parallel but also optimizes memory access and data flow processing through customized hardware logic design. Because the FPGA's processing elements (PEs) have local storage and control logic, they can communicate directly with each other with minimal memory bandwidth requirements, significantly increasing data reuse. This feature is particularly critical for the low-power and efficient computing requirements of portable EEG devices.

In contrast, although GPUs perform well in the field of general computing, in the literature we surveyed, most of them were implemented in high-power PCs. Additionally, their low customization capabilities limit their application on portable devices. ASICs, on the other hand, although capable of delivering extreme performance and efficiency for specific applications, are generally only suitable for specific applications in mass production due to their high development costs and lack of flexibility. Therefore, FPGAs have become the preferred hardware platform for CNN acceleration in portable EEG signal-processing devices due to their low power consumption, flexibility, and sufficient computing power.

Table 3 summarizes the state-of-the-art CNN hardware solutions discussed in this section.

Table 3. List of reviewed CNN hardware techniques.

Method	Authors	Work Description
Weight Pruning	[51,52]	Hessian matrix of the loss function
	[53,54]	Three-step method
	[57]	Frequency-domain approach
	[51]	Variable compression rates
	[58]	GMM
	[59]	Energy reduction with accuracy loss
Structural Pruning	[60–62]	Criterion-based pruning
	[63]	Ranking-based pruning
	[64,65]	Optimization-based pruning
	[66–68]	Whole-layer pruning methods
	[69–71]	Automatically identifies and prunes
Quantization	[72]	Characterized by even step sizes
	[73]	Represented by logarithmic distributions
	[75,76]	Post-training quantization
	[77–79]	Quantization-aware training
	[80–82]	Binary quantization networks
	[72,82–85]	Optimized binarization techniques
LRM Decomposition	[88]	Decomposes the product of weight matrices and input data
	[89]	Sparsity: Maintaining a lower rank for less critical neurons
	[90]	Channel-wise SVD decomposition by dividing kernels into two layers
	[91]	Low-rank approximation without invoking SVD
	[92]	Joint matrix factorization scheme
Tensorized Decomposition	[93]	Tucker decomposition
	[94,95]	CANDECOMP/PARAFAC (CP)
	[96–98]	Tensor train decomposition
	[99]	Tensor ring decomposition
Knowledge Distillation	[101,103–105]	Offline distillation
	[106–108]	Online distillation
	[109,110]	Self-knowledge distillation
NA Search Space	[113]	NASNet
	[114]	ENAS
	[115]	AutoDispNet
	[116]	MnasNet
NA Search Algorithm	[112]	NAS-RL
	[117]	MetaQnn-RL
	[119]	Optimizing neural structures
NA Search Performance	[120]	Progressive NAS
	[121]	ReNAS
	[122]	One-shot NAS
	[123,124]	Hardware-aware NAS

"NA" means neural architecture. "LRM" means low-rank matrix.

4. Challenges and Opportunities

4.1. Standardized Evaluation Paradigms

Due to the different EEG application tasks, different indicators are needed to evaluate their performance. These indicators are diverse and complex. Due to the lack of standardized performance evaluation paradigms, not all EEG signal-processing hardware solutions provide the same indicators. Examples of these indicators include the following:

- **Wearability:** Wearability is an important feature in tasks that require constant, unobtrusive, real-time monitoring, such as sleep tracking or emotion classification. Uncomfortable or bulky devices may discourage users from consistently wearing them, potentially affecting data quality and continuity.
- **Energy Efficiency and Area**: Different hardware architectures cannot usually be directly compared. For example, FPGAs typically consume more power than ASICs. Moreover, the power consumption of an FPGA largely depends on its evaluation board. In other words, even the same FPGA design may lack comparability. For example, Ref. [131] introduced a normalized energy metric for comparing the power consumption of different technologies. The formula is as follows:

$$Normalized_{Energy} = \frac{Power \times ExecTime}{N_{Class}\left(\frac{V_{DD}}{V_{ref}}\right)^2} \tag{6}$$

  where the unit of normalized energy is nJ and N_{Class} is the number of classes.
- **Latency**: Latency is the time it takes for the system to produce classification results after receiving an EEG epoch. For instance, iMotor imagery tasks are highly dynamic, so shorter delays are desirable in such tasks. Subjects are less likely to maintain high levels of attention and exhibit consistent responses over longer time intervals. This contrasts sharply with seizure detection devices based on EEG, where longer detection time windows are more effective, as relevant seizures become sparser over time [132].

Standardized metrics and evaluation paradigms help to fairly compare different hardware solutions, promote healthy competition, and accelerate technological progress. For techniques such as compression, quantization, and hardware acceleration, standardized metrics help to explore the best efficiency for specific applications.

4.2. High-Quality EEG Datasets

Acquiring high-quality EEG datasets is a formidable challenge, significantly impacting the robustness assessment of systems. For DL networks, such as CNNs, that require large and diverse data, the quantity and quality of EEG data affect their training results. Unfortunately, most available EEG datasets lack the approval needed as benchmark resources to validate EEG-based classifiers in medical settings. Therefore, standardization of data collection or involving more subjects in the dataset development process can promote the development of EEG signal-processing technology, both in software and hardware.

Therefore, datasets must provide raw EEG data along with comprehensive data acquisition details. For instance, [133,134] used bandpass Hamming windows as filters. In contrast, [135] employed built-in Sinc filters to eliminate frequency components above specified thresholds, and [131] used a high-pass filter to remove unwanted event codes and noise channels from the raw input data. Equally important is defining stimulus types before EEG data collection, with common choices being auditory, visual, and audiovisual mixed stimuli. However, in a particular task, the choice of stimuli is determined subjectively by the researcher, making it difficult to replicate the classification results. Stimulus selection may become one of the challenges of EEG signal acquisition.

Reporting other subject-related information, such as race, occupation, personality traits, and physiological conditions, as well as comprehensive experimental setup details, is crucial for identifying external influences. Reference sensors in EEG data collection are also vital for anchoring data points. Some datasets [136] use single-modal approaches (only

EEG data), but multimodal datasets [133,137], including ECG, facial video clips, peripheral physiological signals, and infrared spectra, offer a more comprehensive perspective and additional classification strategies.

In particular, CNNs can seamlessly integrate these multi-modal data sources to enhance the model's generalization ability. In conclusion, future EEG-based datasets for standardized benchmarks should encompass multi-modality, universality, consistent labeling, and preprocessing libraries, among other features.

4.3. Embedded Preprocessing

As previously mentioned, the removal of artifacts can directly impact the performance of CNNs. These artifacts often mask meaningful neural activity. For instance, power-line noise can obscure cortical activity around 50 Hz (or 60 Hz), while electrode drift can disrupt slow cortical potentials. When eye- or muscle-related activity occurs simultaneously with external stimulation, these artifacts may be mistaken for cortical activity [138]. This is a common situation in EEG data collection experiments. However, artifact removal can sometimes introduce unintended side effects, such as the multiplicative offset triggered by high-pass filters, which may occasionally be misinterpreted as neural activity [139]. Filter-based preprocessing alone cannot eliminate all artifacts from EEG data. The design of EEG-based hardware systems also needs to be combined with other efficient preprocessing techniques. While the ICA-based techniques mentioned earlier are effective, they are not entirely hardware-friendly and may not comprehensively address all types of artifacts. For instance, ICA cannot distinguish artifacts originating from electrode-specific or sensor-specific sources [140]. Therefore, finding and developing solutions that can effectively handle various artifacts while maximizing the use of hardware characteristics remains a key direction for achieving real-time classification of EEG signals.

4.4. High-Resolution Feature Extraction

The concept of EEG frequency bands, although valuable in engineering applications for enhancing classification accuracy, lacks comprehensive neurophysiological underpinning. Relying on PSD as a proxy for extracting EEG signals representing brain activity is an effective but somewhat rudimentary approach. Integrating CFC features into EEG-based hardware classifiers presents a promising alternative for wearable devices. This approach has the potential to enhance robustness and reduce memory requirements by emphasizing connectivity graphs rather than storing complete raw EEG datasets [141].

In the field of frequency-domain features, another avenue for improvement involves systematically incorporating spatial asymmetries. As discussed in Section 2.2.4, extending this practice to other EEG electrode configurations can further refine our understanding of EEG wave patterns and their impact on CNN-based EEG classifiers in both software and hardware implementations.

5. Conclusions

This review focuses on preprocessing, feature extraction hardware solutions, and acceleration techniques for CNNs analyzing EEG signals. It encompasses a wide range of applications, including emotion classification, motor imagery, seizure detection, and sleep monitoring. This review begins with a detailed analysis of hardware implementation techniques for EEG signal preprocessing and five types of feature extraction: frequency-based, time-based, time-frequency, spatial features, and raw signals. Subsequently, efficient CNN solutions based on EEG signals are elaborated, including pruning, quantization, tensor decomposition, knowledge distillation, and NAS. In the following sections, this review addresses the challenges and opportunities associated with EEG hardware accelerators, such as multi-modal universal datasets, standardized metrics, embedded preprocessing, and high-resolution feature extraction. Therefore, this article summarizes existing research results in the field of EEG signal-processing hardware implementation by reviewing the current literature. It also identifies unresolved problems and research gaps and provides future research directions for other researchers.

Author Contributions: Conceptualization, Y.X. and S.O.; methodology, Y.X.; validation, Y.X.; formal analysis, Y.X.; investigation, Y.X.; writing—original draft preparation, Y.X.; writing—review and editing, Y.X. and S.O.; supervision, S.O. All authors have read and agreed to the published version of the manuscript.

Funding: This research received no external funding.

Conflicts of Interest: The authors declare no conflicts of interest.

Abbreviations

The following abbreviations are used in this manuscript:

ARRC	Auto-Regressive Reflection Coefficient	HOS	Higher-Order Spectral
APoZ	Average Percentage of Zeros	ICA	Independent Component Analysis
ASKI	Approximate Skewness Indicator	KDCL	Collaborative Learning Knowledge Distillation
BCI	Brain–Computer Interface	LFCC	Linear-Frequency Cepstral Coefficient
BNN	Binary Neural Networks	LDA	Linear Discriminant Analysis
CP	CANDECOMP/PARAFAC	LBP	Local Binary Pattern
CFC	Cross-Frequency Coupling	LPP	Late Positive Potentials
CNN	Convolutional Neural Network	LSL	Layer Selective Learning
CSP	Common Spatial Pattern	LSL	Layer Selective Learning
CWT	Continuous Wavelet Transform	MFCC	Mel-Frequency Cepstral Coefficient
CORDIC	Coordinate Rotation Digital Computer	MD	Mahalanobis Distance
DCT	Discrete Cosine Transform	NAS	Neural Architecture Search
DE	Differential Entropy	IMF	Independent Mode Function
DTF	Directed Transfer Function	RMS	Root Mean Square
DWT	Discrete Wavelet Transform	Var	Variance
DNN	Deep Neural Network	SD	Standard Deviation
GPUs	Graphics Processing Units	NTSPP	Neural Time-Series Prediction Processing
GP	Grassberger–Procaccia	OBD	Optimal Brain Damage
GMM	Gaussian Mixture Model	OBS	Optimal Brain Surgeon
EEG	Electroencephalography	PSD	Power Spectral Density
EVD	Eigenvalue Decomposition	RNN	Recurrent Neural Network
EMD	Empirical Mode Decomposition	RL	Reinforcement Learning
FPGA	Field-Programmable Gate Array	RS	Random Search
FRL	Final Response Layer	SK	Skewness
FDSOI	Fully Depleted Silicon on Insulator	SE	Sample Entropy
HFD	Higuchi's Fractal Dimension	SCSSP	Scalp-Coupled Common Spatial Spectral Pattern
HHT	Hilbert–Huang Transform	SVD	Singular-Value Decomposition
HAF	Hybrid Adaptive Filtering	SFP	Soft Filter Pruning
SVM	Support Vector Machine	TT	Tensor Train
STFT	Short-Time Fourier Transform	WT	Wavelet Transform
TR	Tensor Ring		

References

1. Craik, A.; He, Y.; Contreras-Vidal, J.L. Deep learning for electroencephalogram (EEG) classification tasks: A review. *J. Neural Eng.* **2019**, *16*, 031001. [CrossRef] [PubMed]
2. Lotte, F.; Bougrain, L.; Cichocki, A.; Clerc, M.; Congedo, M.; Rakotomamonjy, A.; Yger, F. A review of classification algorithms for EEG-based brain–computer interfaces: A 10 year update. *J. Neural Eng.* **2018**, *15*, 031005. [CrossRef]
3. Wei, Y.; Zhou, J.; Wang, Y.; Liu, Y.; Liu, Q.; Luo, J.; Wang, C.; Ren, F.; Huang, L. A review of algorithm & hardware design for AI-based biomedical applications. *IEEE Trans. Biomed. Circuits Syst.* **2020**, *14*, 145–163.
4. Nunez, P.L.; Srinivasan, R. *Electric Fields of the Brain: The Neurophysics of EEG*; Oxford University Press: Oxford, UK, 2006.
5. Aggarwal, S.; Chugh, N. Review of machine learning techniques for EEG based brain computer interface. *Arch. Comput. Methods Eng.* **2022**, *29*, 3001–3020. [CrossRef]
6. Lu, H.; Wang, M.; Sangaiah, A.K. Human emotion recognition using an EEG cloud computing platform. *Mob. Netw. Appl.* **2020**, *25*, 1023–1032. [CrossRef]
7. Dong, L.; Li, J.; Zou, Q.; Zhang, Y.; Zhao, L.; Wen, X.; Gong, J.; Li, F.; Liu, T.; Evans, A.C.; et al. WeBrain: A web-based brainformatics platform of computational ecosystem for EEG big data analysis. *NeuroImage* **2021**, *245*, 118713. [CrossRef] [PubMed]

8. Ghimire, D.; Kil, D.; Kim, S.H. A survey on efficient convolutional neural networks and hardware acceleration. *Electronics* **2022**, *11*, 945. [CrossRef]
9. Xie, Y.; Majoros, T.; Oniga, S. FPGA-Based Hardware Accelerator on Portable Equipment for EEG Signal Patterns Recognition. *Electronics* **2022**, *11*, 2410. [CrossRef]
10. Johannisson, T. Correlations between personality traits and specific groups of alpha waves in the human EEG. *PeerJ* **2016**, *4*, e2245. [CrossRef]
11. Radüntz, T.; Scouten, J.; Hochmuth, O.; Meffert, B. EEG artifact elimination by extraction of ICA-component features using image processing algorithms. *J. Neurosci. Methods* **2015**, *243*, 84–93. [CrossRef]
12. Narayan, R. *Encyclopedia of Biomedical Engineering*; Elsevier: Amsterdam, The Netherlands, 2018.
13. Stone, D.B.; Tamburro, G.; Fiedler, P.; Haueisen, J.; Comani, S. Automatic removal of physiological artifacts in EEG: The optimized fingerprint method for sports science applications. *Front. Hum. Neurosci.* **2018**, *12*, 96. [CrossRef] [PubMed]
14. Bhuvaneswari, P.; Kumar, J.S. Methods used for identifying EEG signal artifacts. In Proceedings of the International Conference on Computational Intelligence and Information Technology, Coimbatore, Tamilnadu, India, 2–3 March 2012; pp. 375–379.
15. Khatter, A.; Bansal, D.; Mahajan, R. Study of various automatic eeg artifact removal techniques. *Int. J. Res. Appl. Sci. Eng. Technol.* **2017**, *5*, 1027–1037. [CrossRef]
16. Saltuklaroglu, T.; Bowers, A.; Harkrider, A.W.; Casenhiser, D.; Reilly, K.J.; Jenson, D.E.; Thornton, D. EEG mu rhythms: Rich sources of sensorimotor information in speech processing. *Brain Lang.* **2018**, *187*, 41–61. [CrossRef]
17. Yang, C.J.; Fahier, N.; He, C.Y.; Li, W.C.; Fang, W.C. An ai-edge platform with multimodal wearable physiological signals monitoring sensors for affective computing applications. In Proceedings of the 2020 IEEE International Symposium on Circuits and Systems (ISCAS), Online, 10–21 October 2020; pp. 1–5.
18. Aslam, A.R.; Altaf, M.A.B. An 8 channel patient specific neuromorphic processor for the early screening of autistic children through emotion detection. In Proceedings of the 2019 IEEE International Symposium on Circuits and Systems (ISCAS), Sapporo, Japan, 26–29 May; pp. 1–5.
19. Aslam, A.R.; Iqbal, T.; Aftab, M.; Saadeh, W.; Altaf, M.A.B. A10. 13 uJ/classification 2-channel deep neural network-based SoC for emotion detection of autistic children. In Proceedings of the 2020 IEEE Custom Integrated Circuits Conference (CICC), Boston, MA, USA, 22–25 March 2020; pp. 1–4.
20. Yarahuaman, J.C.R.; Huamaní-Navarrete, P.F. Design and Simulation of a Digital Filter in Hardware for EEG Signals Based on FPGA. In Proceedings of the 2020 IEEE ANDESCON, Quito, Ecuador, 13–16 October 2020; pp. 1–6.
21. Aslam, A.R.; Altaf, M.A.B. An on-chip processor for chronic neurological disorders assistance using negative affectivity classification. *IEEE Trans. Biomed. Circuits Syst.* **2020**, *14*, 838–851. [CrossRef]
22. Gonzalez, H.A.; Muzaffar, S.; Yoo, J.; Elfadel, I.A.M. An inference hardware accelerator for EEG-based emotion detection. In Proceedings of the 2020 IEEE International Symposium on Circuits and Systems (ISCAS), Online, 10–21 October 2020; pp. 1–5.
23. Yang, C.H.; Shih, Y.H.; Chiueh, H. An 81.6 μW FastICA Processor for Epileptic Seizure Detection. *IEEE Trans. Biomed. Circuits Syst.* **2014**, *9*, 60–71. [CrossRef]
24. Wu, D.; Han, X.; Yang, Z.; Wang, R. Exploiting transfer learning for emotion recognition under cloud-edge-client collaborations. *IEEE J. Sel. Areas Commun.* **2020**, *39*, 479–490. [CrossRef]
25. Tao, X.; Chen, Z.; Xu, M.; Lu, J. Rebuffering optimization for DASH via pricing and EEG-based QoE modeling. *IEEE J. Sel. Areas Commun.* **2019**, *37*, 1549–1565. [CrossRef]
26. Chen, Y.H.; Chen, S.W.; Wei, M.X. A VLSI implementation of independent component analysis for biomedical signal separation using CORDIC engine. *IEEE Trans. Biomed. Circuits Syst.* **2020**, *14*, 373–381. [CrossRef] [PubMed]
27. Chen, Y.H.; Wang, S.P. Low-cost implementation of independent component analysis for biomedical signal separation using very-large-scale integration. *IEEE Trans. Circuits Syst. II Express Briefs* **2020**, *67*, 3437–3441. [CrossRef]
28. Altaheri, H.; Muhammad, G.; Alsulaiman, M.; Amin, S.U.; Altuwaijri, G.A.; Abdul, W.; Bencherif, M.A.; Faisal, M. Deep learning techniques for classification of electroencephalogram (EEG) motor imagery (MI) signals: A review. *Neural Comput. Appl.* **2023**, *35*, 14681–14722. [CrossRef]
29. Knyazev, G.G.; Savostyanov, A.N.; Bocharov, A.V.; Tamozhnikov, S.S.; Kozlova, E.A.; Leto, I.V.; Slobodskaya, H.R. Cross-frequency coupling in developmental perspective. *Front. Hum. Neurosci.* **2019**, *13*, 158. [CrossRef]
30. Yang, J.; Sawan, M. From seizure detection to smart and fully embedded seizure prediction engine: A review. *IEEE Trans. Biomed. Circuits Syst.* **2020**, *14*, 1008–1023. [CrossRef]
31. Davis, P.; Creusere, C.D.; Tang, W. ASIC implementation of the cross frequency coupling algorithm for EEG signal processing. In Proceedings of the 2014 International Symposium on Integrated Circuits (ISIC), Marina Bay Sands, Singapore, 10–12 December 2014; pp. 248–251.
32. Rizal, A.; Hadiyoso, S.; Ramdani, A.Z. FPGA-Based Implementation for Real-Time Epileptic EEG Classification Using Hjorth Descriptor and KNN. *Electronics* **2022**, *11*, 3026. [CrossRef]
33. Richman, J.S.; Moorman, J.R. Physiological time-series analysis using approximate entropy and sample entropy. *Am. J. Physiol. Heart Circ. Physiol.* **2000**, *278*, H2039–H2049. [CrossRef]
34. Chen, C.; Da Silva, B.; Li, J.; Liu, C. Acceleration of Fast Sample Entropy Towards Biomedical Applications on FPGAs. In Proceedings of the 2022 International Conference on Field-Programmable Technology (ICFPT), Hong Kong, China, 5–9 December 2022; pp. 1–4.

35. Wickramasuriya, D.S.; Wijesinghe, L.P.; Mallawaarachchi, S. Seizure prediction using Hilbert Huang Transform on field programmable gate array. In Proceedings of the 2015 IEEE Global Conference on Signal and Information Processing (GlobalSIP), Orlando, FL, USA, 14–16 December 2015; pp. 933–937.
36. Zhang, L. Real-time feature extraction for multi-channel EEG signals time-frequency analysis. In Proceedings of the 2017 8th International IEEE/EMBS Conference on Neural Engineering (NER), Shanghai, China, 25–28 May 2017; pp. 493–496.
37. Beeraka, S.M.; Kumar, A.; Sameer, M.; Ghosh, S.; Gupta, B. Accuracy enhancement of epileptic seizure detection: A deep learning approach with hardware realization of STFT. *Circuits Syst. Signal Process.* **2022**, *41*, 461–484. [CrossRef]
38. Fang, W.C.; Wang, K.Y.; Fahier, N.; Ho, Y.L.; Huang, Y.D. Development and validation of an EEG-based real-time emotion recognition system using edge AI computing platform with convolutional neural network system-on-chip design. *IEEE J. Emerg. Sel. Top. Circuits Syst.* **2019**, *9*, 645–657. [CrossRef]
39. Ellawala, N.; Thayaparan, S. Hardware implementation of eeg classifier using lda. In Proceedings of the 2019 2nd International Conference on Bioinformatics, Biotechnology and Biomedical Engineering (BioMIC)-Bioinformatics and Biomedical Engineering, Yogyakarta, IN, USA, 12–13 September 2019; Volume 1, pp. 1–5.
40. Qassim, Y.T.; Cutmore, T.; James, D.; Rowlands, D. FPGA implementation of Morlet continuous wavelet transform for EEG analysis. In Proceedings of the 2012 International Conference on Computer and Communication Engineering (ICCCE), Kuala Lumpur, Malaysia, 3–5 July 2012; pp. 59–64.
41. Belwafi, K.; Ghaffari, F.; Djemal, R.; Romain, O. A hardware/software prototype of EEG-based BCI system for home device control. *J. Signal Process. Syst.* **2017**, *89*, 263–279. [CrossRef]
42. Kais, B.; Ghaffari, F.; Romain, O.; Djemal, R. An embedded implementation of home devices control system based on brain computer interface. In Proceedings of the 2014 26th International Conference on Microelectronics (ICM), Doha, Qatar, 14–17 December 2014; pp. 140–143.
43. Malekmohammadi, A.; Mohammadzade, H.; Chamanzar, A.; Shabany, M.; Ghojogh, B. An efficient hardware implementation for a motor imagery brain computer interface system. *Sci. Iran.* **2019**, *26*, 72–94. [CrossRef]
44. Duan, R.N.; Zhu, J.Y.; Lu, B.L. Differential entropy feature for EEG-based emotion classification. In Proceedings of the 2013 6th International IEEE/EMBS Conference on Neural Engineering (NER), San Diego, CA, USA, 6–8 November 2013; pp. 81–84.
45. Kaminski, M.J.; Blinowska, K.J. A new method of the description of the information flow in the brain structures. *Biol. Cybern.* **1991**, *65*, 203–210. [CrossRef]
46. Wang, Y.; Huang, Z.; McCane, B.; Neo, P. EmotioNet: A 3-D convolutional neural network for EEG-based emotion recognition. In Proceedings of the 2018 International Joint Conference on Neural Networks (IJCNN), Rio de Janeiro, Brazil, 8–13 July 2018; pp. 1–7.
47. Salama, E.S.; El-Khoribi, R.A.; Shoman, M.E.; Shalaby, M.A.W. EEG-based emotion recognition using 3D convolutional neural networks. *Int. J. Adv. Comput. Sci. Appl.* **2018**, *9*, *8*. [CrossRef]
48. Singh, K.; de Gyvez, J.P. Twenty years of near/sub-threshold design trends and enablement. *IEEE Trans. Circuits Syst. II Express Briefs* **2020**, *68*, 5–11. [CrossRef]
49. Höppner, S.; Eisenreich, H.; Walter, D.; Scharfe, A.; Oefelein, A.; Schraut, F.; Schreiter, J.; Riedel, T.; Bauer, H.; Niebsch, R. Adaptive body bias aware implementation for ultra-low-voltage designs in 22FDX technology. *IEEE Trans. Circuits Syst. II Express Briefs* **2019**, *67*, 2159–2163. [CrossRef]
50. Gao, Z.; Li, Y.; Yang, Y.; Dong, N.; Yang, X.; Grebogi, C. A coincidence-filtering-based approach for CNNs in EEG-based recognition. *IEEE Trans. Ind. Inform.* **2019**, *16*, 7159–7167. [CrossRef]
51. Ding, X.; Zhou, X.; Guo, Y.; Han, J.; Liu, J. Global sparse momentum sgd for pruning very deep neural networks. *Adv. Neural Inf. Process. Syst.* **2019**, *32*, 573.
52. Hassibi, B.; Stork, D. Second order derivatives for network pruning: Optimal brain surgeon. *Adv. Neural Inf. Process. Syst.* **1992**, *5*.
53. Han, S.; Pool, J.; Tran, J.; Dally, W. Learning both weights and connections for efficient neural network. *Adv. Neural Inf. Process. Syst.* **2015**, *28*.
54. Han, S.; Mao, H.; Dally, W.J. Deep compression: Compressing deep neural networks with pruning, trained quantization and huffman coding. *arXiv* **2015**, arXiv:1510.00149.
55. Zhu, M.; Gupta, S. To prune, or not to prune: Exploring the efficacy of pruning for model compression. *arXiv* **2017**, arXiv:1710.01878.
56. Alford, S.; Robinett, R.; Milechin, L.; Kepner, J. Training behavior of sparse neural network topologies. In Proceedings of the 2019 IEEE High Performance Extreme Computing Conference (HPEC), Westin Hotel, Waltham, MA, USA, 24–26 September 2019; pp. 1–6.
57. Liu, Z.; Xu, J.; Peng, X.; Xiong, R. Frequency-domain dynamic pruning for convolutional neural networks. *Adv. Neural Inf. Process. Syst.* **2018**, *11*, 1051–1061.
58. Lee, E.; Hwang, Y. Layer-Wise Network Compression Using Gaussian Mixture Model. *Electronics* **2021**, *10*, 72. [CrossRef]
59. Yang, T.J.; Chen, Y.H.; Sze, V. Designing energy-efficient convolutional neural networks using energy-aware pruning. In Proceedings of the IEEE Conference on Computer Vision and Pattern Recognition, Honolulu, HI, USA, 21–26 July 2017; pp. 5687–5695.
60. Li, H.; Kadav, A.; Durdanovic, I.; Samet, H.; Graf, H.P. Pruning filters for efficient convnets. *arXiv* **2016**, arXiv:1608.08710.
61. He, Y.; Kang, G.; Dong, X.; Fu, Y.; Yang, Y. Soft filter pruning for accelerating deep convolutional neural networks. *arXiv* **2018**, arXiv:1808.06866.

62. Hu, H.; Peng, R.; Tai, Y.W.; Tang, C.K. Network trimming: A data-driven neuron pruning approach towards efficient deep architectures. *arXiv* **2016**, arXiv:1607.03250.
63. Lin, M.; Ji, R.; Wang, Y.; Zhang, Y.; Zhang, B.; Tian, Y.; Shao, L. Hrank: Filter pruning using high-rank feature map. In Proceedings of the IEEE/CVF Conference on Computer Vision and Pattern Recognition, Seattle, WA, USA, 13–19 June 2020; pp. 1529–1538.
64. Luo, J.H.; Wu, J.; Lin, W. Thinet: A filter level pruning method for deep neural network compression. In Proceedings of the IEEE International Conference on Computer Vision, Venice, Italy, 22–29 October 2017; pp. 5058–5066.
65. Yu, R.; Li, A.; Chen, C.F.; Lai, J.H.; Morariu, V.I.; Han, X.; Gao, M.; Lin, C.Y.; Davis, L.S. Nisp: Pruning networks using neuron importance score propagation. In Proceedings of the IEEE Conference on Computer Vision and Pattern Recognition, Salt Lake City, UT, USA, 18–23 June 2018; pp. 9194–9203.
66. Chen, S.; Zhao, Q. Shallowing deep networks: Layer-wise pruning based on feature representations. *IEEE Trans. Pattern Anal. Mach. Intell.* **2018**, *41*, 3048–3056. [CrossRef] [PubMed]
67. Elkerdawy, S.; Elhoushi, M.; Singh, A.; Zhang, H.; Ray, N. To filter prune, or to layer prune, that is the question. In Proceedings of the Asian Conference on Computer Vision, Kyoto, Japan, 30 November–4 December 2020.
68. Xu, P.; Cao, J.; Shang, F.; Sun, W.; Li, P. Layer pruning via fusible residual convolutional block for deep neural networks. *arXiv* **2020**, arXiv:2011.14356.
69. Liu, Z.; Sun, M.; Zhou, T.; Huang, G.; Darrell, T. Rethinking the value of network pruning. *arXiv* **2018**, arXiv:1810.05270.
70. Liu, Z.; Li, J.; Shen, Z.; Huang, G.; Yan, S.; Zhang, C. Learning efficient convolutional networks through network slimming. In Proceedings of the IEEE International Conference on Computer Vision, Venice, Italy, 22–29 October 2017; pp. 2736–2744.
71. Wang, Y.; Zhang, X.; Xie, L.; Zhou, J.; Su, H.; Zhang, B.; Hu, X. Pruning from scratch. In Proceedings of the AAAI Conference on Artificial Intelligence, New York, NY, USA, 7–12 February 2020; Volume 34, pp. 12273–12280.
72. Zhou, S.; Wu, Y.; Ni, Z.; Zhou, X.; Wen, H.; Zou, Y. Dorefa-net: Training low bitwidth convolutional neural networks with low bitwidth gradients. *arXiv* **2016**, arXiv:1606.06160.
73. Miyashita, D.; Lee, E.H.; Murmann, B. Convolutional neural networks using logarithmic data representation. *arXiv* **2016**, arXiv:1603.01025.
74. Fiesler, E.; Choudry, A.; Caulfield, H.J. Weight discretization paradigm for optical neural networks. In Proceedings of the Optical Interconnections and Networks, Hamburg, Germany, 14–15 March 1990; Volume 1281, pp. 164–173.
75. Wu, H.; Judd, P.; Zhang, X.; Isaev, M.; Micikevicius, P. Integer quantization for deep learning inference: Principles and empirical evaluation. *arXiv* **2020**, arXiv:2004.09602.
76. Banner, R.; Nahshan, Y.; Soudry, D. Post training 4-bit quantization of convolutional networks for rapid-deployment. *Adv. Neural Inf. Process. Syst.* **2019**, *32*.
77. Jacob, B.; Kligys, S.; Chen, B.; Zhu, M.; Tang, M.; Howard, A.; Adam, H.; Kalenichenko, D. Quantization and training of neural networks for efficient integer-arithmetic-only inference. In Proceedings of the IEEE Conference on Computer Vision and Pattern Recognition, Salt Lake City, UT, USA, 18–22 June 2018; pp. 2704–2713.
78. Hubara, I.; Courbariaux, M.; Soudry, D.; El-Yaniv, R.; Bengio, Y. Quantized neural networks: Training neural networks with low precision weights and activations. *J. Mach. Learn. Res.* **2017**, *18*, 6869–6898.
79. Zhou, Q.; Guo, S.; Qu, Z.; Guo, J.; Xu, Z.; Zhang, J.; Guo, T.; Luo, B.; Zhou, J. Octo:INT8 training with loss-aware compensation and backward quantization for tiny on-device learning. In Proceedings of the 2021 USENIX Annual Technical Conference (USENIX ATC 21), Online, 14–16 July 2021; pp. 177–191.
80. Courbariaux, M.; Bengio, Y.; David, J.P. Binaryconnect: Training deep neural networks with binary weights during propagations. *Adv. Neural Inf. Process. Syst.* **2015**, *28*.
81. Courbariaux, M.; Hubara, I.; Soudry, D.; El-Yaniv, R.; Bengio, Y. Binarized neural networks: Training deep neural networks with weights and activations constrained to +1 or −1. *arXiv* **2016**, arXiv:1602.02830.
82. Rastegari, M.; Ordonez, V.; Redmon, J.; Farhadi, A. Xnor-net: Imagenet classification using binary convolutional neural networks. In Proceedings of the European Conference on Computer Vision, Amsterdam, The Netherlands, 8–16 October 2016; pp. 525–542.
83. Hou, L.; Yao, Q.; Kwok, J.T. Loss-aware binarization of deep networks. *arXiv* **2016**, arXiv:1611.01600.
84. Zhou, A.; Yao, A.; Guo, Y.; Xu, L.; Chen, Y. Incremental network quantization: Towards lossless cnns with low-precision weights. *arXiv* **2017**, arXiv:1702.03044.
85. Qin, H.; Gong, R.; Liu, X.; Shen, M.; Wei, Z.; Yu, F.; Song, J. Forward and backward information retention for accurate binary neural networks. In Proceedings of the IEEE/CVF Conference on Computer Vision and Pattern Recognition, Seattle, WA, USA, 13–19 June 2020; pp. 2250–2259.
86. Denil, M.; Shakibi, B.; Dinh, L.; Ranzato, M.; De Freitas, N. Predicting parameters in deep learning. *Adv. Neural Inf. Process. Syst.* **2013**, *26*.
87. Klema, V.; Laub, A. The singular value decomposition: Its computation and some applications. *IEEE Trans. Autom. Control.* **1980**, *25*, 164–176. [CrossRef]
88. Xue, J.; Li, J.; Yu, D.; Seltzer, M.; Gong, Y. Singular value decomposition based low-footprint speaker adaptation and personalization for deep neural network. In Proceedings of the 2014 IEEE International Conference on Acoustics, Speech and Signal Processing (ICASSP), Florence, Italy, 4–9 May 2014; pp. 6359–6363.
89. Swaminathan, S.; Garg, D.; Kannan, R.; Andres, F. Sparse low rank factorization for deep neural network compression. *Neurocomputing* **2020**, *398*, 185–196. [CrossRef]

90. Zhang, X.; Zou, J.; He, K.; Sun, J. Accelerating very deep convolutional networks for classification and detection. *IEEE Trans. Pattern Anal. Mach. Intell.* **2015**, *38*, 1943–1955. [CrossRef]
91. Yang, H.; Tang, M.; Wen, W.; Yan, F.; Hu, D.; Li, A.; Li, H.; Chen, Y. Learning low-rank deep neural networks via singular vector orthogonality regularization and singular value sparsification. In Proceedings of the IEEE/CVF Conference on Computer Vision and Pattern Recognition Workshops, Seattle, WA, USA, 14–19 June 2020; pp. 678–679.
92. Chen, S.; Zhou, J.; Sun, W.; Huang, L. Joint matrix decomposition for deep convolutional neural networks compression. *Neurocomputing* **2023**, *516*, 11–26. [CrossRef]
93. Kim, Y.D.; Park, E.; Yoo, S.; Choi, T.; Yang, L.; Shin, D. Compression of deep convolutional neural networks for fast and low power mobile applications. *arXiv* **2015**, arXiv:1511.06530.
94. Lebedev, V.; Ganin, Y.; Rakhuba, M.; Oseledets, I.; Lempitsky, V. Speeding-up convolutional neural networks using fine-tuned cp-decomposition. *arXiv* **2014**, arXiv:1412.6553.
95. Astrid, M.; Lee, S.I. Cp-decomposition with tensor power method for convolutional neural networks compression. In Proceedings of the 2017 IEEE International Conference on Big Data and Smart Computing (BigComp), Jeju Island, Republic of Korea, 13–16 February 2017; pp. 115–118.
96. Yang, Y.; Krompass, D.; Tresp, V. Tensor-train recurrent neural networks for video classification. In Proceedings of the International Conference on Machine Learning, Sydney, Australia, 6–11 August 2017; pp. 3891–3900.
97. Yin, M.; Sui, Y.; Liao, S.; Yuan, B. Towards efficient tensor decomposition-based dnn model compression with optimization framework. In Proceedings of the IEEE/CVF Conference on Computer Vision and Pattern Recognition, Nashville, TN, USA, 20–25 June 2021; pp. 10674–10683.
98. Wang, D.; Zhao, G.; Li, G.; Deng, L.; Wu, Y. Compressing 3DCNNs based on tensor train decomposition. *Neural Netw.* **2020**, *131*, 215–230. [CrossRef]
99. Li, N.; Pan, Y.; Chen, Y.; Ding, Z.; Zhao, D.; Xu, Z. Heuristic rank selection with progressively searching tensor ring network. In *Complex & Intelligent Systems*; Springer: Berlin/Heidelberg, Germany, 2021; pp. 1–15.
100. Phan, A.H.; Sobolev, K.; Sozykin, K.; Ermilov, D.; Gusak, J.; Tichavský, P.; Glukhov, V.; Oseledets, I.; Cichocki, A. Stable low-rank tensor decomposition for compression of convolutional neural network. In Proceedings of the Computer Vision—ECCV 2020: 16th European Conference, Glasgow, UK, 23–28 August 2020; Proceedings, Part XXIX 16; Springer: Berlin/Heidelberg, Germany, 2020; pp. 522–539.
101. Polino, A.; Pascanu, R.; Alistarh, D. Model compression via distillation and quantization. *arXiv* **2018**, arXiv:1802.05668.
102. Hinton, G.; Vinyals, O.; Dean, J. Distilling the knowledge in a neural network. *arXiv* **2015**, arXiv:1503.02531.
103. Fukuda, T.; Suzuki, M.; Kurata, G.; Thomas, S.; Cui, J.; Ramabhadran, B. Efficient Knowledge Distillation from an Ensemble of Teachers. In Proceedings of the Interspeech, Stockholm, Sweden, 20–24 August 2017; pp. 3697–3701.
104. Nayak, G.K.; Mopuri, K.R.; Shaj, V.; Radhakrishnan, V.B.; Chakraborty, A. Zero-shot knowledge distillation in deep networks. In Proceedings of the International Conference on Machine Learning, Long Beach, CA, USA, 9–15 June 2019; pp. 4743–4751.
105. Li, H.T.; Lin, S.C.; Chen, C.Y.; Chiang, C.K. Layer-level knowledge distillation for deep neural network learning. *Appl. Sci.* **2019**, *9*, 1966. [CrossRef]
106. Jin, X.; Peng, B.; Wu, Y.; Liu, Y.; Liu, J.; Liang, D.; Yan, J.; Hu, X. Knowledge distillation via route constrained optimization. In Proceedings of the IEEE/CVF International Conference on Computer Vision, Seoul, Republic of Korea, 27 October–2 November 2019; pp. 1345–1354.
107. Guo, Q.; Wang, X.; Wu, Y.; Yu, Z.; Liang, D.; Hu, X.; Luo, P. Online knowledge distillation via collaborative learning. In Proceedings of the IEEE/CVF Conference on Computer Vision and Pattern Recognition, Seattle, WA, USA, 13–19 June 2020; pp. 11020–11029.
108. Walawalkar, D.; Shen, Z.; Savvides, M. Online ensemble model compression using knowledge distillation. In Proceedings of the Computer Vision—ECCV 2020: 16th European Conference, Glasgow, UK, 23–28 August 2020; Proceedings, Part XIX 16; Springer: Berlin/Heidelberg, Germany, 2020; pp. 18–35.
109. Ji, M.; Shin, S.; Hwang, S.; Park, G.; Moon, I.C. Refine myself by teaching myself: Feature refinement via self-knowledge distillation. In Proceedings of the IEEE/CVF Conference on Computer Vision and Pattern Recognition, Nashville, TN, USA, 20–25 June 2021; pp. 10664–10673.
110. Vu, D.Q.; Le, N.; Wang, J.C. Teaching yourself: A self-knowledge distillation approach to action recognition. *IEEE Access* **2021**, *9*, 105711–105723. [CrossRef]
111. Stewart, R.; Nowlan, A.; Bacchus, P.; Ducasse, Q.; Komendantskaya, E. Optimising hardware accelerated neural networks with quantisation and a knowledge distillation evolutionary algorithm. *Electronics* **2021**, *10*, 396. [CrossRef]
112. Zoph, B.; Le, Q.V. Neural architecture search with reinforcement learning. *arXiv* **2016**, arXiv:1611.01578.
113. Zoph, B.; Vasudevan, V.; Shlens, J.; Le, Q.V. Learning transferable architectures for scalable image recognition. In Proceedings of the IEEE Conference on Computer Vision and Pattern Recognition, Salt Lake City, UT, USA, 18–23 June 2018; pp. 8697–8710.
114. Pham, H.; Guan, M.; Zoph, B.; Le, Q.; Dean, J. Efficient neural architecture search via parameters sharing. In Proceedings of the International Conference on Machine Learning, Stockholmsmässan, Stockholm Sweden, 10–15 July 2018; pp. 4095–4104.
115. Saikia, T.; Marrakchi, Y.; Zela, A.; Hutter, F.; Brox, T. Autodispnet: Improving disparity estimation with automl. In Proceedings of the IEEE/CVF International Conference on Computer Vision, Seoul, Republic of Korea, 27 October–2 November 2019; pp. 1812–1823.

116. Tan, M.; Chen, B.; Pang, R.; Vasudevan, V.; Sandler, M.; Howard, A.; Le, Q.V. Mnasnet: Platform-aware neural architecture search for mobile. In Proceedings of the IEEE/CVF Conference on Computer Vision and Pattern Recognition, Long Beach, CA, USA, 15–20 June 2019; pp. 2820–2828.
117. Baker, B.; Gupta, O.; Naik, N.; Raskar, R. Designing neural network architectures using reinforcement learning. *arXiv* **2016**, arXiv:1611.02167.
118. Stanley, K.O.; Miikkulainen, R. Evolving neural networks through augmenting topologies. *Evol. Comput.* **2002**, *10*, 99–127. [CrossRef]
119. Real, E.; Aggarwal, A.; Huang, Y.; Le, Q.V. Aging evolution for image classifier architecture search. In Proceedings of the AAAI Conference on Artificial Intelligence, Hilton Hawaiian Village, Honolulu, HI, USA, 27 January–1 February 2019; Volume 2, p. 2.
120. Liu, C.; Zoph, B.; Neumann, M.; Shlens, J.; Hua, W.; Li, L.J.; Fei-Fei, L.; Yuille, A.; Huang, J.; Murphy, K. Progressive neural architecture search. In Proceedings of the European Conference on Computer Vision (ECCV), Munich, Germany, 8–14 September 2018; pp. 19–34.
121. Xu, Y.; Wang, Y.; Han, K.; Tang, Y.; Jui, S.; Xu, C.; Xu, C. Renas: Relativistic evaluation of neural architecture search. In Proceedings of the IEEE/CVF Conference on Computer Vision and Pattern Recognition, Nashville, TN, USA, 20–25 June 2021; pp. 4411–4420.
122. Brock, A.; Lim, T.; Ritchie, J.M.; Weston, N. Smash: One-shot model architecture search through hypernetworks. *arXiv* **2017**, arXiv:1708.05344.
123. Cai, H.; Gan, C.; Wang, T.; Zhang, Z.; Han, S. Once-for-all: Train one network and specialize it for efficient deployment. *arXiv* **2019**, arXiv:1908.09791.
124. Xia, X.; Xiao, X.; Wang, X.; Zheng, M. Progressive automatic design of search space for one-shot neural architecture search. In Proceedings of the IEEE/CVF Winter Conference on Applications of Computer Vision, Waikoloa, HI, USA, 3–8 January 2022; pp. 2455–2464.
125. Zhang, L.L.; Yang, Y.; Jiang, Y.; Zhu, W.; Liu, Y. Fast hardware-aware neural architecture search. In Proceedings of the IEEE/CVF Conference on Computer Vision and Pattern Recognition Workshops, Seattle, WA, USA, 14–19 June 2020; pp. 692–693.
126. Liu, P.; Wu, B.; Ma, H.; Seok, M. MemNAS: Memory-efficient neural architecture search with grow-trim learning. In Proceedings of the IEEE/CVF Conference on Computer Vision and Pattern Recognition, Seattle, WA, USA, 13–19 June 2020; pp. 2108–2116.
127. Pacini, F.; Pacini, T.; Lai, G.; Zocco, A.M.; Fanucci, L. Design and Evaluation of CPU-, GPU-, and FPGA-Based Deployment of a CNN for Motor Imagery Classification in Brain-Computer Interfaces. *Electronics* **2024**, *13*, 1646. [CrossRef]
128. Ma, X.; Zheng, W.; Peng, Z.; Yang, J. Fpga-based rapid electroencephalography signal classification system. In Proceedings of the 2019 IEEE 11th International Conference on Advanced Infocomm Technology (ICAIT), Jinan, China, 18–20 October 2019; pp. 223–227.
129. Lee, S.Y.; Hung, Y.W.; Chang, Y.T.; Lin, C.C.; Shieh, G.S. RISC-V CNN coprocessor for real-time epilepsy detection in wearable application. *IEEE Trans. Biomed. Circuits Syst.* **2021**, *15*, 679–691. [CrossRef]
130. Huang, Y.D.; Wang, K.Y.; Ho, Y.L.; He, C.Y.; Fang, W.C. An edge AI system-on-chip design with customized convolutional-neural-network architecture for real-time EEG-based affective computing system. In Proceedings of the 2019 IEEE Biomedical Circuits and Systems Conference (BioCAS), Nara, Japan, 17–19 October 2019; pp. 1–4.
131. Gonzalez, H.A.; George, R.; Muzaffar, S.; Acevedo, J.; Hoeppner, S.; Mayr, C.; Yoo, J.; Fitzek, F.H.; Elfadel, I.M. Hardware acceleration of EEG-based emotion classification systems: A comprehensive survey. *IEEE Trans. Biomed. Circuits Syst.* **2021**, *15*, 412–442. [CrossRef] [PubMed]
132. Yoo, J.; Yan, L.; El-Damak, D.; Altaf, M.B.; Shoeb, A.; Yoo, H.J.; Chandrakasan, A. An 8-channel scalable EEG acquisition SoC with fully integrated patient-specific seizure classification and recording processor. In Proceedings of the 2012 IEEE International Solid-State Circuits Conference, San Francisco, CA, USA, 19–23 February 2012; pp. 292–294.
133. Koelstra, S.; Muhl, C.; Soleymani, M.; Lee, J.S.; Yazdani, A.; Ebrahimi, T.; Pun, T.; Nijholt, A.; Patras, I. Deap: A database for emotion analysis; using physiological signals. *IEEE Trans. Affect. Comput.* **2011**, *3*, 18–31. [CrossRef]
134. Zhao, K.; Xu, D. Food image-induced discrete emotion recognition using a single-channel scalp-EEG recording. In Proceedings of the 2019 12th International Congress on Image and Signal Processing, BioMedical Engineering and Informatics (CISP-BMEI), Suzhou China, 19–21 October 2019; pp. 1–6.
135. Alakus, T.B.; Gonen, M.; Turkoglu, I. Database for an emotion recognition system based on EEG signals and various computer games–GAMEEMO. *Biomed. Signal Process. Control* **2020**, *60*, 101951. [CrossRef]
136. Goldberger, A.L.; Amaral, L.A.; Glass, L.; Hausdorff, J.M.; Ivanov, P.C.; Mark, R.G.; Mietus, J.E.; Moody, G.B.; Peng, C.K.; Stanley, H.E. PhysioBank, PhysioToolkit, and PhysioNet: Components of a new research resource for complex physiologic signals. *Circulation* **2000**, *101*, e215–e220. [CrossRef]
137. Park, C.Y.; Cha, N.; Kang, S.; Kim, A.; Khandoker, A.H.; Hadjileontiadis, L.; Oh, A.; Jeong, Y.; Lee, U. K-EmoCon, a multimodal sensor dataset for continuous emotion recognition in naturalistic conversations. *Sci. Data* **2020**, *7*, 293. [CrossRef]
138. Yeung, N.; Bogacz, R.; Holroyd, C.B.; Cohen, J.D. Detection of synchronized oscillations in the electroencephalogram: An evaluation of methods. *Psychophysiology* **2004**, *41*, 822–832. [CrossRef] [PubMed]
139. Tanner, D.; Morgan-Short, K.; Luck, S.J. How inappropriate high-pass filters can produce artifactual effects and incorrect conclusions in ERP studies of language and cognition. *Psychophysiology* **2015**, *52*, 997–1009. [CrossRef] [PubMed]

140. Viola, F.C.; Debener, S.; Thorne, J.; Schneider, T.R. Using ICA for the analysis of multi-channel EEG data. In *Simultaneous EEG and fMRI: Recording, Analysis, and Application: Recording, Analysis, and Application*; Oxford Academic: Oxford, UK, 2010; pp. 121–133.
141. Gwon, D.; Ahn, M. Alpha and high gamma phase amplitude coupling during motor imagery and weighted cross-frequency coupling to extract discriminative cross-frequency patterns. *NeuroImage* **2021**, *240*, 118403. [CrossRef] [PubMed]

MDPI AG
Grosspeteranlage 5
4052 Basel
Switzerland
Tel.: +41 61 683 77 34

Sensors Editorial Office
E-mail: sensors@mdpi.com
www.mdpi.com/journal/sensors

www.ingramcontent.com/pod-product-compliance
Lightning Source LLC
Chambersburg PA
CBHW040753250926
51850CB00007BB/1484
* 9 7 8 3 7 2 5 8 7 5 7 1 9 *